LIFE SCIENCES AND SPACE RESEARCH XIV

Organized by

THE COMMITTEE ON SPACE RESEARCH — COSPAR

and

THE BULGARIAN ACADEMY OF SCIENCES

Open Meeting of the Working Group on Space Biology

Sponsored by COSPAR

The Symposium

on Gravitational Physiology

Sponsored by

THE COMMITTEE ON SPACE RESEARCH — COSPAR

THE INTERNATIONAL UNION

OF PHYSIOLOGICAL SCIENCES — IUPS

and

THE INTERNATIONAL ACADEMY OF ASTRONAUTICS — IAA

COSPAR

LIFE SCIENCES AND SPACE RESEARCH XIV

Proceedings of the Open Meeting of the Working Group on Space Biology of the Eighteenth Plenary Meeting of COSPAR

Varna, Bulgaria — 29 May — 7 June 1975

and

Symposium on Gravitational Physiology

Varna, Bulgaria — 30 and 31 May 1975

Edited by

P. H. A. SNEATH

AKADEMIE-VERLAG · BERLIN

1976

Executive Editor:
Dr. A. C. Stickland

Library of Congress Catalog Card Number 63-6132

202 · 100/468/76
Gesamtherstellung: VEB Druckhaus „Maxim Gorki", 74 Altenburg
Bestellnummer: 762 298 2 (3060/XIV) · LSV: 1305, 1495
Printed in GDR

Preface

The increasing development of space technology, with its capability for supporting astronauts in extended space flights, has led to an urgent need to learn more about the long-term effects of weightlessness. This is reflected in the present volume of *Life Sciences and Space Research*, which contains the scientific proceedings in life sciences of the COSPAR meeting at Varna, Bulgaria in 1975, by the inclusion of a symposium on gravitational physiology. Many of the papers in that symposium relate to this problem.

Questions of radiation biology also continue to attract interest. With the advent of well-instrumented probes to the planets, particularly Mars and Jupiter, the study of prebiological processes in planetary environments is making notable progress, and the related questions of planetary quarantine continue to require investigation. The present volume, therefore, contains a broad cross section of reports that should appeal to all who are interested in space biology.

P. H. A. Sneath
Department of Microbiology,
University of Leicester.

Contents List

Radiation Biology

Gravitational Biology

Planetary Quarantine

Exobiology

Biomedical Results of the Skylab Programme

Life Sciences and Space Research XIV — Akademie-Verlag, Berlin 1976

BIOMEDICAL RESULTS OF THE SKYLAB PROGRAM

E. L. Michel, R. S. Johnston and L. F. Dietlein

Lyndon B. Johnson Space Center, National Aeronautics
and Space Administration, Houston, Texas, USA

Skylab, the fourth in a logical sequence of USA manned space flight projects following Mercury, Gemini and Apollo, presented life scientists with their first opportunity for an in-depth study of man's response to the space environment. Extensive medical investigations were undertaken to increase our understanding of man's adaptation to the space environment and his readaptation to gravity upon return to earth. The flight durations of the three Skylab missions were progressively increased from 28 days to 59 days and, finally, 84 days. The results of these investigations of the various body systems clearly demonstrated that man can adapt to zero gravity and perform useful work during long-duration space flight. However, definite changes (some unexpected) in the vestibular, cardiovascular, musculo-skeletal, renal and electrolyte areas were documented. The most significant were: the occurrence of space motion sickness early in the missions; diminished orthostatic tolerance, both in-flight and post-flight; moderate losses of calcium, phosphorus and nitrogen; and decreased tolerance for exercise post-flight. The mechanisms responsible for these physiological responses must be understood and, if necessary, effective countermeasures developed before man can endure unlimited exposure to space flight.

1. Introduction

The United States National Aeronautics and Space Administration (NASA) Skylab Program was designed to expand our knowledge of manned earth orbital operations. Skylab had as its goal the accomplishment of four basic objectives: (i) scientific investigations designed to learn more about the universe, the space environment and the influence of the solar system phenomena on the environment of man on earth; (ii) application experiments involving studies in meteorology, earth resources and communications; (iii) development of technology to provide for the effective and economical development of future long-duration space missions; and (iv) medical investigations leading to the qualification of man for long-duration missions, which are the topic of this report.

Since detailed descriptions of the medical experiment results have been reported previously [1], the emphasis of this paper will be to summarize the significant medical findings noted during the Skylab Missions, and the countermeasures needed or being investigated to overcome these potential problems. Candidate biomedical experiments for the Space Shuttle era will be discussed in relation to knowledge obtained from the Skylab Program.

2. Skylab Medical Experiments Program

The medical experiments were performed in the crew quarters area of the orbital workshop (Fig. 1). This area was approximately 32 meters (20 feet) in diameter totaling 645 meters^3 (400 feet^3) consisting of four compartments: a sleep compartment, a wardroom for food preparation and eating, a waste management section, and an experiment compartment. The major medical subsystems provided in the Skylab orbital workshop to sustain the crewmen, protect their health, and to support medical studies have been reported previously [2].

In general, the medical experiments planned for the Skylab Program were based on physiological alterations observed post-flight in Gemini and Apollo astronauts [3—6]. The Skylab Program offered the first opportunity to follow the time course of these physiological alterations during space flight. The medical experiments performed in Skylab can be separated by body system into six different categories as in Fig. 2. The overall purpose of the M 070 series of experiments was to assess the effect of space flight on nutritional and musculo-skeletal systems. There were four experiments in this series:

a. M 071 — Mineral balance
b. M 073 — Bio-assay of body fluids
c. M 074 — Small mass measurement device
d. M 078 — Bone mineral measurement

These experiments were intended to give a composite picture of any skeletal and/or muscular alterations; evaluate mineral, electrolyte, and hormonal changes; and provide nutritional relationships. In addition to providing the means for evaluating the effects of long-term space flight, this group of experiments was valuable as a real-time operational tool in that body weight changes and accurate food and water intake records were maintained on a daily basis as part of the experiment requirements.

The M 090 series of experiments was designed to assess the effect of space flight on cardiovascular adaptive processes. There were two experiments in this series:

a. M 092 — Lower body negative pressure
b. M 093 — Vectorcardiogram

Since a loss of "orthostatic tolerance" was anticipated as a likely consequence of exposure to the weightlessness during space flight, the M 092, lower body negative pressure experiment was considered one of the prime experiments necessary to qualify man for longer duration space flights. The M 093 vectorcardiogram (VCG) experiment was designed to detect changes in the electrical activity of the heart during space flight and to correlate these changes with anatomical shifts of heart position and body fluids. The VCG was also measured during the performance of M 092 and M 171 experiments.

The M 110 series consisted of five experiments:

a. M 111 — Cytogenetic studies of blood
b. M 112 — Man's immunity — in vitro aspects
c. M 113 — Blood volume and red cell life span
d. M 114 — Red blood cell metabolism
e. M 115 — Special hematologic effects

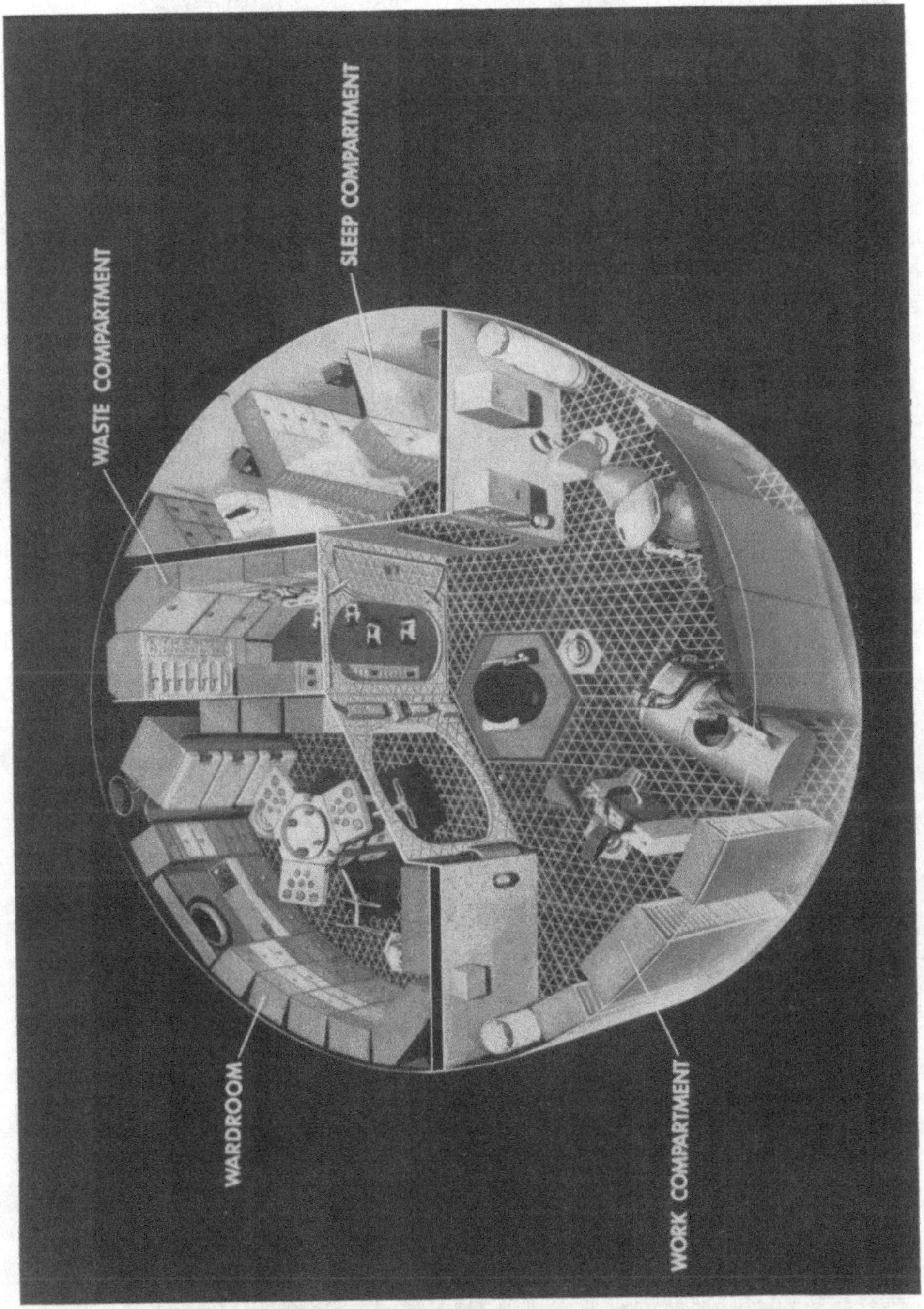

Fig. 1. Skylab crew quarters layout.

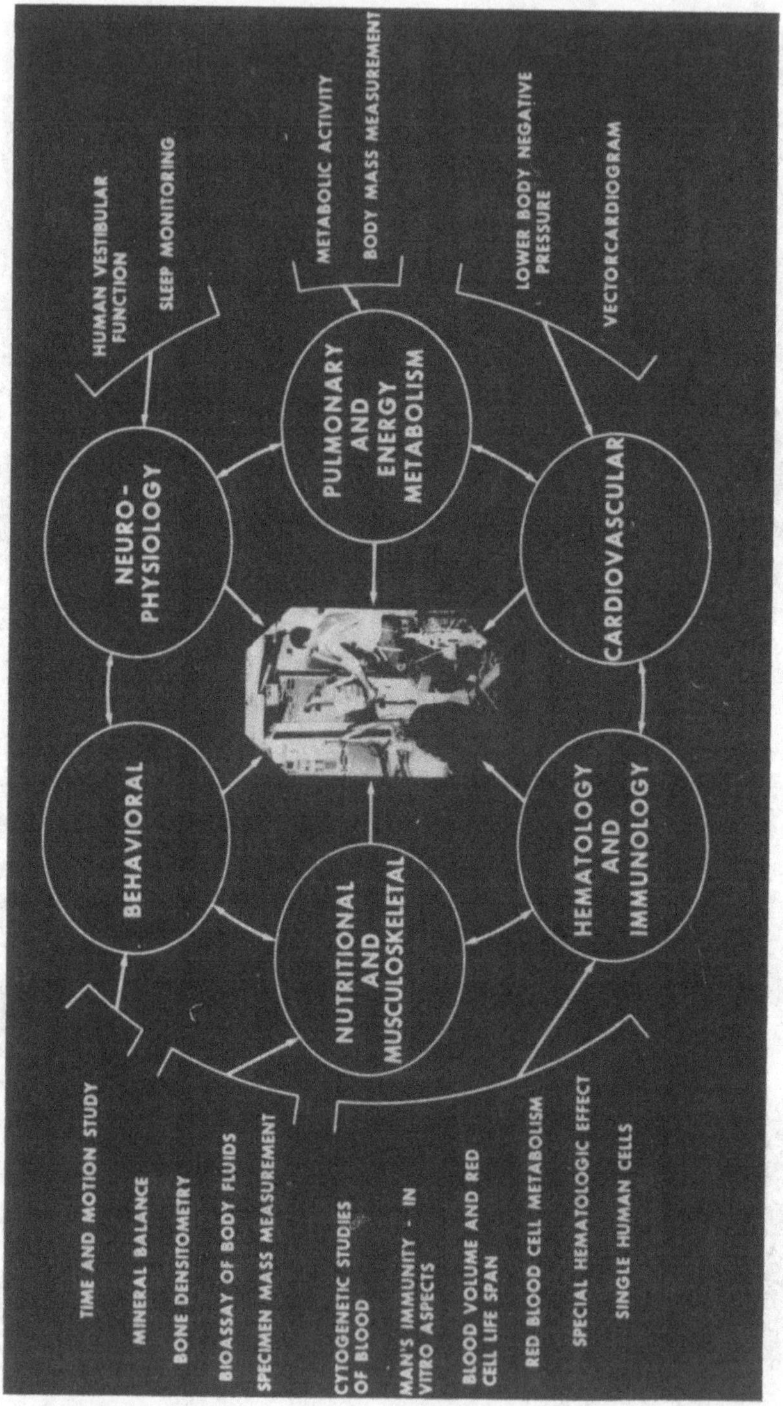

Fig. 2. Life sciences experiment interrelationships.

These experiments were designed to study certain hematologic and cellular anomalies observed post-flight in earlier programs as well as other possible immunological changes that might occur during space flight exposures. Data for this whole group of experiments were obtained from blood samples taken pre-flight, in-flight and post-flight and provided analytical information to support other experiments.

The M 130 Neurophysiology series consisted of two experiments:

a. M 131 – Human vestibular function
b. M 133 – Sleep monitoring

The purpose of the M 131 experiment was to test the crew's susceptibility to motion sickness in the Skylab environment and to acquire data fundamental to understanding the function of human gravity receptors during prolonged absence of gravity. The M 133 experiment evaluated the quantity and quality of in-flight sleep through analyses of electroencephalographic (EEG) and electro-oculographic (EOG) activity.

M 151, Time in motion study, was the only experiment in the M 150 behavioral series. The purpose of this experiment was to evaluate, through cinematographic records, the differences, correlation, and relative consistency between ground-based and in-flight task performance of the crewmen.

The M 170 series consists of two experiments:

a. M 171 – Metabolic activity
b. M 172 – Body mass measurement (BMMD)

M 171 "Metabolic activity" was another of the priority medical experiments in the qualification of man for extended space flight. The experiment was conducted to establish whether man's ability to perform mechanical work would be progressively altered as a result of exposure to zero gravity.

The purpose of the M 174 experiment was to demonstrate body mass measurement in a zero gravity environment and to validate the design of the BMMD. Basically, the device utilized the inertial property of mass in lieu of a gravity field to determine mass. The M 172 experiment supported those experiments for which body mass measurements were a requirement, in particular, M 071, M 073 and M 171.

Generalizing, the Skylab Medical Experiments Program consisted of 16 experiments whose data were to provide a scientific basis for making predictions as to the capability and limitations of man in long-duration space flight missions.

3. Results

The medical experiment results demonstrated that man could adapt and function effectively in the weightless environment of space for extended periods of time. Crew health and well-being were maintained by adequate dietary intake,

Table 1

Skylab Medical Experiment Findings

Headward body fluid shifts
Space motion sickness
Bone and muscle alteration
Cardiovascular deconditioning
Fluid and electrolyte changes
Reduced red cell mass and plasma volume

programmed daily in-flight personal exercise, and appropriate sleep, work and recreation periods. Table 1 lists those findings which require additional research to understand causative mechanisms and in some cases may require the development of remedial or preventative measures for those missions exceding durations of 9—12 months.

3.1. Headward Body Fluid Shifts

Anthropometric measurements made before, during, and after flight, supplemented by subjective comments (full feeling of the head, nasal and ocular congestion), photographs and center of mass determinations documented an expected headward fluid shift. This fluid shift remained throughout flight, but was ameliorated by cycle ergometer exercise in-flight, and was immediately reversible (2—3 hours) upon re-exposure to the $1g$ environment. It is possible that this observed headward fluid shift upon exposure to the weightless environment may be responsible for, or at least contributed to, other physiological alterations noted in the Skylab Program and which will now be reviewed.

3.2. Space Motion Sickness

Table 2 summarizes the incidence of space motion sickness in Skylab. Under operational conditions, six of the nine crewmen experienced motion sickness. However, the administration of anti-motion sickness drugs makes it difficult

Table 2

Skylab Space Motion Sickness Summary

Six of the nine Skylab crewmen had motion sickness of varying severity:

Mild (stomach awareness) — 2 crewmen
Moderate (stomach awareness and nausea) — 2 crewmen
Severe (nausea and vomiting) — 2 crewmen

to determine the individual susceptibility and severity of incidents. With this in mind, it appears that the Skylab 2 crewmen did not experience clear-cut symptoms in-flight and only the scientist pilot experienced seasickness after recovery. Among the Skylab 3 crew, the pilot experienced motion sickness shortly after orbital insertion. The remaining Skylab 3 crewmen experienced motion sickness only after entering the workshop. In Skylab 4, both the commander and pilot exhibited symptoms even with doses of anti-motion-sickness medication. In general, Skylab motion sickness symptoms persisted for 3—4 days with complete recovery by Mission Day 7. This finding will be discussed in detail subsequently because of its potential impact on future space missions.

3.3. Bone and Muscle Alterations

Moderate in-flight losses of calcium, phosphorus and nitrogen were observed during the Skylab Program [7]. The observed losses of nitrogen and phosphorus appear to be associated with reduction in leg muscle tissue. Reference to Fig. 3 shows that typically in-flight urinary calcium excretion rapidly doubled its ground-based level and levelled off at levels analogous to the losses observed in bed rest. There was no indication of a decrease in calcium excretion toward the end of the 84-days Skylab 4 mission. Even though these losses are reversible upon return to 1*g*, nominal musculo-skeletal functions might be impaired if mineral and muscle loss were allowed to continue for long periods of time such as during extended space flight. This finding will also be addressed further in a subsequent section of this report.

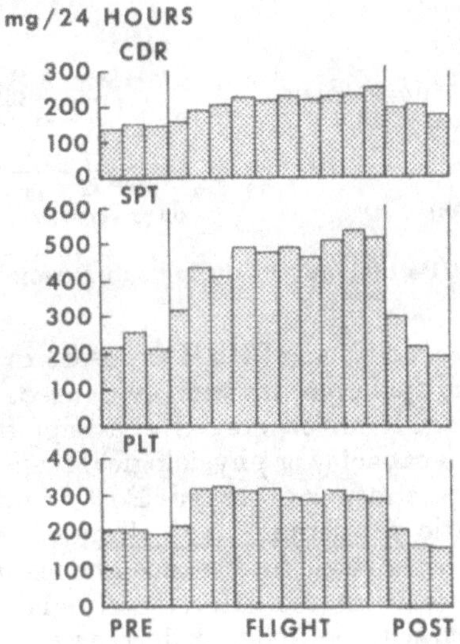

Fig. 3. Skylab 3 urinary calcium. Each bar represents a 6-day average.

3.4. Cardiovascular Deconditioning

Provocative lower body negative pressure (LBNP) tests conducted in-flight have shown that the Skylab crewmen exhibited cardiovascular deconditioning which was adaptive in nature and which tended to stabilize after 4—6 weeks. These changes, however, did not impair crew health or the ability to function effectively in the zero-*g* environment. Table 3 summarizes the incidents of

Table 3
Skylab 2, 3 and 4
M 092, Lower Body Negative Pressure

Mission days	No. of early terminations of LBNP tests
4—14	5
15—28	4[a]
29—42	1
43—56	1
57—70	1
71—83	1

[a] Three of these were between days 15—20

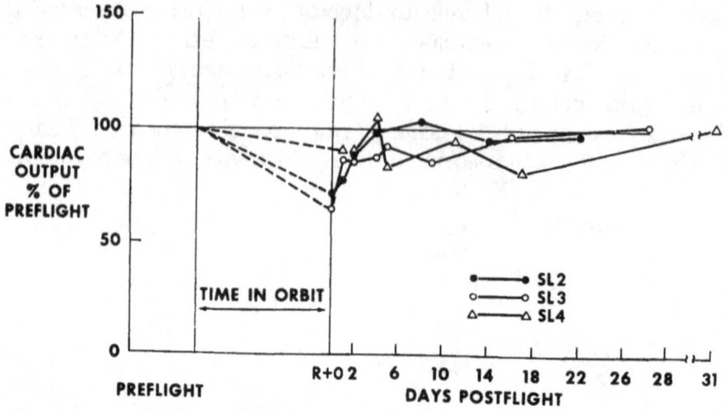

Fig. 4. Cardiac output during submaximal exercise.

presyncopal episodes observed during LBNP exposure in three Skylab missions. A total of 13 in-flight presyncopal episodes were evidenced. They generally occurred during tests conducted seven hours or more after arising from sleep. No significant in-flight decrement in work capacity or physiological response to both submaximal and maximal levels of exercise was observed. Post flight, however, all crewmen evidenced both orthostatic intolerance and diminished response to exercise. The etiology for both the orthostatic intolerance and reduced response to exercise appears to be related to a decreased effective circulating blood volume (post-flight relative to pre-flight) with consequent decreased venous return and cardiac output. Fig. 4 summarizes the mean cardiac output obtained during exercise for each of the three Skylab crews. The values are expressed as a percentage of pre-

Table 4

In-Flight Quantitative Personal Exercise Summary

	[1]Total (watt-min)	[2]Daily average (watt-min)	[3]Daily average (watt-min/kg body weight)	
SL-2				
CDR	62810	2855	47	31.3 avg
SPT	45307	1618	21	
PLT	55795	1993	26	
SL-3				
CDR	228581	3874	58	65 avg
SPT	214645	3638	62	
PLT	386193	6545	75	
SL-4				
CDR	349210	4108	62	72.3 avg
SPT	469420	5523	80	
PLT	414760	4879	75	

[1] Includes M 171 experiment tests and personal exercise.
[2] Based on 28-day SL-2 mission, 59-day SL-3 mission and 84-day SL-4 mission.
[3] Based on mean in-flight body weight.

flight values and demonstrate the time course of return to pre-flight values. As can be seen, the Skylab crews exhibited an approximate 30% decrease in cardiac output immediately post-flight [8]. Recovery was for the most part rapid, requiring 4—5 days for the Skylab 3 and 4 crews and approximately 21 days for the Skylab 2 crew. Increased in-flight exercise by Skylab 3 and 4 crewmen is thought to be a factor in their improved recovery rate [9, 10]. Based on crew comments after each successive flight, time allocated for daily personal exercise was extended from 0.5 hour per crewman in Skylab 2 to 1.0 hour per crewman in Skylab 3 and 1.5 hours per crewman in Skylab 4. The cycle ergometer, the prime mode of exercise, proved to be an excellent stressor of the cardiovascular system. Additional exercise devices were provided for each successive mission in order to attempt

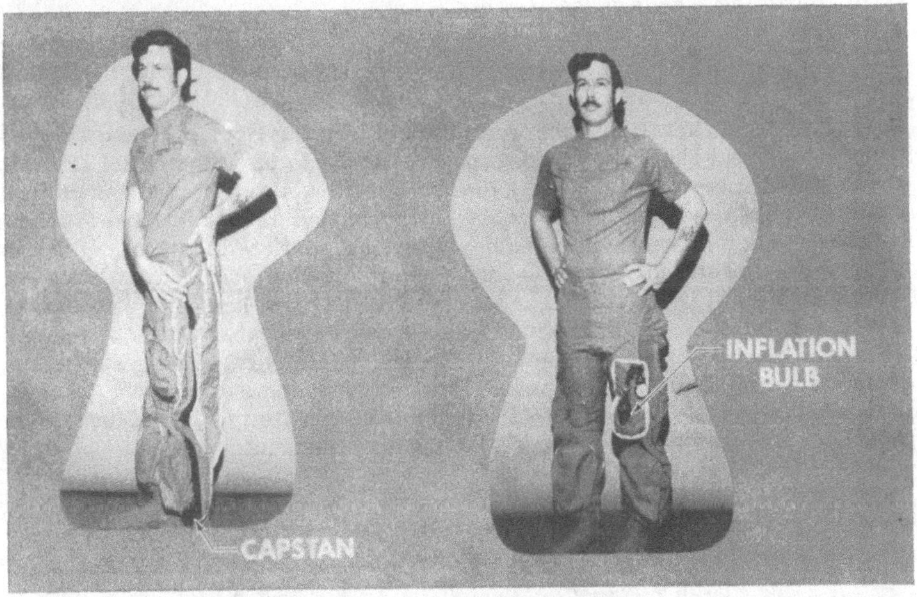

Fig. 5. Skylab cardiovascular counter pressure garment.

to maintain strength in those anti-gravity muscles not adequately exercised by the cycle ergometer. Reference to Table 4, which compares the personal exercise performed on the cycle ergometer by each of the crewmen. The far right column, showing these data normalized to the crewman's body weight indicates that the Skylab 3 crew exercised about 100% more than the Skylab 2 crew while the Skylab 4 crew performed about 130% more exercise than did the Skylab 2 crew.

In an attempt to provide the Skylab crewmen with some protection against anticipated decreased effective circulating blood volumes upon return to the $1g$ environment, individually sized cardiovascular counterpressure garments were provided and were donned prior to re-entry. As seen in Fig. 5, the garment had built-in lateral capstans in each leg which, when inflated to a gage pressure of 170—180 mmHg, produced a pressure gradient which ranged from 85 to 90 mmHg at the ankles to 10 mmHg at the waist. The efficacy of these garments in

Table 5
Efficacy of Cardiovascular Counter Pressure Garment (five-minute means)

	Standing inflated	Standing deflated
Heart rate	103 ± 4	121 ± 5
Systolic blood pressure	124 ± 11	113 ± 8
Diastolic blood pressure	70 ± 6	66 ± 10
Pulse pressure	54 ± 9	43 ± 8

reducing post-flight postural hypotension is demonstrated in Table 5. The pressurized garments afforded a 10—20 beats/min reduction in heart rate together with a 5—15 mmHg increase in pulse pressure.

3.5. Fluid and Electrolytes

Significant biochemical changes were observed during Skylab. These changes returned to normal shortly after re-exposure to the 1*g* environment. Basically, the observed changes were indicative of a successful adaptation to the combined stresses of weightlessness. This adaptation permitted re-establishment of the homeostasis of the internal milieu to new levels appropriate to the zero-*g* condition. However, as stated previously, no such adaptation occurred with respect to the catabolism of the musculo-skeletal system. The new homeostatic set point exerts a cost on the organism which is reflected in elevated levels of cortisol (Fig. 6) and changes in other endocrinological parameters. Changes in these parameters clearly indicate that the astronauts were stressed by their exposure to weightlessness. The overall effect on fluid and electrolyte homeostasis over a prolonged period of time remains conjectural at this point.

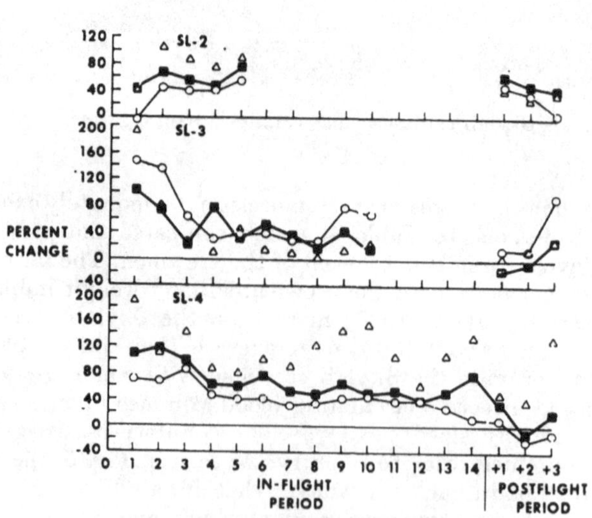

Fig. 6. Urinary cortisol (Skylab data). Each point is a 6-day average.

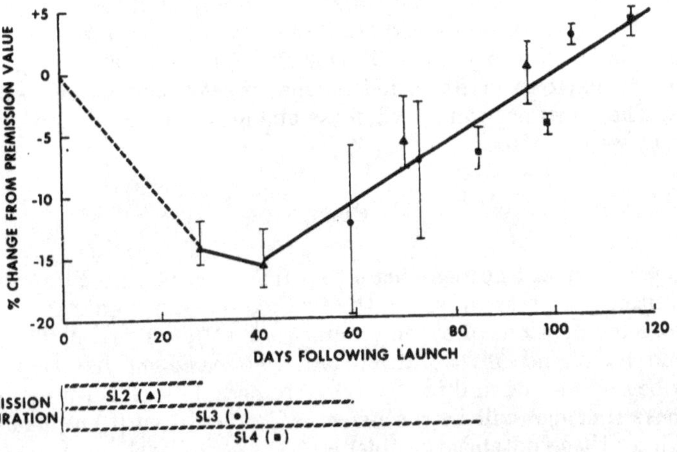

Fig. 7. Red cell mass (Skylab data).

3.6. Reduced Red Cell Mass and Plasma Volume

Red cell mass decreases averaging approximately 10% have been measured regularly among crewmen returning from space flight. The Skylab Missions afforded us our first opportunity to rule out the hyperoxic hypothesis of the red cell mass decrease while permitting us to follow the time course of these changes. Analyses of data from the Skylab red cell mass determinations (Fig. 7) indicate that the red cell mass drops occurred within the first 30 days of flight and that a gradual recovery of red cell mass deficits began approximately 60 days after launch. The plasma volume changes shown in Table 6 are expressed as the mean percentage change from the pre-mission value observed for each of the three Skylab crews. As can be seen, the plasma volumes decrease and do not return to

Table 6
Plasma Volume
(percentage change from pre-mission value)

Skylab 2		
CDR		− 2.5
SPT		−10.3
PLT		−12.3
	Mean	− 8.4
Skylab 3		
CDR		−18.4
SPT		− 9.1
PLT		−11.8
	Mean	−13.1
Skylab 4		
CDR		−15.7
SPT		−19.2
PLT		−12.9
	Mean	−15.9

normal values with extension of the mission. The Skylab 2 (28-day) plasma volume decrease was less than those for Skylab 3 (59-day) and Skylab 4 (84-day) but still greater than that during Apollo (−5% mean). For the most part, plasma volume values reverted back to or exceeded normal pre-mission values within 14 days after flight. The transient nature of these changes appears to indicate successful adaptation to weightlessness.

4. Discussion

Of the physiological findings emanating from the Skylab Program, two stand out as potential problems in either the forthcoming short duration Shuttle Program or plans for missions of extended duration ($1^1/_2$—3 years). These are motion sickness and bone and muscle alteration. The basic mechanisms which evoke these must be understood and protective or remedial measures must be developed. Each of these findings will be discussed individually with emphasis on how we plan to address these potential problems.

4.1. Space Motion Sickness

Prior to Skylab there were eleven reported incidents of motion sickness in orbital flight [11]. With a few exceptions, the symptoms experienced were mild. In-flight motion sickness was not reported prior to the Apollo missions. It is believed that the Apollo command module provided conditions far more favorable for eliciting motion sickness than either the Mercury or Gemini spacecrafts. The incidence of motion sickness during the Skylab orbital flights (6 of 9 crewmen) was greater than had been anticipated. The Apollo and Skylab experiences have underscored the importance of vestibular physiological research in future manned space flight programs.

Accordingly, NASA has attached a high priority to broad-based investigations of neurosensory and related physiological processes believed to be associated with the space flight motion sickness syndrome. These investigations may be divided into the four principal categories of mechanisms, prevention, treatment, and prediction.

a. Mechanisms — Available information is inadequate to explain the causes or underlying mechanisms of the space motion sickness syndrome. Studies are being conducted to (i) investigate the semicircular canal-otolith interaction process; (ii) determine the influence of body fluid shifts on the symptoms expressed; (iii) investigate the role of vision; (iv) investigate the role of biochemical alterations; and (v) evaluate possible neurological alterations.

b. Prevention — The total effectiveness of anti-motion sickness drugs used in Skylab is presently not clear. Studies must, therefore, be continued not only to evaluate the absolute effectiveness of such drugs in preventing motion sickness, but also to determine their possible deleterious side effects after both single or repeated doses.

c. Treatment — In order to insure that effective treatment procedures will be available, should they be required, additional research must be performed. An attempt should be made to identify drugs which would virtually eliminate symptoms even after their onset. Consideration must also be given to the development of operationally acceptable procedures to hasten the rate of adaptation to the zero-*g* environment if indeed an individual is experiencing difficulties. In this regard, the feasibility of using drugs to aid the adaptation process should also be considered.

d. Prediction (required for selection) — Techniques for reliably identifying individuals who are susceptible to space motion sickness are presently lacking. These must be developed to permit the selection of individuals who have a low probability of experiencing motion sickness in zero gravity.

It is intuitively obvious that the development of predictive techniques or selection criteria must be largely dependent on a firm understanding of underlying mechanisms. That is, we must know what we are looking for and why. This dependence on better understanding of the underlying mechanisms is equally important for the logical and expeditious development of effective methods for prevention and treatment. Regardless of the new knowledge that can be acquired from our ground-based studies, the final validation of techniques for the prediction, prevention and treatment of the space motion sickness syndrome must await the forthcoming Shuttle Program.

Several candidate Shuttle in-flight experiments have already emerged. One of the most prominent and well-defined among these at the present time is an experiment designed to employ electronystagmography to measure alterations in semicircular canal input as a function of modifications in otolith behavior following transition into weightlessness. Information from this type experiment would be invaluable in establishing the role of intralabyrinthine sensory conflict as a primary underlying mechanism in zero-*g* motion sickness.

Two other candidate Shuttle experiments are aimed toward defining mechanisms and involve the use of small animals such as guinea pigs or monkeys. The first is designed to measure possible intralabyrinthine fluid pressure changes in response to body fluid shifts in zero gravity. Information from this type of experiment might allow a quantitative determination of the contribution of body fluid shifts to space sickness symptomatology. A second animal experiment under consideration would be designed to evaluate sensory conflict in zero-*g* sickness. Both visual-vestibular and canal-otolith conflict would be investigated in normal and surgically treated animals.

4.2. Bone and Muscle Alteration

As stated previously, the moderate losses of calcium, phosphorus and nitrogen observed in the Skylab crewmen are comparable with those seen in bed-rested subjects. There has been much controversy as to whether bed rest is a valid analog of weightlessness. It appears that bed rest is a very good analog, at least in the area of musculo-skeletal function. NASA has sponsored bed rest studies since 1967 in order to assess the magnitude and duration of bone mineral loss and to study the efficacy of various countermeasures. The first studies involved the exposure of three subjects to 30—36 weeks of continuous bed rest in order to assess the magnitude and duration of bone mineral loss [12]. It was found that during bed rest, bone mineral is lost at a rate of 6 grams of calcium per month or an estimated 0.5% of total body pool of calcium per month, and that this process persists for at least 36 weeks. Additionally, transmission scans of the central calcaneus, ulna and radius suggested that mineral loss during bed rest occurs preferentially from bones which are normally weight-bearing. The observed demineralization was found to be reversible and mineral returned to baseline levels at about the same rate as it was lost. This study suggested two potential hazards: (a) that astronauts who undertake prolonged space flights face an increased risk of bone fractures, (b) that the higher urinary calcium levels could increase the risk of developing kidney stones. Subsequent studies were planned to develop countermeasures which would be effective in reducing these hazards. Specifically, as seen in Table 6, the following potential protective measures were assessed before, during and after 24 to 30 weeks of continuous bed rest.

a. Horizontal isotonic/isometric exercise with the Exer-Genie 80 minutes per day. There was no reduction in bone mineral loss as assessed by either metabolic balance techniques or gamma ray transmission scanning of the calcaneus.

b. Oral phosphate supplements, 1.3 g of phosphorus daily in divided doses as the neutral potassium salt. This regimen, as would be expected, significantly reduced the hypercalciuria in five subjects so treated; however, negative calcium balance was only partially reduced and calcaneal mineral loss was not prevented by phosphate supplementation [13].

c. Longitudinal compression applied with a gravitational acceleration simulating suit (GASS). A force equalling 80% of body weight was applied 4 hours daily. The results showed at best only slight reduction in negative mineral balances and no prevention of mineral loss from the calcaneus [14].

d. Synthetic salmon calcitonin, 100 MRC units I.M. daily in order to reduce the accelerated bone resorption which is thought to be responsible for disuse osteoporosis. Based on this very preliminary study, this countermeasure appeared to be ineffective in preventing bone mineral loss [14].

e. Combination of calcium and phosphorus supplements; 1.2 g of calcium, 1.3 g of phosphorus daily. These supplements appeared to be partially effective in protecting the skeleton during the first 3 months of bed rest after which there was apparently an escape from the beneficial effect [14].

f. EHDP (disodium etidronate), a diphosphonate which inhibits bone turnover and prevents disuse osteoporosis in animals was studied in subjects receiving either 5 mg kg^{-1} day^{-1} or 20 mg kg^{-1} day^{-1} prior to bed rest. The low dose of EHDP had no apparent effects. The high dose of this drug reduced bone turnover but did not prevent the negative calcium balance or loss of calcaneal mineral which attends bed rest.

g. Lower body negative pressure (LBNP) at −30 mmHg for either (1) 4 hours daily; (2) same, but delivered intermittently with a cycle of 2 minutes on and 1 minute off; (3) LBNP at −30 mmHg delivered continuously for 1 week only during the fourth week of bed rest. Results of these studies indicate that LBNP at −30 mmHg is not effective in preventing bone mineral loss during bed rest.

h. Horizontal impact loading: The importance of direct physical stress (weight-bearing) on maintaining the integrity of bone is being investigated. Studies are underway to assess the protection afforded by impact loading during prolonged bed rest in preventing bone demineralization. At this time, it is not possible to draw any conclusions.

In summarizing our bed rest experience, of the potential countermeasures investigated to date only the calcium and phosphorus supplements appear to have some effectiveness in minimizing bone demineralization. It is recognized, however, that other nutrient combinations may have therapeutic efficacy. These will no doubt be the subject of future research programs. One concept which has evolved from retrospective analyses of the patterns of mineral loss is that it may be possible through the use of a "predictive equation" of Vogel et al. [15] to determine which crewmen are likely to resist the demineralizing influences of space flight.

If this equation, which is based upon the baseline urinary hydroxyproline content and the baseline calcaneal mineral content is in fact correct, one could predict whether a given individual is either extremely resistant to mineral loss or will lose mineral at a rapid rate early in space flight.

At the present time, it is obvious that we do not have an effective countermeasure for the bone demineralization observed in space flight. It is also evident that we must continue our ground-based research to provide protective techniques and/or procedures which will give reasonable assurance of healthy functioning astronaut skeletal systems in astronauts during and after extremely long flights. Attention must be given to the selection of individuals who are most resistant to mineral loss. Shuttle era experiments should be designed to define absolute catabolic change in the musculo-skeletal system of animals during zero-*g* exposure. Finally, new and improved methods should be sought to monitor the body's elemental constituents. Considering the difficulties and extreme detail of classic metabolic balance studies, various isotope techniques may have an advantage where patterns of change in metabolic performance need not be known. One of the most promising future approaches for very long flights appears to be the pre- and post-flight measurement of total body calcium by neutron activation analysis.

5. Summary and Conclusions

The Skylab medical experiment results indicate that man adapts well and can function effectively in space flight for 84 days. Although no physiological alterations were observed that would preclude longer duration flights, research is required to understand the mechanisms responsible for the numerous changes observed both in-flight and immediately post-flight. These alterations appear to result from the human body seeking a new homeostatic set point which is in concert with the zero gravity environment and which, in turn, is further altered when the body is forced to readapt to the earth-based environment on return from space. Two findings, those of space motion sickness and bone and muscle alterations, require further research to understand causative mechanisms and do appear at this point in time to require development of protective mechanisms. The short duration Shuttle era missions should provide a test bed for potential preventive or treatment measures developed through ground-based research. Additionally, further observations on man in earth orbit for up to 6 months should logically precede a Mars-type extended space mission. In addition to insuring that previously observed findings do not impair man's performance even on missions of twice the duration of Skylab, this would also afford the opportunity to evaluate the efficacy of bone demineralization countermeasures developed through bed rest studies which may be necessary for extremely long flights.

References

[1] Proceedings of the Skylab Life Sciences Symposium, Vols. 1 and 2, NASA TMX-58154 (1974).
[2] R. S. Johnston and L. F. Dietlein, Int. Fed. Astronaut. XXVth Congr. Rep. 74–100 (1974).
[3] Gemini Midprogram Conference, (including Experiment Results) NASA SP-121 (1966).

[4] Gemini Summary Conference, NASA SP-138 (1967).
[5] C. A. Berry, Aerospace Med. **45**, 9 (1974).
[6] Biomedical Results of Apollo, NASA SP-368 (1975).
[7] G. D. Whedon et al., Proc. Skylab Life Sci. Symp. Vol. 1, p. 353, NASA TMX-58154 (1974).
[8] M. C. Buderer et al., Aviation, Space and Environ. Med. **46**, 6 (1975).
[9] J. A. Rummel et. al., J. Am. Med. Women's Assoc. **30**, 4, (1975).
[10] C. B. Sawin et al., Aviat., Space and Environ. Med. **46**, 4 (1975).
[11] J. L. Homick and E. F. Miller, in: Biomedical Results of Apollo, NASA SP-368, Chap. 8 (1975).
[12] C. L. Donaldson et al., Metabolism **19**, 1071 (1970).
[13] S. Hulley, et. al., J. Clin. Investig. **50**, 2506 (1971).
[14] D. Hamtman et al., J. Clin. Endocrin. and Metab. **36**, 845 (1973).
[15] J. M. Vogel and M. W. Whittle, Proc. Skylab Life Sciences Symp., Volume **1**, 387 (1974).

Symposium on Gravitational Physiology

Life Sciences and Space Research XIV — Akademie-Verlag, Berlin 1976

CONSIDERATIONS OF GEOTROPISM IN PLANTS

H. Kaldewey

Botany Section, University of the Saarland, Saarbrücken, FRG

Four periods of research in the reaction of plants to gravity are distinguished: (1) An early period of purely mechanistic interpretation of the phenomena observed. This period, starting in 1700, is characterized by studies of French and British plant physiologists. They demonstrated that gravity influenced the growth direction of plant parts. (2) In 1868 Frank introduced the term geotropism and postulated an active, peculiar force within the plant which causes a physiological reaction of the plant organ. This is the beginning of the second period, directed by consideration of the physiology of irritability. The elucidation of the *stimulus-reaction chain*, a quantitative description of the phenomena observed, including terminological questions, and the search for geosensors (statolith starch) dominate this "German" period. (3) With the detection of plant hormones to regulate cell elongation in plants at the beginning of the 20th century, the third period begins with the *lateral-auxin-distribution theory* of Cholodny and Went (1926). (4) After a "silent" period of 25 years research in geotropism of plants was revived with modern, critical studies using refined methods. It was found that there might be different principles of geo-reactions in unicellular geotropic plant organs, in roots, and shoots, respectively. In shoots the asymmetric distribution of auxin might be caused by differences in the longitudinal auxin transport, possibly as a result of an asymmetry of active *Golgi* vesicles. This hypothesis is discussed in connection with some recent papers on this subject.

1. Early Period of Studies in Geotropism

It is precisely 275 years ago that the first scientific paper was presented at the "Académie royale des Sciences" at Paris [1] dealing with a problem of plant growth or with a problem of plant movements which we, nowadays, call geotropism. In those early days of research in plant physiology, however, this term geotropism was not yet known. It took more than 150 years until this child of the Frenchman Dodart was named by the German plant physiologist Frank, who introduced the term geotropism in 1868 [2]. In fact, Dodart, although he used the word "la perpendiculaire" in the title of his paper, obviously did not recognize that the growth reactions of his plants might be reactions to gravity. He observed that roots and shoots of his bean seedlings always grew downwards and upwards, respectively, regardless of the orientation of the seeds while germinating. He stated that the roots of inverted oak seedlings made a sharp downward curvature and he noted the upward bending in shoots of different plants which had been brought out of the perpendicular.

Dodart was honest enough to admit that he had been unable to detect the cause of the plant movements observed in his experiments. With the utmost caution, almost hesitation, he pointed to the hypothesis that moistening of the lower side might lead to an extension, and desiccation of the upper side to a contraction of fibres within the opposite sides of aerial plant parts; root fibres might react in a reverse manner, thus leading to the differing behaviour of roots and shoots. There was no word of gravity as a force able to bring about the movements! This idea was presented eight years later, again to the Royal Academy of Sciences at Paris. Astruc [3] suggested that the plant sap, due to its density, was mainly transported in the lower side when shoots were placed horizontally. Thus, by getting more food than the upper side, the lower side would grow faster than the former and thereby make the shoot bend up.

Research in gravity effects on plant growth appears to have been a *French domain* during the first century after Dodart's publication: the names of de la Hire, Bazin, Duhamel de Monceau may be mentioned [cf. 4]. Their investigations confirmed Astruc's idea of gravity as being involved in the growth direction of shoots and roots. The crucial experiment, however, which proved that mass acceleration, i.e. gravity under natural conditions, really *is* the essential force was made by Thomas Andrew Knight who presented his famous paper "On the direction of the radicle and germen during the vegetation of seeds" at the Royal Society of London in 1806 [5]. Knight considered that, if the assumptions of his French colleagues were true, "a centrifugal force should have the same effect as gravity" [4, p. 35].

The experiments confirmed his consideration. Knight was overjoyed to observe that the radicles of bean seedlings, when rotated around a horizontal axis, grew in a right angle away from the axis no matter what the orientation of the seed had originally been, whereas the shoot tips after a few days of germinating joined in the middle of the rotating wheel (Fig. 1). Even more important were Knight's observations on the growth directions of roots and shoots in seedlings while being rotated in a horizontal plane. In this case the roots and shoots pointed obliquely

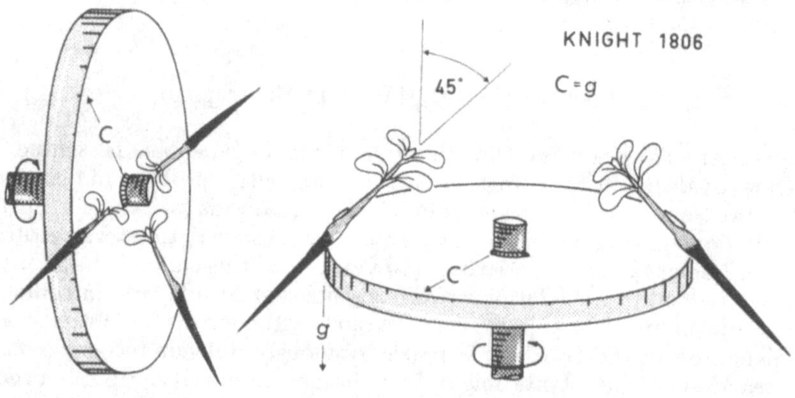

Fig. 1. Knight's centrifuge experiments of 1806. Negatively and positively orthotropic growth in shoots (dotted) and roots (black), respectively, of seedlings germinating under a centrifugal acceleration (C) while rotating around a horizontal axis (left) and oblique growth in the direction of the resultant fo gravitational (g) and centrifugal (C) vectors of equal strength (right) while rotating around a vertical axis.

downward and upward, respectively. Finally if the speed of the wheel was adjusted to put a centrifugal force on the seedlings of close to $1g$, roots and shoots formed an angle to the perpendicular line, estimated by Knight to be 45° [5].

By these two experiments, (i) eliminating the unilateral effect of gravity and exerting centrifugal forces alone, and (ii) combining gravity and centrifugal forces, it was clearly shown not only that a centrifugal force can replace gravity but also that plants react in the same manner to either force, with regard to quality as well as to quantity.

As far as the interpretation of the phenomena is concerned, Knight still followed the traditional mechanistic view of an asymmetric redistribution of the dense plant sap due to acceleration. This purely mechanistic view of a direct physical action of acceleration on plant parts — still remaining predominant during the first half of the 19th century — was overcome when Frank in 1868 [2] postulated that gravity releases an active, peculiar force within the plant organ which is only stimulated to respond physiologically after it has *suscepted* or *percepted* the physical stimulus.

2. Geotropism as a Problem of Physiology of Irritability

Frank's statements signalized the end of the "French- and British-dominated" period. Moreover, they announced the second period of this field of botanical investigations, a period which was directed by aspects and methods for studying the physiology of irritability. Botanical research during this period was very strongly influenced by discoveries in animal physiology. It was the era of Johannes Müller's "*doctrine of specific nervous energy*", the time of exciting findings in neurophysiology and of the detection of statocysts as geosensors in aquatic animals.

The phytophysiological work of this predominantly "german-dominated" period (as examples the names of Sachs, Noll, Pfeffer, Vöchting, Pringsheim, v. Guttenberg and Fitting may be mentioned) can be subdivided into three parts: (i) attempts to elucidate the reaction chain or the chain of stimuli, including terminological questions, (ii) the quantitative description of the phenomena observed, and (iii) the search for geosensors.

2.1. Technical Problems

First, one serious handicap must be mentioned. There is a fundamental difficulty of any quantitative research in geotropism, not only in plants: the stimulus cannot be eliminated as long as experiments are conducted on earth — of course with the exception of free fall, the available time period of which is too short to be useful for experiments with living plants. This handicap was unavoidable until very recently, when a few observations could be made in space [6].

What could be done under terrestrial conditions was to eliminate unilateral gravity stimulation. This had been done already by Knight when he used a centrifuge with a horizontal axis [5], and rotating plants around a horizontal axis became *the* method for quantitative geotropism research. Extensive efforts have been made to construct useful apparatus for this purpose, instruments which were called *clinostats* by Sachs in 1879 [see 7].

The prototype of clinostats is known to every plant physiologist from textbook figures of the spring-powered so-called Pfeffer's clinostat. One main disadvantage of this type of clinostat was that any imbalance in "loading" the machine caused irregularities in the rotary velocity, and therefore unilateral geo-stimulation was not really eliminated. This difficulty was largely overcome by the use of strong motor-powered clinostats. But even if a machine is used as sophisticated, and expensive, as the clinostat constructed by Solon Gordon in the sixties of our century [8], an instrument which guarantees uniform rotation at almost any wanted speed and avoids almost completely the serious implication of vibrations; even then, the question still remains as to how a plant rotating around a horizontal axis "*feels*" weightlessness. Gordon therefore preferred to talk about a *compensation of the unilateral gravity force vector* rather than of simulated weightlessness.

Instead of going into further details, especially with regard to the problem of finding the "right" rotary motion which is, on the one hand, fast enough to prevent the plant from perceiving a unilateral geo-stimulus and, on the other hand, slow enough to avoid centrifugal acceleration, it may be mentioned that the results of the few experiments with plants in earth orbit appear to confirm the results obtained with clinostats, at least in principle [cf. 6]. Thus, the clinostat seems to be a useful tool for research in geotropism of plants and, in fact, the clinostat was and still is a prerequisite for any quantitative work in this field.

Coming back to the aforementioned three directions of the second period of botanical geotropism research, some highlights will be given very briefly to illustrate the discussions and the findings of this period which was dominated by the point of view of plant geotropism being a problem of physiology of irritability.

2.2. Reaction Chain and Terminology

Within the reaction chain a physical phase was distinguished from three important successive physiological phases, the latter being reactions of the living systems of the plant (Fig. 2). As known from neurophysiology, it turned out that the physical phase, called *susception*, which is the exposure of the plant to gravity or to centrifugal forces, had to reach a certain level — or better to pass a certain threshold — in order to cause a visible *reaction*. A sub-threshold stimulus, however, could be demonstrated to be far from having no effect at all. Repeated exposures of the plant to sub-threshold stimuli would bring about a reaction when the threshold had been reached by the sum of the stimuli, and it was Rutten-Pekel-

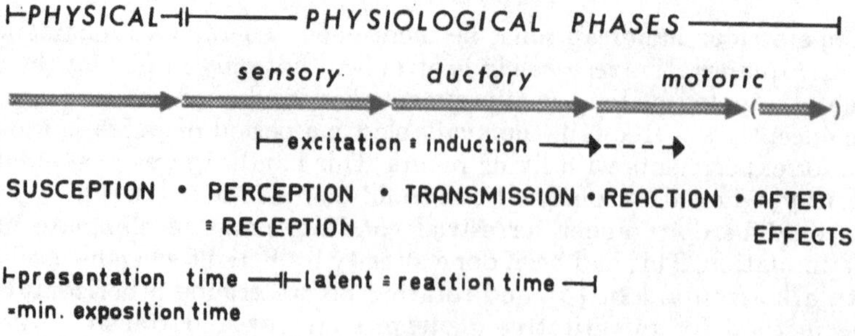

Fig. 2. Stimulus-reaction chain of geotropism in plants. Terminology as suggested by plant physiologists at the beginning of our century.

haring who noticed in 1909 [9] in coleoptiles of oat (*Avena*) — the most famous plant organ used for this kind of experiments not only during the second period of geotropism research — that a summation even of different sub-threshold stimuli, for instance gravitational and centrifugal, could take place. The threshold value, so far discussed, was and is still described by the so-called "*presentation time*", i.e. the minimum duration of unilateral geo-stimulation to cause a visible geotropic reaction. However, the observation of a possible summation of sub-threshold stimuli clearly indicated that the presentation time did not describe the critical level of susception which is appropriate to cause *perception* (or *reception*) as the first physiological phase of the reaction chain. The threshold value for perception is obviously smaller than the presentation time. The discrimination of the threshold values for the perception and for the reaction, respectively, already called for a link between the two physiological phases and this call for a *transmission*, for a *ductory phase*, became necessity when (as in many cases) perception and reaction were observed to be located at different places of the plant organ.

As far as the motoric phase is concerned, it was established by extensive investigations that the geotropic bending reaction at least in higher plants, was due to differences in cell elongation between the opposite sides of the bending zone, without changing the overall growth of the reacting organ. Apart from the fact that the proper reaction may be followed by *after-effects*, for instance by so-called *reverse* or *overcurvatures* (Rück-, Überkrümmung), the geo-reaction is limited to a re-orientation of the displaced plant organ to the *normal* or *liminal direction*, to form a definite angle with the plumb line. This is in accord with the definition of any tropistic reaction, the direction of which is determined by the direction of the driving force.

The liminal direction, which is maintained for a prolonged time period without the plant performing any geo-reactions and which was therefore also called the *resting position* (Ruhelage), turned out to be characteristic, although by no means uniform, for different plants and plant organs, supposing that interferences

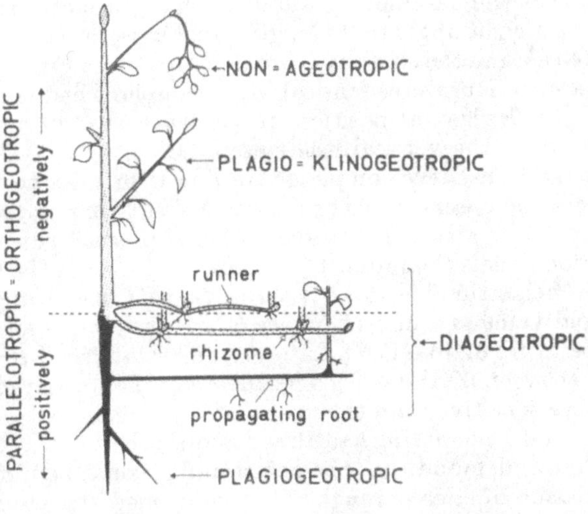

Fig. 3. Normal or liminal growth directions of different plant parts as influenced by gravitational acceleration. Shoots and shoot derivatives dotted; roots black.

with other directional factors, such as light, chemicals, touch, humidity, or temperature gradients are excluded (Fig. 3).

A normal growth direction parallel or close to parallel to the plumb line, which is found in most main roots and main shoots, was called *parallelotropic* or *orthogeotropic*. Any displacement of an orthogeotropic plant part from the vertical causes either a *negatively geotropic* reaction, if the curvature is opposite to the direction of the gravity force (as in shoots) or a *positively geotropic* reaction, if the plant organ bends physically downwards (as in roots). Lateral roots and shoots usually grow at an angle to the direction of the gravity force. This behaviour was called *plagiogeotropism* or *clinogeotropism*; and if the liminal growth direction is 90° or near to a right angle, as in many organs developed for propagation of the plant (some roots, rhizomes and runners) the term *diageotropism* was used. Finally there are plant parts which do not react at all to gravity: they are *non-geotropic* or *ageotropic*, although in certain cases (many inflorescences) they bend downward due to their own weight.

Finally, the liminal growth direction may change during development, if it undergoes a "*change of disposition*" ("geotropische Umstimmung") as observed in many flower and fruit stalks [cf. 10] and further, plagiotropic behaviour especially was often observed to be the result of *correlations* between different plant parts. A branch may become erect if the tip of the main shoot is excised and, similarly, a plagiogeotropic lateral root may become positively geotropic if the main root is damaged.

2.3. Quantitative Description of Geotropic Phenomena

Quantitative studies during the second period of research in geotropism were usually undertaken with orthogeotropic organs and mostly with young primary roots or seedlings or, above all, with cereal (mainly oat) coleoptiles. The main reason for doing so are technical problems, but the disadvantage is that *most of the statements and conclusions cannot be generalized.*

Extensive work was done at the beginning of our century to confirm two rules which should hold if the assumption was true that mass acceleration, affecting the plant organ at a right angle to its longitudinal axis, is the stimulus for a geo-reaction: the *sine-rule* and the *stimulus-quantity law*. It was Fitting who showed in 1905 [11] that geo-reactions were identical whether a plant had been stimulated for a definite time in a horizontal position or for twice that duration at an angle deviating by 30° from the vertical, the sine of 30° being 0.5. The results of his systematic investigations were completely in line with the two rules mentioned above and so were the observations of Rutten-Pekelharing in 1909 [9] who used centrifugal forces. Their results led to the so-called product rule which says that the geo-stimulation equals the product $g \cdot t \cdot \sin \alpha$, where g is the force of gravity, t is the duration of the stimulus and α is the angle with the vertical. It was found that the threshold value of stimulation (see § 2.2) was 300 g s for the *Avena* coleoptile at a temperature of 16—17 °C. Later, it was established that the threshold value may vary between 20 and 600 g s for different plants and plant parts, roots usually being more sensitive than shoots.

Fitting and Rutten-Pekelharing had already noted the temperature dependence of geo-stimulation and found that the geo-stimulus could be nullified by an antagonistic stimulation of equal strengh. They confirmed the product rule over a wide range of force between about 60 and $0.1 \times g$ with corresponding times between about 5 and 4000 seconds. On the other hand they showed th atthe product rule holds only within a limited range of stimuli and that the efficiency of prolonged stimulation in increasing the geo-reaction decreases after a critical level. This observation points to a saturation of at least one of the physiological systems involved in the reaction chain. As mentioned above (§ 2.2) it had been shown that intermittent stimulations at sub-threshold levels could accumulate to induce a geo-reaction. Fitting's observations, however, indicated that the time interval between the successive stimulations was important. He found that the efficiency of intermittent stimulation decreased if the duration of the interval exceeded seven times the duration of the stimulation. Obviously, after that time, a *relaxation* of the stimulus occurred, thus leading to an increase in the time (taken as a measure for the efficiency of stimulation) necessary to induce a visible curvature. However, intermittent stimulation within the relaxation time (the latter appearing to be of variable duration in different species and in different circumstances) was shown to be more effective than a continuous stimulation by Günther-Massias [12].

The last mentioned observations of a decrease in the efficiency of prolonged stimulation, of summation of intermittent sub-threshold stimuli, of relaxation of the stimulus, and finally of intermittent stimulation being more efficient than a continuous one were, at that period of studies of geotropism, mainly discussed under the point of view of similarities to the findings in animal and human physiology. In fact, they were interpreted to support of the validity of Talbot's and Weber-Fechner's laws, respectively, also in geotropism. In a retrospective view many of the discussions of this kind appear partly to be a reaction to the older "purely mechanistic" conceptions; partly, however, they manifest the trend in plant physiology toward an "exact" science.

2.4. The Search for Geosensors

The detection of the *otocysts* as the equilibrium organs of many aquatic animals at the end of the last century gave rise to the search for similar organs in plants. Noll [13, 14] appears to be the first botanist to put forward hypotheses that statocysts and statoliths are present in plants. He postulated particles of submicroscopic dimensions acting as statoliths within a "*centrosphere*" with sensitive "walls"; a complete "micro-parallel" to the corresponding organs of animals. It may be that Noll's ideas inspired Haberlandt [15] and Němec [16]. In any case, both of them found, independently, gravity-displaced particles within plant cells, although not of submicroscopic dimensions: *starch grains* within the *starch sheath* which is characteristic for many plant organs (Fig. 4). Both reported this in the *Transactions of the German Botanical Society* of 1900; Němec, who mainly worked with roots, reported a few pages ahead of Haberlandt's paper dealing with shoots (mainly with nodes of *Tradescantia*). The *statolith-starch theory of geotropism* became, one may say, the most famous theory of geotropic perception or induction in plants. Plenty of papers on this subject appeared during the following years, mostly confirming the hypothesis. Starch grains were also supposed to carry electrical charge and therefore to cause the electrical potential observed between the upper and lower side of a plant organ placed horizontally, this being the mediator for the geo-electric effect (GEE), first mentioned by Bose in 1907 [17] and intensively studied by Brauner in the twenties of this century [cf. 18]. There were, however, also observations of geo-reactions in the absence of starch grains,

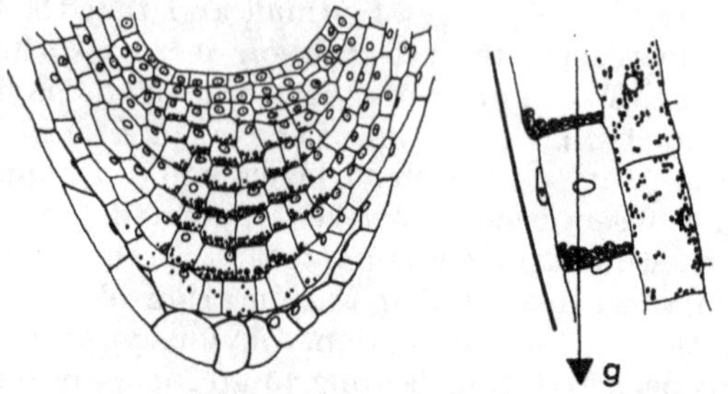

Fig. 4. Gravity-induced asymmetric distribution of statolith starch (black dots) within the statocysts of a root tip and a shoot node which gave rise to the statolith starch theory of geotropism in plants, emphasized by Němec and Haberlandt [15, 16].

a fact which had already been mentioned by the founders of the statolith starch theory. The question of whether starch grains really are the geosensors still remains open in spite of extensive research up to the present time [19, 20].

2.5. Conclusions of the Second Period of Geotropic Research

The easiest way to survey the second period of phytophysiological research in geotropism would be to refer to the extensive reviews of Zimmermann [21, 22] and to Rawitscher's book of 1932 [23] *Der Geotropismus der Pflanzen*. Both publications reflect the efforts which had been made to confirm Frank's postulate that geo-reactions in plants really were physiological answers to an irritation caused by gravity, and hence to elucidate the different phases of a stimulus-reaction chain and to find the geosensors.

To a great extent the studies were designed by analogy to animal physiology and the search for analogies, especially with neurophysiological findings, appears to be an attitude widely assumed by plant physiologists of that period. Moreover, the five decades after Frank's hypothesis in 1868 were characterised by minute observations of the different kinds of geo-reactions. The Darwins' famous book „*The Power of Movement in Plants*" published in 1880 in London [24] may be mentioned as one typical example. As a by-product of these observations, a catalogue of terms was produced, often leading to discussions which were sometimes more of a polemic than of a scientific nature. Lundegårdh [25], Zimmermann and Rawitscher may be mentioned in this field.

3. Auxin Theory of Geotropism

Similar to the almost simultaneous and, in fact, independent detection by Němec [16] and Haberlandt [15] of statolith starch as being possibly involved in the geostimulation of plants, the idea of a lateral transport of the growth-hormone *auxin* (brought about by gravity and thus causing the differences in extension growth between the physically upper and lower sides of an organ placed horizontally) was, so to speak, born twice about 50 years ago by Cholodny [26] and Went [27]. No doubt, there had been forerunners of the so-called Cholodny-Went theory: the participation of some hormone-like substances in tropisms had already been postulated at the end of the last century by Sachs [28] and Darwin [24] and more precise predictions were given by the initiators of the hormone concept of plant growth, such as Boysen-Jensen [29, 30], Paál [31], Söding [32, 33], Seubert [34] and, interestingly enough, by the zoophysiologist J. Loeb [35]. It remained for Cholodny and Went, however, to formulate the *lateral transport* (or *distribution*) *theory* for geotropism. By this theory the enormous complexity of results and speculations in geotropic research could be brought back to "the much simpler problem of a chemical control of growth" [26, p. 457] and could thus be coordinated into a single principle. It was then Dolk [36] who showed that, when *Avena* coleoptiles were placed horizontally, either coleoptile tips or subapical cylinders apically supplied with auxin, more auxin diffused out of their lower halves than from the upper halves into basal agar receiver blocks (Fig. 5) without any alteration of the total amount delivered by the up-

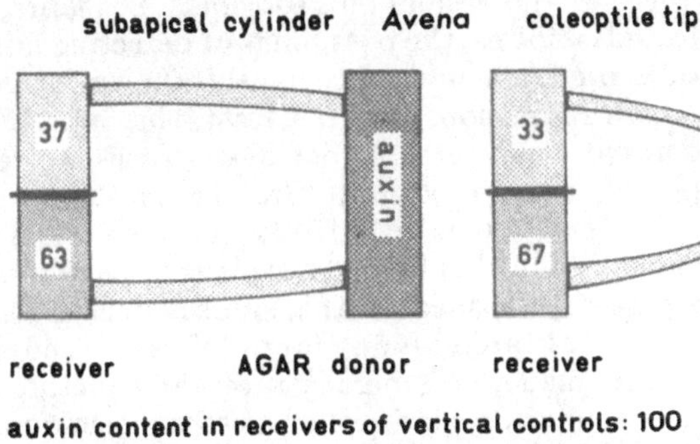

Fig. 5. Dolk's experiments of basipetal transport of auxin in horizontally exposed coleoptile tips of *Avena sativa* and subapical coleoptile cylinders, apically supplied with an agar donor block containing auxin.

right controls. Dolk's results were taken as proof of the validity of lateral auxin transport induced by unilateral gravity stimulation (see however § 4.2). The resulting accumulation of auxin in the physical lower side could explain both the negative geotropism of shoots and the positively geotropic behaviour of roots. In roots, being more sensitive to auxin than shoots, an accumulation led to a supra-optimal, inhibitory auxin concentration whereas cell elongation was promoted in the lower halves of the less sensitive shoots. Dolk's observations in coleoptiles were confirmed in the following years for roots and shoots of many species [cf. 37, 38]. The figures of his doctor thesis entered the textbooks and lateral auxin transport was generally accepted to explain the geo-reactions of plants. In the middle of the thirties, as was enthusiastically emphasised by Cholodny [26] only ten years before, the problem of geotropism was thought to be solved.

4. Modern Critical Period

It was probably the seemingly almost perfect conformity of theory, experimental results and observations which led to a nearly "sterile" period of some 20 years in this field of plant physiology until research in geotropism of plants was revived in the late fifties, especially by some critical papers: Reisener [39] believed he had observed a uniform distribution of the auxin indoleacetic acid (IAA) in geotropically stimulated *Avena* coleoptiles fed with radioactive IAA; Brauner and Hager stated in 1957 [40, 41] that after geo-stimulation auxin-free hypocotyls of *Helianthus* showed a geo-curvature if they were then supplied with IAA and Kaldewey [42] failed to detect a downward movement of IAA across the stem of horizontally oriented flower stalks of *Fritillaria;* instead, differences in the velocity of the *longitudinal* auxin transport were observed between upper and lower halves.

4.1. Lateral Asymmetries

These observations did not fit with the theory and although Reisener's results very soon turned out to be wrong [43] the time appears to have been ripe for a critical evaluation of theories and results on geotropism in plants. Above all, the availability of sophisticated facilities, the possibility of detecting minute quantities of labelled auxin with a precision never previously known, of getting electron microscope pictures of high resolution, and, not least, the opportunity of experiments in space, inaugurated a new era of (not always critical) research of geotropism, not only in plants. The symposium "Gravity and the Organism", organized by S. A. Gordon and M. J. Cohen in 1967 [44], may be taken as a sign of the increasing interest of biologists in this field. There is not space to mention, much less to discuss, all the papers published on this subject during the last 15 years (see the reviews of Wilkins [45], Anker [46], Haupt [47, 48], Schönbohm [49]). In spite of the enormous efforts made, no generally acceptable picture of the primary effect of a gravitational stimulus on plant cells has so far been found.

All the differences established between the upper and the lower sides of a geo-stimulated plant organ (such as differences in pH values, in potassium and calcium ions, in sugar content, in the permeability of membranes, in respiration, in redox potentials, in osmotic potentials (cf. [18]), even the quite recent findings of a gradient in sulphhydryls [50] or an asymmetric distribution of growth hormones other than auxins [51] or of the endoplasmic reticulum (cf. [52, 53])) are probably secondary effects. In other words, these effects may rather be the consequences of a unilateral growth reaction, the latter caused by some as yet unknown asymmetry.

What the recent papers do indicate is that there are probably different principles in the perception as well as in the transmission of the geotropic stimulus in shoots, roots, and — maybe as special cases of unicellular geotropic plant organs — in the rhizoids of the alga *Chara* [54—56] and the sporangiophores of the fungus *Phycomyces* [57] respectively. In the two latter cases auxins are very probably not involved at all in the geo-reaction [19]. In roots, the simple picture of a supra-optimal auxin concentration in the lower side after geo-stimulation appears not to hold (cf. [58—60]); rather, an asymmetric distribution of other growth hormones, possibly of gibberellins [61] and especially of inhibitors [51, 62—64] seems to be part of the reaction chain. Finally, in shoots, it has become more and more unlikely that lateral auxin transport — if it occurs at all in an intact plant — is the *only* reason for an unequal auxin distribution which, in fact, seems to cause geo-curvatures in stems.

4.2. Asymmetries in Longitudinal Auxin Transport

Dolk's experiments (see § 3) by no means proved the correctness of the lateral auxin transport hypothesis as it had been accepted earlier, obviously under the strong influence of the Cholodny-Went theory. The differences in the auxin contents of upper and lower agar receivers (Fig. 5) could be just as well caused by differences between the opposite halves in longitudinal auxin transport. There is still no exactly convincing experiment with intact plants to demonstrate lateral auxin transport after horizontal exposure, perhaps with the exception of lateral transport in the very tip of *Avena* coleoptiles as shown by Brauner and Appel [65]. There is, however, increasing evidence that differences in longitudinal auxin

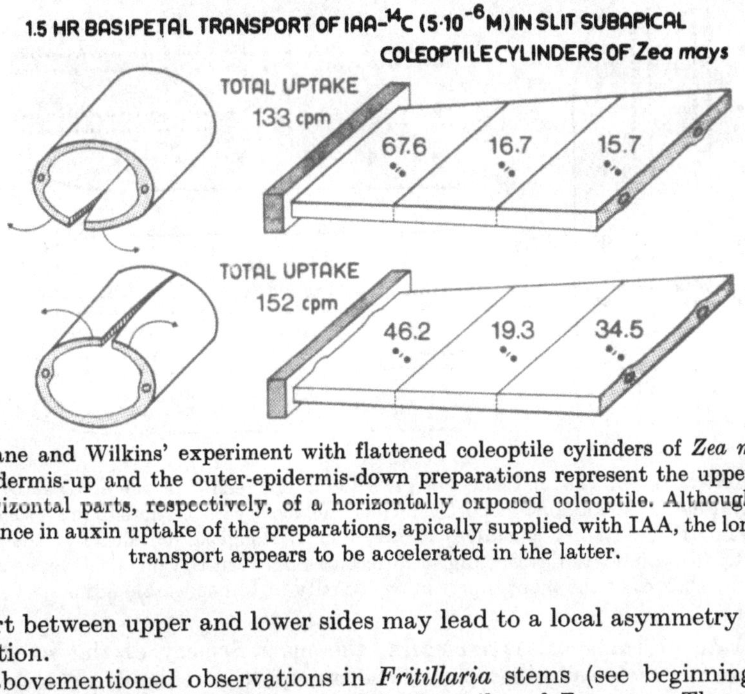

Fig. 6. Cane and Wilkins' experiment with flattened coleoptile cylinders of *Zea mays*. The outer-epidermis-up and the outer-epidermis-down preparations represent the upper and the lower horizontal parts, respectively, of a horizontally exposed coleoptile. Although there is no difference in auxin uptake of the preparations, apically supplied with IAA, the longitudinal transport appears to be accelerated in the latter.

transport between upper and lower sides may lead to a local asymmetry in auxin distribution.

The abovementioned observations in *Fritillaria* stems (see beginning of § 4) induced Naqvi and Gordon [66] to study coleoptiles of *Zea mays*. These authors found that longitudinal auxin transport is enhanced in the lower halves of horizontal coleoptiles as compared with the upper halves. More recently, in 1969, the results of interesting experiments were published by Cane and Wilkins [67]. They flattened longitudinally slit subapical coleoptile cylinders of *Zea mays* and supplied the apical cut surfaces with IAA-^{14}C agar donor blocks (Fig. 6). In vertically oriented preparations there were indications of a lateral downward movement of IAA. Even more interesting were the results in flattened segments placed either with the outer epidermis down (corresponding to the horizontal part of the lower half of a horizontal coleoptile segment) or with the outer epidermis up. No significant differences in the total uptake of radiocarbon were detected between the two types of preparations. However, in the *outer-epidermis-down sheets*, which represent the physically lower side of a horizontal coleoptile, a higher percentage of radiocarbon moved to the basal part of the flattened segment as compared with the *outer-epidermis-up sheets* (Fig. 6). This result points to a speeding up of basipetal auxin transport in the lower side of a horizontal coleoptile. A further highly important observation was reported three years ago by J. Shen-Miller [68]. She showed that different portions of corn coleoptiles differ in their rates of auxin transport. These findings of Naqvi and Gordon, of Cane and Wilkins and especially of Shen-Miller seem to support not only the observations on horizontally exposed *Fritillaria* stems [42], but also the as yet hypothetical deductions drawn by the author. The main idea of these deductions was that alterations in the transport velocity of the auxin transport system must lead to differences in the *density* of the auxin stream. If in a system with a definite velocity, or a definite mass trans-

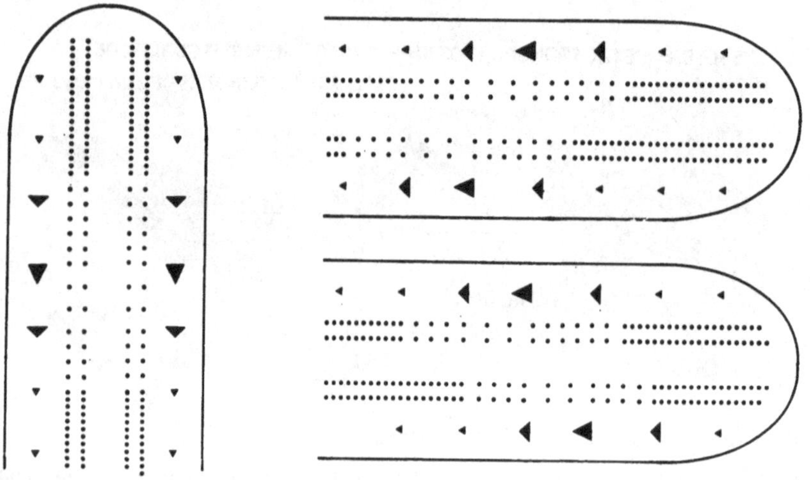

Fig. 7. "Barrier" hypothesis of Kaldewey [42]. Left: The length of the arrows are symbolic of the magnitude of the auxin transport velocity; the density of the dots indicates the density of the auxin stream. Right: Examples for asymmetric longitudinal shifts of the transport pattern after geo-stimulation, resulting in differences between the physical upper and lower halves of the plant organ in the density of transportable auxin.

fer, the rate of transport is *increased*, the mass density of the stream must *decrease*. In contrast, a slowing down of the transport rate should act as a *barrier*, causing an *accumulation* or an increase in the density of the substance transported. There are reports of different auxin transport velocities in differently aged plant tissues. These reports indicate slow rates in very young, scarcely elongating cells, a speeding up of auxin transport in the more basal region of the main region of elongation and a decrease of the transport velocity in the yet older region [69]. Therefore, in the view of the author, there should be a U-shaped distribution of transportable auxin along the axis with high amounts near the apex, a basipetal decrease and an accumulation toward the region of almost completely elongated cells. Then, any unilateral longitudinal shift in the transport pattern along the axis, for instance in a geo-stimulated organ, should cause local differences in the amounts of transportable auxin between opposite sides (Fig. 7).

There is one more point which needs to be mentioned. Too little attention has usually been paid to the developmental stage of the geo-reacting shoot region in

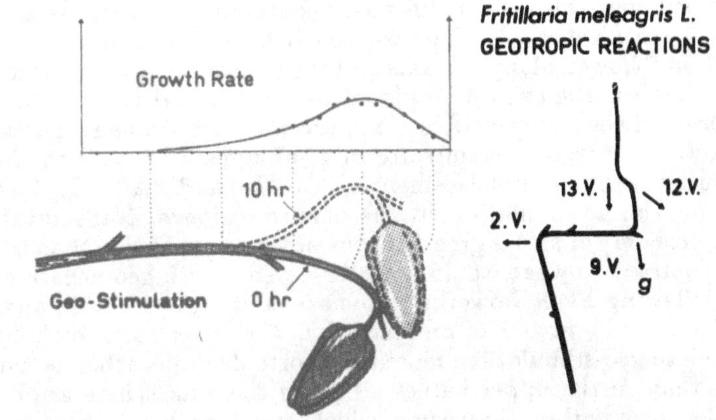

Fig. 8. Negatively geotropic reaction after 10 hr of horizontal exposure of a geo-epinastic flower stalk of *Fritillaria meleagris* and the growth rate pattern along the axis (left). View of a mature plant which had been placed in a horizontal position repeatedly at the dates indicated (right). [After 10, 42, 70].

plants other than coleoptiles or seedlings. Looking at the geo-negative curvature of a horizontally exposed flower stalk of *Fritillaria* (Fig. 8) it is evident that the bending occurs far away from the apex, and growth measurements along the axis show that this is a region at the base of the zone of main elongation [70]. If a plant is repeatedly geo-stimulated, the curvature will always appear in this region, which moves acropetally during development. Similar observations, presented in Fig. 9, were made for flower stalks of *Tulipa*, which had been decapitated

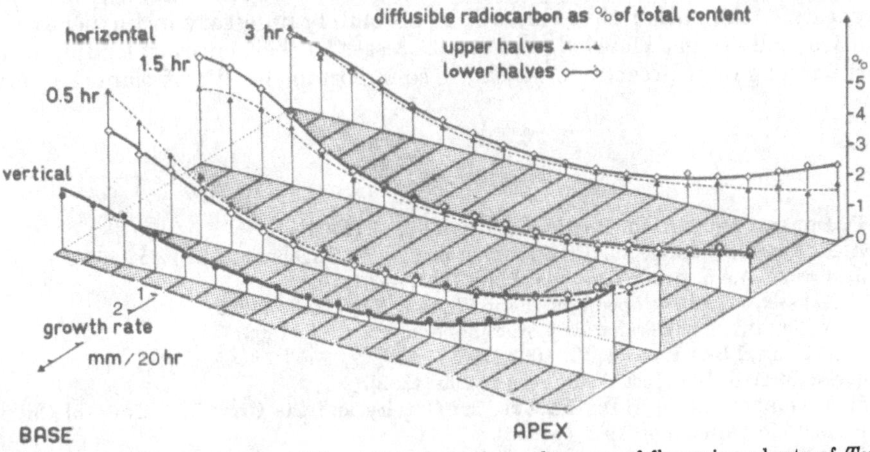

Fig. 9. Distribution of transportable radiocarbon along the axes of flowering plants of *Tulipa Fosteriana* which had been decapitated and supplied with a lanolin paste containing IAA-^{14}C at the apical cut surfaces, 18—20 hr before the experiment. The figures of the erect plant are divided into halves; for the horizontally exposed plants, values for upper and lower half sections are given. The growth rate is indicated as increment in length of 5-mm sections within 20 hr.

and apically supplied with a lanolin paste containing IAA-^{14}C as a substitute for the natural auxin source. In this case, the distribution of transportable radiocarbon has been followed along the axis and the results confirm the picture developed above. Further, the axes were placed in a horizontal position for different periods of time and the transportable auxin was then determined separately in the upper and lower halves. The results are in good agreement with the hypothesis of a geo-induced acropetal displacement of the "*barrier*". After half an hour of geo-stimulation the accumulation in the barrier region of transportable radiocarbon (very probably auxin) is greater in the physical upper side than in the lower side. This is contrary to what would be necessary to give a geo-negative reaction. During the following hour, however, the surplus of transportable auxin in the upper side decreases in favour of an increase in the lower half, so that after one hour and a half of geo-stimulation, more transportable radiocarbon is found in the lower halves than in the upper halves all along the axis. There are only slight absolute differences within the region which starts to bend three hours after geo-stimulation, the proportion between lower and upper halves being 55:45 on an average. The growth rate is, however, small within the bending region so that even a small absolute surplus of elongation in the lower side is sufficient to cause the differences in length necessary to produce the negative geo-curvature observed.

The question arises, of course, as to the reason for the geo-stimulated displacement of the longitudinal auxin transport pattern. Since auxin transport is an active metabolism dependent process a geo-induced lateral asymmetry in metabolic activity would give an appropriate explanation. A suspected candidate for differences in metabolic activity and, thus for the actual geosensor, may have been found in the Golgi apparatus. In fact, Shen-Miller observed an asymmetry of active Golgi apparatus within cells of geo-stimulated coleoptiles [71]. This asymmetry might well influence auxin transport and also auxin immobilization, a process which has as yet scarcely received attention in experiments in geotropism. It appears to the author that this will be absolutely necessary in further studies if we are really to get knowledge of the processes in their entirety leading to the understanding of geo-reactions which are so important for life of plants on earth.

References

[1] D. Dodart, Mém. Acad. Roy. Sci. Paris **1700**, 47 (1703).
[2] A. B. Frank, Beiträge zur Pflanzenphysiologie, W. Engelmann, Leipzig 1868.
[3] J. Astruc, Mém. Acad. Roy. Sci. Paris **1708**, 463 (1709).
[4] P. Larsen, in: Encyclopedia of Plant Physiol. XVII/2, 34 (1962).
[5] T. A. Knight, Phil. Trans. Roy. Soc. London, Pt. I, 99 (1806).
[6] C. J. Lyon, Plant Physiol. **43**, 1002 (1968).
[7] J. Sachs, Arb. Bot. Inst. Würzburg **2**, 209 (1882).
[8] S. A. Gordon and J. Shen-Miller, in: Gravity and the Organism, Univ. of Chicago Press, Chicago, London 1971 (p. 415).
[9] C. J. Rutten-Pekelharing, Rec. Trav. bot. néerl. **7**, 241 (1910).
[10] H. Kaldewey, in: Encyclopedia of Plant Physiol. XVII/2, 200 (1962).
[11] H. Fitting, Jb. wiss. Bot. **41**, 221 (1905).
[12] M. Günther-Massias, Z. Bot. **21**, 129 (1928).
[13] F. Noll, Flora **76**, 265 (1892).
[14] F. Noll, Jb. wiss. Bot. **34**, 457 (1900).

[15] G. HABERLANDT, Ber. dtsch. Bot. Ges. **18**, 261 (1900).
[16] B. NĚMEC, Ber. dtsch. bot. Ges. **18**, 241 (1900).
[17] J. C. BOSE, Comparative Electrophysiology, Longmans, Green, London 1907.
[18] L. BRAUNER, in: Encyclopedia of Plant Physiol. XVII/2, 74 (1962).
[19] K. V. THIMANN, in: Hormonal Regulation in Plant Growth and Development. Verlag Chemie, Weinheim 1972 (p. 155).
[20] R. HERTEL, R. K. DE LA FUENTE and A. C. LEOPOLD, Planta (Berl.) **88**, 204 (1969).
[21] W. ZIMMERMANN, Ergebn. Biol. **2**, 116 (1927).
[22] W. ZIMMERMANN, Jb. wiss. Bot. **66**, 631 (1927).
[23] F. RAWITSCHER, Der Geotropismus der Pflanzen, Fischer, Jena 1932.
[24] CH. and FR. DARWIN, The Power of Movements in Plants, Appleton, London 1880.
[25] H. LUNDEGÅRDH, Planta (Berl.) **2**, 152 (1926).
[26] N. CHOLODNY, Jb. wiss. Bot. **65**, 447 (1926).
[27] F. W. WENT, Proc. Kon. Akad. Wetensch. Amsterdam **30**, 10 (1926).
[28] J. SACHS, Arb. bot. Inst. Würzburg **2**, 226 (1879).
[29] P. BOYSEN-JENSEN, Bull. Acad. Roy Danmark **Nr. 1**, 3 (1911).
[30] P. BOYSEN-JENSEN, Ber. dtsch. bot. Ges. **31**, 559 (1913).
[31] Á. PAÁL, Jb. wiss. Bot. **58**, 406 (1919).
[32] H. SÖDING, Ber. dtsch. bot. Ges. **41**, 396 (1923).
[33] H. SÖDING, Jb. wiss. Bot. **65**, 611 (1926).
[34] E. SEUBERT, Z. Bot. **17**, 49 (1925).
[35] J. LOEB, Bot. Gaz. **63**, 25 (1917).
[36] E. DOLK, Geotropie en groeistof, Diss. Utrecht 1930, Engl. transl.: Rec. Trav. bot. néerl. **33**, 509 (1936).
[37] F. W. WENT and K. V. THIMANN, Phytohormones, Macmillan, New York 1937.
[38] L. ANKER, in: Encyclopedia of Plant Physiol. XVII/2, 103 (1962).
[39] H. J. REISENER, Naturwissenschaften **44**, 120 (1957).
[40] L. BRAUNER and A. HAGER, Naturwissenschaften **44**, 429 (1957).
[41] L. BRAUNER and A. HAGER, Planta (Berl.) **51**, 115 (1958).
[42] H. KALDEWEY, Planta (Berl.) **60**, 178 (1963).
[43] B. GILLESPIE and K. V. THIMANN, Experientia (Basel) **17**, 126 (1961).
[44] S. A. GORDON and M. J. COHEN (eds.), Gravity and the Organism, Univ. of Chicago Press, Chicago, London 1971.
[45] M. B. WILKINS, Ann. Rev. Plant Physiol. **17**, 379 (1966).
[46] L. ANKER, Acta Bot. Neerl. **17**, 385 (1968).
[47] W. HAUPT, Fortschr. Bot. **33**, 167 (1971).
[48] W. HAUPT, in: Hormonal Regulation in Plant Growth and Development, Verlag Chemie, Weinheim 1972 (p. 349).
[49] E. SCHÖNBOHM, Fortschr. Bot. **36**, 167 (1974).
[50] A. FABIAN, Studia Univ. Babes-Bolyai, Ser. Biol. **1**, 71 (1971).
[51] H. M. M. EL-ANTABLY, Z. Pflanzenphysiol. **75**, 17 (1975).
[52] B. E. JUNIPER and A. FRENCH, Planta (Berl.) **109**, 211 (1973).
[53] D. VOLKMANN, Protoplasma **79**, 159 (1974).
[54] A. SIEVERS and K. SCHRÖTER, Planta (Berl.) **96**, 339 (1970).
[55] K. SCHRÖTER, M. I. RODRIGUEZ-GARCIA and A. SIEVERS, Protoplasma **76**, 435 (1973).
[56] U. FRIEDRICH and R. HERTEL, Z. Pflanzenphysiol. **70**, 173 (1973).
[57] D. S. DENNIS, in: Gravity and the Organism, Univ. of Chicago Press, Chicago, London 1971 (p. 65).
[58] H. KONINGS, Acta Bot. Neerl. **16**, 161 (1967).
[59] H. KONINGS, Acta Bot. Neerl. **17**, 203 (1968).
[60] H. G. BURSTRÖM: Amer. J. Bot. **56**, 679 (1969).
[61] I. D. J. PHILLIPS, in: Hormonal Regulation of Plant Growth and Development, Verlag Chemie, Weinheim 1972 (p. 1).
[62] S. SHAW and M. B. WILKINS, Planta (Berl.) **109**, 11 (1973).
[63] P.-E. PILET, Planta (Berl.) **111**, 275 (1973).

[64] W. HARTUNG and R. BEHL, Planta (Berl.) **122**, 53 (1975).
[65] L. BRAUNER and E. APPEL, Planta (Berl.) **55**, 226 (1960).
[66] S. M. NAQVI and S. A. GORDON, Plant Physiol. **41**, 1113 (1966).
[67] A. R. CANE and M. B. WILKINS, Plant Physiol. **44**, 1481 (1969).
[68] J. SHEN-MILLER, in: Plant Growth Substances 1970, Springer, Berlin, Heidelberg, New York 1972 (p. 710).
[69] H. KALDEWEY, Wiss. Z. Univ. Rostock, Math.-Nat. Reihe **16**, 487 (1967).
[70] H. KALDEWEY, in: Gravity and the Organism, Univ. of Chicago Press, Chicago, London 1971 (p. 333).
[71] J. SHEN-MILLER, in: Plant Growth Substances 1970, Springer, Berlin, Heidelberg, New York 1972 (p. 738).

Life Sciences and Space Research XIV — Akademie-Verlag, Berlin 1976

HORMONES AND THE GROWTH OF PLANTS IN RESPONSE TO GRAVITY

D. J. Osborne

Agricultural Research Council Unit of Developmental Botany, Cambridge, England

Plants are remarkable amongst living things in that all their parts (leaves, stems, roots and flowers) can orientate their position in response to gravity. Roots normally grow downwards towards the stimulus, shoots upwards away from the stimulus, while leaves and special kinds of stems like underground rhizomes or aerial runners have adaptations to grow horizontally, approximately at right angles to the gravitational field. This directional positioning of the organ in relation to gravity is achieved by a regulation of the growth of enlarging cells below the meristems or within the organ. Mature tissues in which further cell growth is precluded are therefore unable to reorientate if their position in the field is changed.

Under conditions of zero gravity, roots and shoots continue to grow in opposite directions and the orientation of laterals with respect to the apex is essentially normal. This demonstrates the inherent polarity of plant cells and the internal correlative growth regulation that each organ exerts upon its neighbours.

The perception of gravity, involving statoliths, membranes and "wound" ethylene, is discussed, together with the mechanisms by which the subsequent growth responses can be mediated by changes in endogenous hormones. Evidence for how such hormonal changes can lead to modifications of the rate, extent and reorientation of cell growth is reviewed for several geotropically responding systems.

At the germination of every seed there is a reiteration of a relationship of the plant kingdom to gravity; the root grows downwards towards the gravitational force and the shoot grows in the opposite direction, upwards (Fig. 1). The

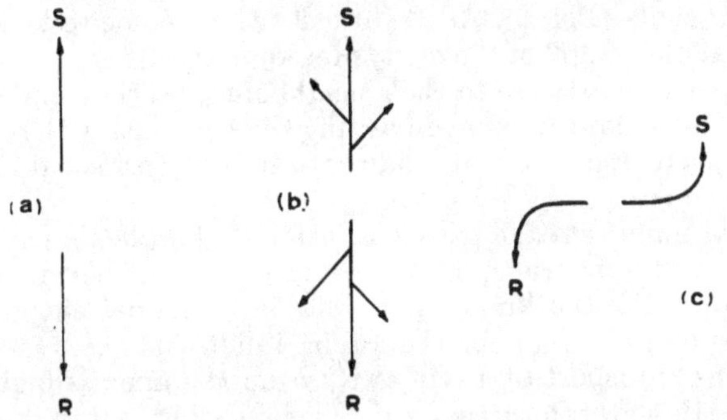

Fig. 1. Normal orientation of main root (R) and shoot (S) (a) and laterals (b) in response to gravity. Curvature occurs in elongating regions (c) when plant placed horizontal.

modification in orientation that may occur in later stages of development in specialized organs of certain plants is no reason for this basic rule to be discounted.

When plants are displaced from their normal position with respect to gravity, the roots and shoots reorientate to bring the growing parts back to their normal alignment to the gravitational force. This occurs, in the short term, by a differential growth of cells on opposite sides of the extending organ, such that a curvature is induced in the growing region (Fig. 2). Normally the presentation time required to initiate a geotropic response is 2 to 3 minutes, and the time for the growth curvature to develop is of the order of 25 minutes [1].

Fig. 2. Tropical response to gravity. Ink marks on root and shoot apices show that curvature responses result from differential growth on the two sides of the organ. The response is the result of enhanced cell elongation*.

Experiments with centrifuges have shown that for inductors of curvature, the intensity of the gravitational force $\times$ the presentation time is a constant over a wide range of forces; for *Avena* coleoptiles the range is 0.08—58.4 g [2], although there is probably a minimum value of $10^{-3}g$ [3] below which a geotropic growth response is not evoked.

On a clinostat with horizontally or vertically compensated gravity of 1g (on earth) or at 0g (in Biosatellite 2) the normal relationship of root to shoot is essentially maintained; roots and shoots extend in opposite directions, and the orientation of laterals with respect to the main organs is normal [4]. There is therefore an inherent factor independent of a continual gravitational stimulus that determines the growth pattern of a plant. This inherent factor is polarity. Each cell of a plant exhibits polarity, in that it possesses a physiological apical and basal end with respect to gravity. This is well exemplified morphologically when a young plant is cut into a series of segments and kept under appropriate and favourable conditions; the "top end" of each will develop buds and the lower end will produce roots (Fig. 3). Adjacent cells which come to be at the "base" of one segment or at the "top" of the next, subsequently differentiate therefore with respect to their position relative to their neighbours, i.e. they undergo a positional differentiation. This is confirmed by inverting the segments with respect to gravity immediately after cutting. The buds and roots are still formed at the physiological apical and basal ends (Fig. 3).

A physiological manifestation of cell polarity is the phenomenon of preferential basipetal or acropetal transport within the tissue of hormonal regulators of cell growth (Fig. 4). All the known non-volatile hormones show some degree of polarity of transport, but that for the auxin, indole-3yl-acetic acid (IAA) is the most marked. The transport of auxin away from the apex is highly basipolar in shoot cells [5], particularly those that are still elongating, but in roots the direction is still predominantly downwards towards the gravitational force, therefore movement is *towards* the root tip rather than away from it [6], i.e. is acropetal (Fig. 4).

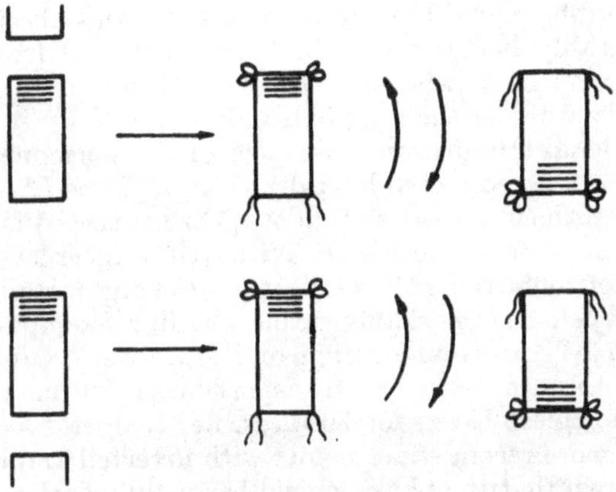

Fig. 3. Segments of plant stem when cultivated show growth of buds and roots at the physiologically apical and basal ends only, irrespective of whether the tissue is inverted with respect to gravity.

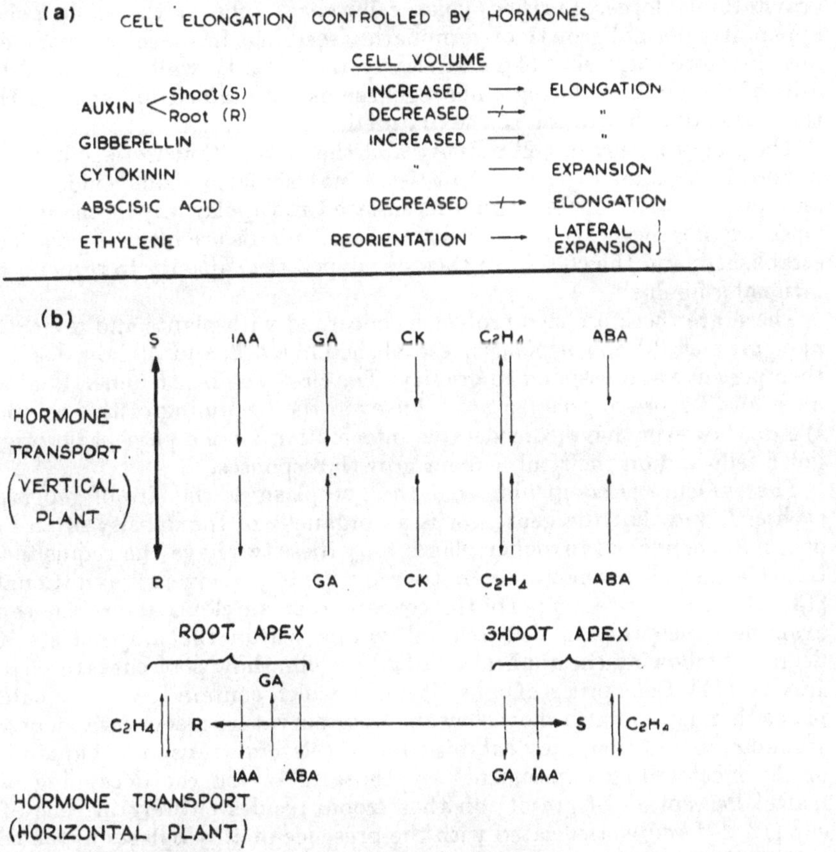

Fig. 4. The major plant hormones and (a) their effects on cell growth and (b) their direction of transport in vertical and horizontally placed roots and shoots.

Other hormones [gibberellins (GA) cytokinins (CK) and abscisic acid ABA] are transported, preferably backwards from both shoot and from root apices, i.e. are basipetally transported, whereas ethylene is believed to move freely in all directions, certainly when in the gas phase.

During gravitational stimulation, transport of the hormone, whilst remaining essentially polar, can be redirected laterally (Fig. 4). Thus IAA and GA preferentially accumulate on the lower sides of shoots [7] whereas GA is transported preferentially to the upper sides of roots [8, 9]. With ABA, higher levels can be recovered in the lower sides of roots and shoots [10] when the organ is placed horizontal.

That polarity of cells is remarkably stable was first exemplified by the experiments of Went [11]. When stem cuttings of *Tagetes* were rooted in the soil in an inverted position, they continued to transport auxin within the stem (but only towards the physiological base), for 36 days, i.e. basipetal transport was maintained. Similarly, more recent experiments with inverted and rooted tomato cuttings [12] have shown that the physiological basipetal polarity of auxin transport is maintained without change in the region between the shoot and rooted part for as long as 4 months, despite cell division and new growth. Clearly, polarity once established would seem to be maintained despite a change in the direction of the gravitational force. Further, polar cells beget cells of the same polarity. The apparently normal growth of germinating seedlings in space or upon a clinostat is not therefore surprising [4], and clearly the polarity which is established in the cells of the embryo during embryogenesis is maintained unimpaired throughout the period of dehydration in the dry seed.

The precise nature of cell polarity and the factors that initiate it are still quite unknown. A callus culture of plant tissue may show no visible evidence of polarity and continue to divide in an undifferentiated and apolar way for many generations. Once organogenesis occurs and root and bud initials are formed, polarity has been established, and the cells have then developed the capacity to respond to a gravitational stimulus.

There are therefore two problems concerned with plants and their dependence upon gravity; (i) how is polarity established in a cell, and (ii) how does a polar cell then perceive and respond to gravity. The first we do not understand at all, and must await more experimentation for even the beginning of this comprehension. We do, however, have considerable information on the perception of gravity by polar cells and on their subsequent growth responses.

The evidence is compelling that the cytoplasm is the site of geoperception of each cell, and that the geosensor is an organelle of the density of an amyloplast or a mitochondrion. In higher plants, only these two have the requisite density to traverse the cell in the 2—3 minutes required to perceive a gravitational stimulus [13]. Much evidence supports the concept that amyloplasts are the sensors. For example, coleoptiles in which starch grains within the amyloplasts have been degraded following the application of gibberellin show poor curvature responses to gravity [14]. Coleoptiles of mutant maize which contain few and smaller amyloplasts than normal also show slow and poor curvature responses to a gravitational stimulus; in addition, they exhibit a reduced lateral transport of auxin [15]. Some of the most telling experiments are those involving the decapping of roots of maize. Perception of gravity in these roots resides entirely in cells of the root cap [16, 17] and is associated with the presence of amyloplasts in the central cap region. Although when the cap is removed a new cap regenerates, recovery of geotropic perception precedes this by many hours, and coincides with the first divisions of cells in the Quiescent Centre (QC) and with the development of amyloplasts within, and in cells immediately surrounding, the cells of the QC [18]. These amyloplasts must be of sufficient number and develop to a minimum size before they act as geosensors, and once the new root cap is regenerated and has formed amyloplasts, those around the QC and the root apex again degenerate [18].

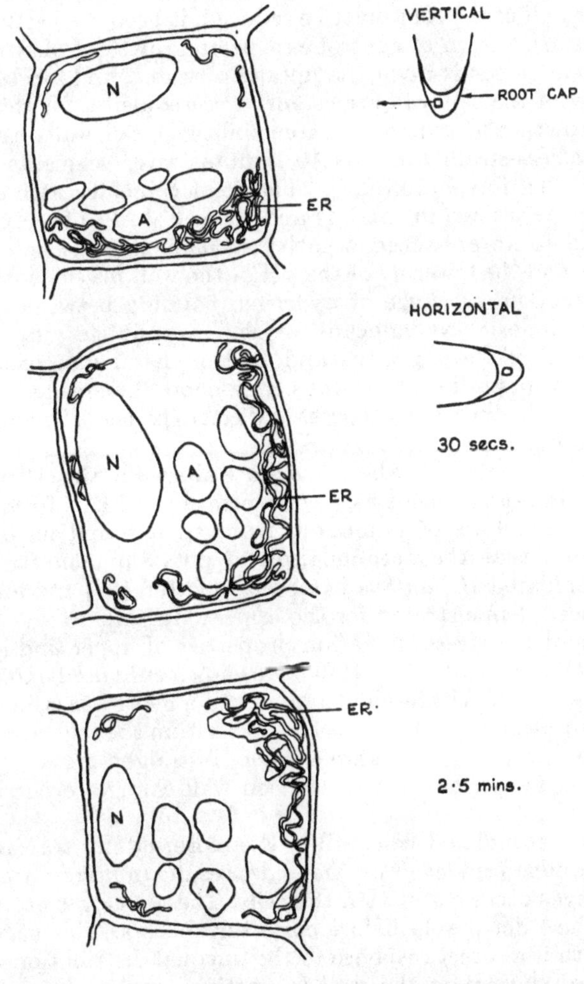

Fig. 5. Redistribution of amyloplasts (A) and endoplasmic reticulum-complex (ER) in root cap cells following geotropic stimulation. N = nucleus (after Volkmann [43]).

More intriguing, however, is what happens in a cell when an amyloplast is displaced. There is now considerable evidence [19, 20] that as the amyloplasts are displaced in root cap cells there is a reorientation of endoplasmic reticulum from the base of a cell towards the upper periclinal wall (Fig. 5). In coleoptiles there is also good evidence [21] that gravitational stimulation results in an increase in the number and activity of dictyosomes in the lower half. But whereas changes in endoplasmic reticulum can be observed by the end of the minimum presentation time for geotropic response (2—3 minutes) the changes in dictyosomes are not visible until 6—9 minutes after stimulation, though still considerably before the start of geotropic growth curvature at 25 minutes.

When considering that a curvature response to a gravitational force requires that in some way growth of cells on the lower half of a shoot must be enhanced, and on the lower half of the root must be reduced, it becomes pertinent to summarize what factors are known to control expansion growth of plant cells.

Cell enlargement depends upon the uptake of water, and a relaxation and enhanced plasticity of the cell wall are essential prerequisites for this to occur [22]. An enhanced growth and enhanced extensibility of cell walls can be measured by mechanical stress-strain analysis 10 minutes after segments of tissue have been exposed to IAA in solution [23]. This event coincides with an extrusion of protons through the plasma membrane and the cell wall to the external ambient solution [24] and to an enhanced negative trans-membrane potential [25]. The efflux of protons and the lowering of the pH of the wall may be intimately related to wall relaxation. The breakage of hydrogen bonding between cellulose microfibrils and hemicellulosic xylanglucans, or the enzymic cleaving of O-glycosidic co-valent bondings between pectins and hydroxylated aminoacids in protein, may be primary events in loosening the components that make up the matrix of the wall, permitting further water ingress and cell expansion. These events may all be favoured at low pH.

Protons are not the only ions whose levels are altered in the cell as a result of an auxin treatment. In *Avena* coleoptiles, concentrations of IAA from 10^{-7} to 10^{-5} M cause an enhanced uptake of potassium from an external medium [26]. It is relevant, therefore, that the accumulation of potassium on the lower side of geotropically stimulated *Helianthus* has been reported [27] and an accumulation of calcium has been demonstrated for the upper side [28].

Careful studies of the stress relaxation properties of upper and lower halves of geotropically stimulated coleoptiles [29] have shown enhanced extensibility of the tissues of the lower half within eight minutes of a geotropic stimulus. This coincides with the initiation of dictyosome changes within the cells resulting from geotropic stimulation [21]. During auxin-enhanced straight growth [30] activation of the dictyosomes is associated with the cell wall changes occurring during cell expansion.

In geotropically stimulated coleoptiles, therefore, lateral transport of IAA to the lower side (or enhanced levels on that side) results in rapid ultrastructural and biochemical changes occurring within the cells. These changes are the prelude to cell growth and are detectable before curvature occurs. This suggests that the differential growth is a direct response to the unequal distribution of auxin.

The lowering of the pH of the wall is clearly a major component of auxin-induced growth. It has been shown, however, that lowering of the pH by other means (buffers at pH 3—4 or CO_2-saturated water) elicit many similar, although temporary, responses. A promotion of water uptake occurs within 10 seconds when sunflower segments are exposed for 3 minutes to CO_2-saturated water, and growth is then temporarily enhanced 8—16 fold [31]. Furthermore, the negative geotropic curvature is increased in these segments by the low pH treatments [32]. Evidence that acid-growth curvatures can be obtained in the absence of an enhanced level of auxin on the lower side of the tissue is afforded by experiments with inhibitors of growth and respiration. In the presence of cycloheximide or arsenite, auxin-induced growth and curvature is precluded, but the growth and curvature at low pH is unimpaired [33]. The application of filter-paper strips at pH 3—4 to one side of vertical or horizontal cucumber hypocotyls leads to increased growth on that side and to negative curvatures. Using sensitive Instron stress-strain relaxation techniques, enhanced cell wall plasticity has been demonstrated in the epidermal tissue stripped from the treated side 15 minutes after treatment, and before bending is observed [34].

The growth during the geotropic response of shoots therefore resembles normal cell growth. The concept has been developed that during cell extension, IAA could function as an effector molecule for a plasma membrane-bound ATP-ase proton pump [35]. Enhanced levels of auxin on one side of a tissue would therefore preferentially activate the pumps on that side. Whereas the low pH or CO_2-induced curvature response is only of limited duration, the auxin-induced response in geotropism continues until the organ again adjusts to gravity, and until the lateral redistribution of auxin has ceased.

In roots, the cells of the root cap have been shown to contain an inhibitor of cell growth [36]. Geotropic curvature can be related to the redistribution of the inhibitor to the lower side [37], which can be simulated by ABA, and may indeed, be ABA [38]. One might expect therefore that removal of the cap and the preclusion of gravity perception would be associated with a temporarily enhanced rate of root growth. This situation has been reported for maize roots [39, 40]. Because geoperception is reinstated on development of amyloplasts in the apex or in the regenerated root cap, the possibility arises that the amyloplast itself might be a source of the growth inhibitor, and be the vehicle of transfer of inhibitor to the lower side of the cell [18]. The idea is attractive, particularly since amyloplasts are derived from plastids which (when they develop into chloroplasts) have been demonstrated to synthesize ABA [41]. For such a concept to be of general application, however, the amyloplast would have to transport IAA in shoot cells. Furthermore, such possibilities are precluded in the rhizoids of the alga *Chara*, where amyloplasts do not occur, and where the geosensors (Glanzkörper) are crystals of barium sulphate [42]. Unlike the membrane-bounded amyloplasts, statoliths of this inert kind are very unlikely to function in the transport of hormones. There could, however, be another function for statoliths in geoperception, which depends solely upon the mass of the particle.

It is possible that as the statoliths are displaced in the cytoplasm they brush against, and perturbate, the membranes of the endoplasmic reticulum and stimulate the release of factors that activate a proton pump at the plasma lemma. Such a stimulation, though mechanical, is an energy-requiring process, and it has been calculated that the pressure exerted by one amyloplast upon the endoplasmic reticulum complex would be of the order of 4 dyn cm^{-2} [43]. Volkmann [43] has observed that in root cap cells of *Lepidium* the timing of the fall of amyloplasts and the displacement of the endoplasmic reticulum to the upper periclinal wall are closely synchronous events, and are first evident within 30 seconds of geostimulation, and are complete by 2.5 minutes. Both amyloplasts and endoplasmic reticulum return to their normal orientation by 60 minutes, after a positive geocurvature has been achieved. Other experiments [44] have indicated that statoliths mechanically activate a cytoplasmic membrane system. When rhizoids of *Chara* were subjected to different centrifugal accelerations for the same time in the horizontal position, different degrees of growth curvature were obtained. For accelerations of 5 minutes, $1g$ gave 7.2°, $5g$, 13.3°, and $10g$ resulted in 17.0°. This points to some pressure mechanism being involved in evoking the georesponse.

Plant cells are unique in possessing a remarkable reaction to small pressures, wounding or any stress component to which they are subjected. Within the shortest intervals of time that have been tested, they produce enhanced levels of ethylene. The stroking of a leaf with a camel's hair brush [45] or the contact of a tendril with a solid surface are sufficient to elicit this response [46]. Ethylene is a potent hormonal regulator of plant growth, and the levels produced by plant are sufficient to modify cell growth and development [47].

The perturbation produced in cells when plants are rotated on a clinostat is sufficient to increase the ethylene production by 100% in 60 minutes [48] and this may account for the petiole epinasty that is commonly observed in these conditions. Similar increases in production occur when an excised leaf is inverted [49].

The temporary doubling in ethylene production within 30 minutes, which results when a box of pea seedlings is handled during transfer from one place to another, leads to a temporary change in membrane synthesis. It has been shown [50] that the incorporation of radioactive glycerol or choline is reduced $2^1/_2$ hours later by 50% and the duration of this reduction mirrors the period of enhanced ethylene production. It could well be, therefore, that the pressure exerted by a moving statolith (amyloplast or Glanzkörper) upon the membranes of the endoplasmic reticulum is sufficient temporarily to alter their permeability characteristics. It has been shown that plants contain substances that function as activators (or co-factors) of ethylene biosynthesis [51], and one of these [52] can be obtained by diffusion into water when cell membranes are impaired at senescence or following stress conditions. It has been proposed [52] that this substance is normally compartmentalized within an organelle, but that any agitation of the tissue is sufficient temporarily to alter the permeability of the organelle membrane, and release the substance of the cytoplasm. The resulting increase in ethylene production is thus a manifestation of an internal "wounding" response.

Ethylene functions as an inhibitor of growth in most plant tissues, and the elongation of shoots and roots can be arrested within 5 minutes of enclosing plants in an atmosphere of the gas. However although a mechanical cytoplasmic stimulation could be invoked for the reduction in growth that occurs on the lower sides of roots when they are exposed to a gravitational force, the mechanism seems at first sight not to be applicable to the negatively geotropic curvatures of shoots. Ethylene is, however, very probably involved in curvature responses of shoots when they are placed horizontal, at right angles to the gravitational field. During upward curvature in the nodes of a horizontal flowering stalk of the grass *Echinochloa colonum* there occurs a lateral redistribution of auxin to the lower side, together with an enhanced ethylene production [53]. Although much of the ethylene produced may result from an auxin-induced ethylene production the following lateral redistribution of auxin, an initial rise may be attributable to an internal "wounding" by the amyloplast geosensors. In terms of cellular ecology, there may also be an important difference between the reaction of geosensors and membranes adjacent to *internally* facing (inner periclinal) and *externally* facing (outer periclinal) cell walls. This may lead to a difference in the auxin transport between the two sides when the organ is horizontal [54]. Certainly, ethylene reduces auxin transport in shoots [55], and a small "wound" ethylene production on the lower side may be sufficient to cause an initial build-up of auxin as in the "barrier" proposal of Kaldewey [56] and to initiate the train of biochemical events that lead to enhanced cell growth on the lower side. Since ethylene is freely diffusible (Fig. 4) few cells only of a tissue may need to function in geoperception. Indeed, in roots, cells containing statoliths (amyloplasts) are limited to the root tip and in the nodal leaf sheath of *Echinochloa* are confined to a few cells around the vascular bundles.

In *Echinochloa* it has been shown that despite the high levels of ethylene produced by the lower side the growth response is primarily regulated by the auxin level, and is therefore seen as an enhancement of cell growth. The presence of auxin essentially overcomes the growth retarding effect of ethylene. In the upper tissues of the horizontal organ, however, auxin is continuously depleted, and ethylene diffusing freely from the lower side to the upper side elicits typical ethylene-type responses in these auxin-depleted cells. The combined action of these two hormones (auxin and ethylene), in concert with other plant growth regulators, is believed to exert an important regulation of the growth response of grass nodes following geostimulation [53], and it may be that such an interaction is generally implicated in plant geotropic and responses.

References

[1] L. J. Audus, Physiology of Plant Growth and Development, McGraw Hill, London 1969, p. 205.
[2] C. J. Rutten-Pekelharing, Rec. Trav. bot. neerl. **7**, 241 (1910).
[3] S. A. Gordon and J. Shen-Miller, in: Life Sciences and Space Research IV, (1966).
[4] C. J. Lyon, Bioscience **18**, 633 (1968).
[5] M. H. M. Goldsmith and P. M. Ray, Planta (Berl.) **111**, 297 (1973).
[6] S. Shaw and M. B. Wilkins, J. Exp. Bot. **84**, 199 (1974).
[7] I. D. Railton and I. D. J. Phillips, Planta (Berl.) **109**, 121 (1973).
[8] H. El-Antably and P. Larsen, Nature **250**, 76 (1974).
[9] J. H. Webster and M. B. Wilkins, Planta (Berl.) **121**, 303 (1974).
[10] H. M. M. El-Antably and P. Larsen, Physiol. Plant **32**, 322 (1974).
[11] F. W. Went, Bot. Gaz. **103**, 386 (1941).
[12] A. R. Sheldrake, New Phytol. **73**, 637 (1974).
[13] L. J. Audus, Symp. Soc. Exp. Biol. **16**, 197 (1962).
[14] B. Pickard and K. V. Thimann, J. Gen. Physiol. **49**, 1065 (1966).
[15] R. Hertel, K. K. de la Fuente and A. C. Leopold, Planta (Berl.) **88**, 204 (1969).
[16] B. E. Juniper, S. Groves, B. Landau-Schachar and L. J. Audus, Nature **209**, 93 (1966).
[17] P. E. Pilet, Ber. schweiz. bot. Ges. **81**, 52 (1971).
[18] P. W. Barlow, J. Exp. Bot. **25**, 1137 (1974).
[19] H. J. Griffiths and L. J. Audus, New Phytol. **63**, 319 (1964).
[20] B. Juniper and A. French, Planta (Berl.) **109**, 211 (1973).
[21] J. Shen-Miller and C. Miller, Plant Physiol. **49**, 634 (1972).
[22] H. G. Burstrom, I. Uhrstrom and B. Olavsson, Physiol. Plant **23**, 1223 (1970).
[23] M. L. Evans, Ann. Rev. Plant. Physiol. **25**, 195 (1974).
[24] R. Cleland, Proc. Nat. Acad. Sci. Washington **70**, 3092 (1973).
[25] E. Marré, P. Lado, A. Ferroni and A. B. Denti, Plant Sci. Lett. **2**, 257 (1974).
[26] H. P. Haschke and U. Lüttge, Z. Naturforsch. **28c**, 555 (1973).
[27] H. R. Bode, Planta (Berl.) **54**, 15 (1960).
[28] N. Arslan-Cerim, J. Exp. Bot. **17**, 236 (1966).
[29] J. Shen-Miller and Y. Masuda, Plant Physiol. **51**, 464 (1973).
[30] S. R. Gawlik and J. Shen-Miller, Plant Physiol. **54**, 217 (1974).
[31] L. Reinhold and Z. Glinka, Plant Physiol. **41**, 39 (1966).

[32] D. Ganot and L. Reinhold, Planta (Berl.) **95**, 65 (1970).
[33] M. L. Evans, P. M. Ray and L. Reinhold, Plant Physiol. **47**, 335 (1971).
[34] S. Iwami and Y. Masuda, Plant and Cell Physiol. **15**, 121 (1974).
[35] R. Hertel, K. Thomson and V. E. A. Russo, Planta (Berl.) **107**, 325 (1972).
[36] G. S. B. Gibbons and M. B. Wilkins, Nature **226**, 558 (1970).
[37] P. E. Pilet, Planta (Berl.) **111**, 275 (1973).
[38] K. K. Kundu and L. J. Audus, Planta (Berl.) **117**, 183 (1974).
[39] P. E. Pilet, Planta (Berl.) **106**, 169 (1972).
[40] J. Bednara, in: Structure and Function of Primary Root Tissues, Bratislava Veda Publ. House of Slovak Acad. Sci. 1974 (p. 23).
[41] B. V. Milborrow and M. Garmston, Phytochemistry **12**, 1597 (1973).
[42] K. Schröter, A. Läuchli and A. Sievers, Planta (Berl.) **122**, 213 (1975).
[43] D. Volkmann, Protoplasma **79**, 159 (1974).
[44] U. Friedrich and R. Hertel, Z. Pflanzenphysiol. **70**, 173 (1973).
[45] D. J. Osborne, in: Crop Processes in Controlled Environments, Academic Press, New York, London 1972 (p. 251).
[46] M. J. Jaffe, Plant Physiol. **46**, 631 (1970).
[47] F. B. Abeles, in: Ethylene in Plant Biology, Academic Press, New York, London 1973.
[48] G. R. Leather, L. E. Forrence and F. B. Abeles, Plant Physiol. **49**, 183 (1972).
[49] J. H. Palmer, Physiol. Plant **28**, 188 (1973).
[50] R. F. Irvine and D. J. Osborne, Biochem. J. **136**, 1133 (1973).
[51] L. W. Mapson, Biol. Rev. **44**, 155 (1969).
[52] D. J. Osborne, M. B. Jackson and B. V. Milborrow, Nature New Biol. **240**, 98 (1972).
[53] D. J. Osborne, in: Mechanisms of Regulation of Plant Growth, Bulletin 12, Royal Society N.Z., Wellington 1974 (p. 645).
[54] S. M. Naqvi and S. A. Gordon, Plant Physiol. **41**, 1113 (1966).
[55] E. M. Beyer and P. W. Morgan, Plant Physiol. **44**, 1690 (1969).
[56] H. Kaldewey, Planta (Berl.) **60**, 178 (1963).

Life Sciences and Space Research XIV — Akademie-Verlag, Berlin 1976

THE EVOLUTIONARY ROLE OF GRAVITY

N. P. Dubinin and E. N. Vaulina

Institute of General Genetics, USSR Academy of Sciences, Moscow, USSR

Analysis of the part played by gravity in development in the organic world shows that this factor has had an impact on evolution. All terrestrial organisms, including man, have adapted themselves to gravity by developing a number of important features of their composition and functions. Variations of gravitational field in any direction bring about numerous changes in organisms, ranging from metabolism to changes in more conservative systems which also include hereditary structures. Gravitational forces determine the form and the size of organisms, the development of skeletal supporting organs, and energetics. The study of the role of gravity in the variability of the organic world will be of great importance for long-term systems of life support and for work on space orbital stations or at bases on the moon and planets where gravitational forces may differ greatly those from on the earth.

It is customary to divide the factors determining evolution into two major groups: (a) geophysical factors which include climate, changes in distribution of continents and seas, and tectonic processes; and (b) ecological factors which include changes in the numbers of organisms and the effect of the immediate environment. At the present time we distinguish (c) a third group of factors which ought to be referred to as "space factors of organismic evolution". This group of evolutionary factors includes: (i) force fields (gravitational, magnetic, electric), (ii) electromagnetic waves (including light, heat and UV rays, radio-waves and ionizing radiation). These evolutionary factors are distinguished by their all-pervading character and relative constancy over time.

In his classical investigations V. I. Vernadsky [1] for the first time considered evolution in connection with space influences on the earth's biosphere. Nowadays ever-growing consideration is given to this problem. Recently there have been made attempts to study differentially the effect of space factors as a moving force of evolution in the organic world. Thus, this question was discussed at the symposium "Space and evolution of organisms" held in Moscow in 1974 [2]. This symposium reviewed a great number of facts from paleontology, geophysics, biology, cosmology and other sciences, which demonstrate the existence of relations between changes in space factors over the period of development of the biosphere and the evolution of the organic world. The time is ripe to determine their importance in evolution.

These effects of space factors have two aspects. They affect all organic forms and are very constant. Such factors should produce adaptive evolution. During vast periods of time they have been playing a part in evolution. However, the constancy of space factors can be disturbed by temporal changes. Environmental conditions sometimes change catastrophically. Such changes, some authors [2—4] believe, could lead to strong shifts in the evolution of the organic world and could determine certain critical points in evolution, such as the invasion by plants of the terrestrial biosphere, the appearance of flying animals, the death of some big taxonomic groups, etc.

The gravitational field — a force field — is one of the most constant and comprehensive factors acting on evolution. In classical physics gravity is identified as gravitational force acting between any bodies and determined by body masses. Gravity is one of the most important characteristics of nature. According to Newton's law the gravitational force depends on the mass m_a, m_b of bodies and the distance (r) between them:

$$F = -K \frac{m_a m_b}{r^3}.$$

Consequently the part of gravity increases when transferred to larger objects. Gravitational forces acting between two electrons are insignificant and about 5×10^{42} times less than the force of their electrostatic repulsion. The gravitational relationship of large bodies with the earth ("terrestrial gravity") is very great and plays an important part in all processes on the earth. As a result of the constancy of the earth mass the value of terrestrial gravity was considered for a long time to be invariable. Variations in gravitational field on the earth surface are small and restricted to a range of 5%. At present the problem of changes of gravitational field during the course of the evolution of the earth is being discussed. Thus, there are estimates of the change of the earth's gravitational field during the movement of the solar system round the galactic nucleus which is accounted for by the heterogeneity of masses in different parts of galaxy [5]. The reconstruction of vital forms of some plants [6], and analysis of the granulometric composition of subaqual sediments [7] suggest the possibility of periodic changes of the earth's gravitational field in the geological past.

The study of the evolutionary role of gravity is particularly urgent at the present time in connection with the development of astronautics. It has practical consequences when gravity either decreases until it disappears (e.g. under conditions of orbital flight) or increases (e.g. for a short period at the start and landing of space vehicles or for long periods, e.g. on other planets which have a mass exceeding that of the earth). Such studies require the prediction both of its effect on individual organisms and their communities and its influence on the evolutional process in the case of prolonged existence in altered gravitational fields.

A great number of works have appeared since the beginning of this century devoted to the different relations between organic nature and terrestrial gravity and some interesting information has been gained on the role of gravity in evolution and on the development and orientation of organisms. Gravitational forces are known to determine the form and the size of organisms, the development of supporting, skeletal organs and to affect their energetics [4,8—21]. From the evolutionary aspect the effect of gravity is well observed in organisms that entered the terrestrial habitats from the aqueous habitat where gravitational forces were largely counteracted by buoyancy.

The invasion of terrestrial habitats by animals required a series of new features: diminution of body size, increase of the relative weight of skeleton, development of extremities, appearance of a double blood circulation, increase in the heart weight due to the fact that energetic requirements are larger, appearance of centres of hemoglobin synthesis and increase in the life-span of red blood cells, all these forming a complex of adjustments assuring the stability of the internal environment [4, 8, 9, 14, 15, 19, 20].

The invasion of the terrestrial habitat by plants resulted in a similar series of changes: reduction of size in herbs, development of strong supporting formations in trees, elaboration of the system of conducting and skeletal tissue, appearance of a complex of adjustments supplying plants with water, oxygen and mineral salts, thus assuring preservation of moisture and protection against light and heat [18, 21].

Responses to gravity permit organisms to have a certain spatial position with respect to the gravitational vector, to form supporting structures which determine the form of the organism and the position of its organs, to develop energetic and biochemical bases allowing vital functions under conditions of the gravitational field.

There are no organisms on the earth which are unaffected by the action of gravity. Evolution in the gravitational field resulted in formation of specific organelles, organs and organ systems responsible for the spatial positioning of organisms in the gravitational field. In protozoa this function is performed by certain organelles of the cell. Thus it is known that amyloplasts perform the function of spatoliths, and mitochondria in cells are distributed according to the action of gravitational forces [22]. Infusoria have an organelle which looks like a bubble filled with otocomia, in paramecia this role is performed by food vacuoles [23, 26]. With further complexity of organisms one can distinguish two lines of development: (i) animals — organisms leading a mobile life; (ii) plants — organisms leading a sedentary life.

In animals the organ of statistic equilibrium developed from a protozoan organelle (statocyst) into a perfect organ in vertebrates which have a complicated labyrinth (including ampules, utriculus, sacculus and Corti's organ). Evolution of animals led to formation of a sensorimotor system which not only determines the position of the organism and its parts with respect to the gravitational vector, and controls the deviation of their position from the normal position, but also corrects the position of the organism, keeping it in the range required for life, and can register the magnitude and direction of linear accelerations attending the movement of animals [27].

Plants have subcellular "sensors of gravity" (statoliths), which move in consequence of the re-orientation of the organ with respect to the gravitational vector and thus bring about a response of a cell or an organ to this re-orientation. In apical cells of Hara rhizoids the statoliths regulate the growth of cell walls, locally opening or closing the supply of cells with substances which build up cellular membrane [22]. In multicellular structures the statoliths are identified with amyloplasts [28]. When these move in the cell due to the action of gravity they create polarity in the organ which may change the growth region. There is also a suggestion that the Golgi apparatus in cells is a geosensor and is responsible for a response to gravity [28]. Orientation of plants with respect to gravity manifests itself by the presence of orthotropic organs (in the vertical position) and plagiotropic organs (in the horizontal position). The growth of plants with respect to gravity is expressed as negative geotropism (for stems) and positive geotropism (for roots). The mechanism of its action was worked out by Holodnyi [29] and Went and Thiemann [30] who developed the hormonal theory of tropisms.

The formation of supporting structures determining the shape of the organism and disposition of organs manifested itself in the development of the skeletal-muscular system in animals, and in the development of the conducting system of skeletal tissues and specific supporting structures in plants. The necessity for an energy supply led to the formation and perfection of the blood circulation system. Simulation experiments on the earth with chickens and rats exposed to long-term centrifugation has demonstrated inhibition of their growth with diminution of size, body weight and fat content [31]. This shows that the organism responds to increasing gravitational forces by an increase of energy consumption. Gravity also affects embryonic development, where gravitational dependence of some stages and processes is observed.

Some knowledge of how organisms function in the absence of a gravitational field was gained in numerous experiments on space vehicles under conditions of weightlessness. Changes in gas exchange and hydrodynamics, and in heat and mass exchange cause changes in metabolism which disturb physiological functions. The higher the evolutionary position of the organism the more significant are these disturbances.

The greater part of the most reliable data from experiments concerns human physiology [32]. An essential feature of these characteristics is that they corroborated the understanding of the role of gravitational effects which was based on analyses of evolutional processes in organisms on the earth. The absence of a gravitational field which serves as an orientator in the environment caused some disturbances of the sensorimotor apparatus during space flight. These disturbances included disturbances in movement coordination, in the vestibular and nerve-muscle apparatus, changes in regulation of the vertical position, etc. [33—37]. Disturbances were also observed in the supporting motor system and in water-salt exchange. Disturbances in water-salt exchange in their turn brought about changes in mineral content of bones and consequently some thinning of skeletal bones. The tone and perimeters of lower extremities decrease, and their power indices become lower. These changes are apt to increase with increasing flight duration [35—38]. Significant changes are also observed in the energy balance. These affect respiration and blood circulation. For example oxygen consumption by V. N. Tereshkova and V. Bykovsky was 30% higher for 8—10 hours [38]. A decrease in the volume of blood plasma, mass of red blood cells, blood volume and a number of other changes were also noted [34—36]. The mode of movement in space also changes. In spaceships under the condition of weightlessness the most suitable appears to be flotation; on the moon where gravity is reduced to 1/6 g jumping on two legs like a kangaroo is best [39]. Staying on the moon resulted in decrease of elasticity of those muscles which are of great importance in running.

Similar changes are observed in other vertebrates [40—42]. Changes in some physiological functions were also noted in plants. Thus, experiments with wheat sprouts demonstrated an increase of respiration, changes in orientation of sprout organs which seem to be brought about by replacement of auxins, increase of the volume of interphase nuclei in cells of primary and lateral roots of sprouts and significant lengthening (by 13.5%) of their cells [43]. Rearrangement of cell organelles by weightlessness was also observed. The most significant changes were noted in the distribution of starch grains. In the weightless condition they are distributed all over the cell whereas in the earth control they are grouped along the lower side of the cell. Changes in insect metabolism are provided by increase in the lifetime of *Drosophila* females exposed in space [44], by data on the reduction of xanthindehydrogenase in *Habrobracone* [45] and reduction of growth rate in amoebas [46]. Apparently, redistribution of cellular energy sources under weightlessness resulted in an increase of growth rate in bacteria [47]. The authors believe that cells of the bacterium *Salmonella typhimurium* randomly distribute in a population in the weightless condition where the nutritional gradient observed in a sedimenting population of high density on the earth is absent. Cell energy under these conditions is used for reproduction and not for movement of the cell to a region with better nutritional conditions.

It has already been mentioned that both reduction and increase of gravitational forces bring about varied changes in organisms. Aviation and space physiology and medicine have gained much information on the effect of accelerations on organisms and from this the following conclusions are made: as a result of acceleration, changes in the organism may be expressed by hardly noticeable functional disturbances or by extremely grave conditions associated with disorders of the respiratory, cardiovascular, nervous and other systems which can result in the loss of consciousness and sometimes in serious anatomic injuries [48, p. 109]. In addition to these, significant changes are also observed in the hereditary apparatus [49].

The major feature of organism structure determined by gravitation is the polarity of organisms. It is expressed in morphology, embryology, development and growth. The basis of the polarizing role of gravity is the gravitational dependence of physical and chemical processes on the earth. All earth organisms from unicellular organisms to man have upper, lower, front, back and lateral sides. The structure of organisms is determined by their mode of life and gravitational vector. Thus, the attached, sedentary, life results in the position of body along the gravity vector with the oral cavity situated in the upper part of the organism in animals. Such organisms are characterized either by radial-symmetric or spiral disposition of organs or organelles. Most organisms possessing independent means of mobility have bilateral symmetry. Terrestrial creeping modes of life cause vertical flattening of organisms. Actively moving, swimming, running or flying animals usually have a spindle-shaped body with flat sides. This shape developed under selection through a set of complicated interrelations between the effect of gravity and physiological processes. Organisms which do not possess the ability of active movement, like many aquatic unicellular organisms which adapted themselves to floating, developed a complicated flat form of body structure as a result of their struggle against sedimentation of cells by gravity [50]. The relation of life mode to gravity and body shape is clearly observed in organisms which changed their mode of life and passed from an active swimming to an attached sedentary state.

The second gravitation-dependent feature of organismic structure is the formation of skeletal structures. The third structural feature is the presence of gravitation receptors which become more elaborate at higher levels of organization. The fourth structural and functional feature of organisms under gravitation is correlated with heat and mass exchange in the organism. On the one hand, the overcoming of gravity requires certain energy losses which affect the processes of heat and mass exchange in the organism and related functional systems which ensure blood circulation, respiration, digestion and excretion. On the other hand, the processes of heat and mass exchange of the organism with the environment also depend on gravity and can significantly change with a change of gravity.

The higher the organization of an organism the greater is its specialization to gravity. The most primitive unicellular organisms and bacteria have weakly polarized cells in which it is very difficult to distinguish the upper and lower sides, gravity receptors are absent, and the effect of gravity is too small to affect it and its structures [51]. More highly organized protozoa, e.g. *Vorticella,* have a complicated differentiated structure of the cell: there is a front or upper end of the cell and its distinctive structure corresponds to their mode of life. Some features of their internal structures respond to the earth's gravitational field, e.g. the form of the cell, the presence of gravitational receptors [27]. In higher vertebrates we find the most elaborate adaptations to the effects of gravity.

The question arises: to what extent has stabilization of organisms to the action of gravitational forces turned into specialization? It is known that particular specialization of organisms is not always a progressive phenomenon. Shmalhousen notes that usually all ways of evolution except those allowing further specialization are closed for specialists [52, p. 147]. If specialization has gone too far it can result in extinction of organisms due to rapid changes in the environment. In the choice between the alternative of death or change (adaptation to the new environmental conditions) the decisive factor is the rate of alteration of environmental conditions and the tempo of evolution of organism.

This problem of the degree of specialization of organisms to the earth gravity is one of the most important in estimating the effect of reduced gravity and weightlessness on organisms. The available facts suggest that absence of gravitational forces does not lead to mass catastrophical death of organisms of the kind which takes place, for example, in the case of specialization of a parasite when the absence of its host causes it to die out. It is possible that a number of other factors are able to produce a polarizing effect, e.g. magnetic and electric fields and light, and can serve as substitutes for gravity in some vital activities. For instance, light in the absence of gravity can determine the direction of plant growth. However, changes in environmental conditions must increase variability, both adaptive and mutational. The increase of variability in its turn intensifies the evolutionary process. The effect of gravity on the rate of evolution raises interesting questions, though we have not enough data to answer them, and can only put forward some speculative hypotheses based on general laws of evolution.

The history of evolution shows that often the progressive and adaptive modification of an organism involves unfavourable changes in some of its parts. A general rapid increase of growth in many phylogenetic branches had negative results in addition to a number of advantages promoting the vital activity of the organism. The proportional enlargement of the whole organism requires the preservation of a system of correlations between its organs and functions. In its turn this requires a long development of the organism under these conditions, or a slow gradual alteration of environmental conditions.

Transference of organisms into reduced or zero gravity will require much time for adaptation to these conditions. The time and results of this adaptation will depend on the organization level of the organism and this process in micro-organisms will proceed in a different way from those in highly organized animals. If energy losses decrease under weightlessness then the first result may be equated with an expansion of nutrition sources and the second one with their contraction. The evalution will take different forms in these cases. The reduction of gravity down to its absence will provide the organism with redundant energy (on account of the loss of necessity to resist the gravitational field) in the case of stable correlation between the structure of organs and their functions. Under these conditions in the evolutionary aspect the appearance of hypermorphose is to be expected as a consequence of disturbances in correlation of organisms with the environment. If such disturbances are absent and the redundant energy acts as an additional source of vital activity, evolution should be regarded in the direction of appearance of aromorphose. In the case where the energy redundancy is compensated by increase of energy losses required for processes bound up with obtaining this energy, allomorphous modifications will be observed with which the organism preserves the previous interrelations of limited adaptation with the environment. Some connections with the environment are substituted by others and the energy of vital activity remains on the same level.

Other trends of evolution are to be expected under increased gravity, i.e. when energy losses of the organism increase. This would cause evolution towards diminution of body sizes. However, diminution of body size beneath existing limits results in simplification of organization which is characterized by underdevelopment [52]. In such cases one expects a selection of mutations correlated with delayed metamorphosis or with arrest of development of some organs at the larval stage of development.

Under weightlessness some organs and related functions are not used, i.e. gravity receptors and their correlations with other systems. This may result in accumulation of mutations (neutral mutations in this case) which will bring about an underdevelopment of the organ. In the evolutionary process this leads to degeneration of the organ, which may be followed by the phenomenon of despecialization. Despecialization always stimulates a new progress. In our case the regression of specialized organs (i.e. the gravity receptors) will lead to despecialization and under new environmental conditions it will certainly afford new opportunities for adaptation.

The knowledge of the evolutionary role of gravity has also a practical significance. It is necessary for solution of problems of construction and functioning of biological life-support systems aboard space ships and stations in the absence of gravitational forces. On the basis of experimental data and analysis of the role of gravity in evolution of organisms on the earth we suggest that higher plants and animals will not be able to ensure the safe working of life-support systems under prolonged absence of gravity. Since the rate of evolution depends mainly on the size of populations, the rate of mutational process and the pressure of selection, the small number of these organisms in the life-support system does not offer hope of spontaneous evolutionary processes under these conditions. The restriction of territory and habitable conditions in ecological systems, as well as gravitational specialization of higher organisms, increase the danger of their extinction during long-term space flights. Work on the development of forms capable of living under weightlessness and during rotation of a closed ecological system is needed. The development of such forms will require much time. Therefore at the initial stage of developing a life-support system for space flights artificial gravity is required. The degree of this should be determined experimentally.

As for lower organisms (bacteria, algae) as components of life-support systems, their higher resistance to the action of gravity, enormous numbers and rapid multiplication should permit in a time the selection of organisms capable of vital activity both under weightlessness and in small gravitational fields.

References

[1] V. I. VERNADSKY, Biogeochemical Essays, Izdat. Akad. Nauk SSSR, Moscow—Leningrad 1940.
[2] Space and evolution of organisms, Proc. Conf. on Space Factors and Evolution of Organic World, Moscow 1974.
[3] A. S. PRESSMAN, Electromagnetic Fields and Animate Nature, Nauka, Moscow 1968.
[4] P. A. KORZHUEV, Evolution, Gravity, Weightlessness, Nauka, Moscow 1971.
[5] G. P. TAMRAZYAN, Izv. VUZ'ov, II (1967).
[6] V. A. KRASILOV, Paleontol. 1 (1969).
[7] L. S. SMIRNOVA and YU. N. LYUBINA, Dokl. Akad. Nauk SSSR, Moscow (1961).
[8] K. E. TSIOLKOVSKY, in: Path to the Stars, Izd. Akad. Nauk SSSR, Moscow 1961.
[9] A. N. SEVERTSEV, Main Trends in the Evolutional Process, Biomedgiz, Moscow 1934.
[10] A. N. SEVERTSEV, Morphological Laws of Evolution, Izd. Akad. Nauk SSSR, Moscow 1939.
[11] I. I. SHMALHOUSEN, Trends and Laws of the Evolutional Process, Izd. Akad. Nauk SSSR, Moscow 1940.
[12] I. I. SHMALHOUSEN, Origin of Ground Vertebrates, Nauka, Moscow 1964.
[13] H. SZARSKI, Pochodzenie plazow Warszawa, 1961.
[14] D. THOMPSON, Growth and Form. Cambridge University Press, Cambridge 1917.
[15] V. YA. BROVAR, Forces of Gravity and Morphology of Animals, Izd. Akad. Nauk SSSR, Moscow 1960.
[16] V. S. IVLEV, Zh. Obschei Biologii, 2nd edn (1959) (p. 20).
[17] V. S. KORZHUEV, Aviation and Space Medicine, Moscow 1963.
[18] V. F. RAZDORSKY, Architectonics of Plants. Sov. Nauka, Moscow 1955.
[19] L. A. ZENKEVITCH, Zh. Obschei Biologii, 5, 3rd edn (1944).
[20] G. FOXON, Bul. Rev., 30, (2) (1955).
[21] G. H. MOLOTKOVSKY, Dokl. Akad. Nauk, 78, 3 (1951).
[22] P. LARSEN, Life Sciences and Space Research XI, 141 (1973).
[23] O. KOEHLER, Arch. Protistenkunde, 45, 37 (1922).
[24] O. KOEHLER, in: Handbuch Norm. Pathol. Physiol., II, Berlin 1926 (p. 1027).
[25] Y. DEMBROWSKI, Arch. Protistenkunde, 66, 104 (1929).
[26] H. MERTON, Arch. Protistenkunde, 85, 33 (1935).
[27] YA. A. VINNIKOV et al., Receptor of Gravity. Evolution of structural cytochemical and functional organization, Nauka, Leningrad 1971.
[28] J. SHEN-MILLER, paper presented at XV COSPAR Meeting, Madrid, Spain, 1972.
[29] N. G. HOLODNYI, Growth Hormones and Dronisms in Plants, 1927.
[30] F. W. WENT and K. V. THIMANN, Phytohormones, Macmillan Co, London, New York 1937.
[31] G. C. PITTS, Life Sciences and Space Research XI, 171 (1973).
[32] Man in Space, Nauka, Moscow 1974.
[33] C. A. BERRY, Life Sciences and Space Research XI, 89 (1973).
[34] C. A. BERRY et al., NASA, SP-121, Washington, D.C., 1966.
[35] N. N. GUROVSKY, Life Sciences and Space Research XI, 77 (1973).
[36] M. A. CHEREPAKHIN et al., Life Sciences and Space Research XI, 117 (1973).
[37] R. MARGARIA, Life Sciences and Space Research XI, 177 (1973).
[38] V. V. PARIN, Space Science Symp., Florence 1964.
[39] C. L. FISCHER et al., J. Amer. Med. Assoc. 200, 99 (1967).
[40] E. H. KASS, paper presented at XV COSPAR Meeting, Madrid, Spain, 1972.
[41] J. HANLEY et al., Life Sciences and Space Research XI, 123 (1973).
[42] N. A. GAIDAMAKIN et al., Kosm Issled. 7, 931 (1969).
[43] S. W. GRAY and B. F. EDWARDS, in: The Experiments of Biosatellite II, NASA, Wash., D.C., 1971 (p. 123).
[44] I. I. OSTER, in: The Experiments of Biosatellite II, NASA, Washington D.C., 1971 (p. 17).

[45] R. C. VON BORSTEL et al., in: The Experiments of Biosatellite II, NASA, Washington, D.C., 1971 (p. 17).
[46] J. H. ABEL et al., in: The Experiments of Biosatellite II, NASA, Washington, D.C., 1971 (p. 291).
[47] R. H. T. MATTONI et al., in: The Experiments of Biosatellite II, NASA, Washington D.C., 1971 (p. 309).
[48] P. V. VASILYEV and A. R. KOTOVSKAYA, in: Space Biology and Medicine, Nauka, Moscow 1966 (p. 109).
[49] N. P. DUBININ and E. N. VAULINA, Life Sciences and Space Research XII, 93 (1974).
[50] M. M. GOLLERBACH and V. I. POLYANSKY, Freshwater Algae and their Study, Sov. Nauka, Moscow 1951.
[51] S. KONDO, Japan J. Gen., **43**, 468 (1968).
[52] I. I. SHMALHOUSEN, Trends and Laws of the Evolutionary Process, Izd. Akad. Nauk, Moscow—Leningrad 1939.

Life Sciences and Space Research XIV — Akademie-Verlag, Berlin 1976

WEIGHT AND SHAPE

S. J. Gould

Museum of Comparative Zoology, Harvard University, Cambridge, Mass., USA

In his *Discorsi* of 1638, Galileo argued that large terrestrial vertebrates would have to grow very thick legs in order to avoid the decreasing cross-sectional area/weight ratios of large objects geometrically similar to small prototypes. Since then, scaling theory (especially the principle of decreasing relative surface area in large forms) has been one of the most widely appreciated and, paradoxically, least studied areas of biology. The two major themes of scaling theory have been: (1) That small and large animals live in different adaptive "worlds" regulated by forces dominant at their size (surface forces for insects, gravity for large organisms). These forces place limits on the size of organic designs (breathing through external tracheae and presence of exoskeleton constrain insects to be small; buckling strength limits the height of trees; scaling of kinetic energy at something close to l^5 sets maximum height of terrestrial bipeds). (2) That small and large animals have characteristic differences in form and function conditioned by the scaling of surfaces and volumes (large animals have relatively smaller brains, thicker legs, lower metabolism, longer life, more convoluted internal surfaces for gas exchange, digestion and circulation).

We know that organic form is adapted to body size, but a major issue for space research involves the degree of purely genetic determination for such adaptation. If, as D'Arcy Thompson argued for the trabeculae of the human femur, adaptations to large size require the immediate action of gravitational forces, then prolonged weightless flight will provoke reversion by removal of the necessary stimulus. In any case, we must know what scaling theory predicts (and what animals actually display) in order to predict the potential problems of prolonged weightlessness.

1. Introduction

In the second dialogue of Galileo's *Discorsi* [1], the work he wrote under house arrest in 1638, Professor Salviati asks his Aristotelian fall-guy Simplicio how to build a cylinder twice as large as its prototype but equal in strength. Simplicio responds with the common-sense proposal that the required cylinder of equal strength should retain the same proportions as its prototype; but Galileo (masquerading as Salviati) proves that the larger cylinder of equal strength must have a different shape from its smaller prototype: it must be relatively wider in diameter (Fig. 1). Supporting strength is measured as cross-sectional area, but the weight to be supported increases as volume (or length (l) cubed). Since volumes (l^3) scale so much more rapidly than surfaces (l^2), equal strength in the larger cylinder can only be maintained by changing the object's shape to increase the cross-sectional area differentially. Galileo also recognized that this physical principle would apply equally well to organisms; he even sketched (Fig. 2) the relatively thick leg bone of a large animal in comparison with the slender bone of its smaller prototype. Indeed, the principle works in nature: Fig. 3 shows the femora of three fossil pelycosaurian reptiles — all drawn to the same size, but scaled at 1:2:4 in life; note the relatively thicker bones of the larger animals.

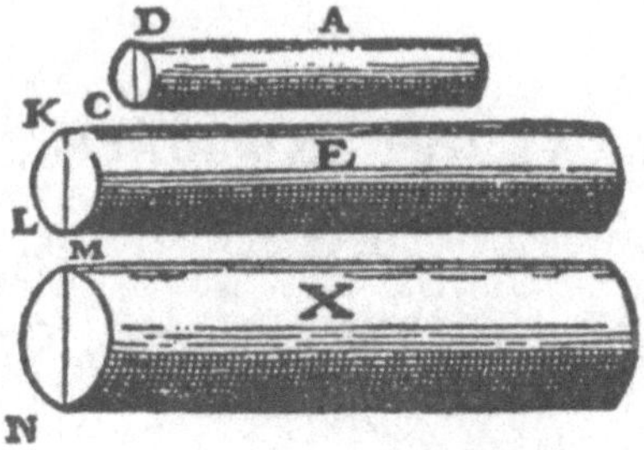

Fig. 1. Galileo's illustration of gravitational scaling (from his *Discorsi* of 1638). The small prototype is on top. Simplicio's solution for equal strength is below it and Galileo's correct solution (with relatively increased width) is at the bottom.

It is a general rule in the design of organisms that larger models must be of radically different shape than smaller prototypes in order to function in the same way. Many physical properties that organisms need to keep constant will change systematically with size if shape is not altered. If shape and density of building material do not change, for example, then Reynolds number will scale as length

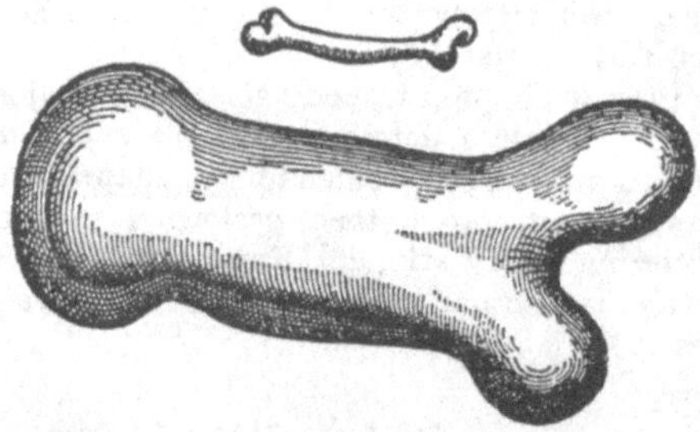

Fig. 2. From Galileo's *Discorsi* of 1638. Galileo recognized that large animals would have relatively thicker weight supporting bones than smaller prototypes.

and regimes of flow around larger objects will change drastically. Kinetic energy scales, depending upon the problem, between l^4 and l^5 [2]; the head of a falling child will hit the ground with, at most, 1/16 the kinetic energy of a parent twice as tall. A child's near imperviousness to head injury is a function of its small size, not its "soft" bones.

Yet, of all the differential scalings that affect the forms of organisms, none is so pervasive in its influence as the pheonomenon that Galileo recognized — the decline, as size increases, of the surface/volume ratio of similarly shaped objects. And, to turn finally to our subject, most of these effects arise because gravitational force increases as the surface/volume ratio drops. Some of the famous surface/

volume effects do not involve gravity *per se* — namely those in which the surfaces of large animals must increase differentially in order to support a growing body *volume* (not weight). Thus we, as large mammals, have villi on our small intestine in order to increase the absorptive area of a gut that must feed an entire body volume; small mammals do not have villi. And we, again as large vertebrates, possess lungs — little more than a convoluted bag of large surface area to process blood that must serve the entire body volume.

But the best recognized effects of the surface-volume principle are those related directly to gravity, i.e. those in which volume has its effect via increase in *weight*. Galileo's case of relatively thick legs in large terrestrial vertebrates remains the standard example; the necessary flightlessness of large birds is another. In fact, I want to argue that the main differences in design of small vs large organisms (particularly of external shape) are a direct response by natural selection to the rigid mechanical requirements of gravity's increasing influence upon large objects.

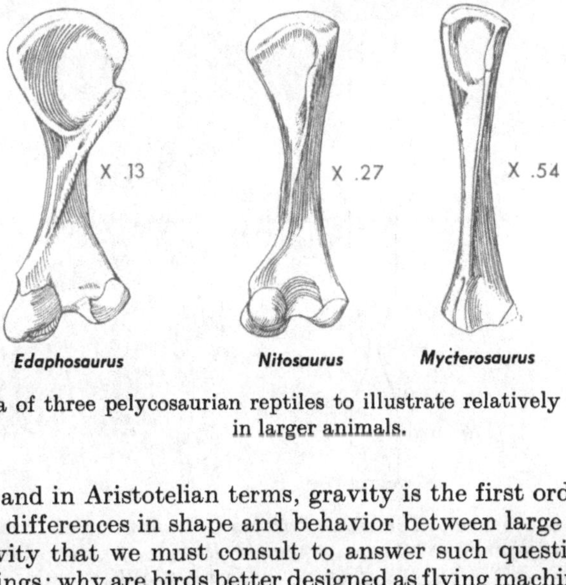

Fig. 3. Femora of three pelycosaurian reptiles to illustrate relatively thicker bones in larger animals.

In other words, and in Aristotelian terms, gravity is the first order formal cause of characteristic differences in shape and behavior between large and small organisms. It is gravity that we must consult to answer such questions as: why do flies walk on ceilings; why are birds better designed as flying machines than bumble bees; why can nothing as large as an elephant be shaped like a dachshund. And the influence of gravity in explaining the difference between small and large extends well beyond the organic; for it tells us why the small inner planets are cratered while the earth is not (retention of an eroding atmosphere by greater gravity), and it can explain, as I will argue later, why large medieval cathedrals are relatively narrow.

2. The First Theme: D'Arcy Thompson's Theory of Form: Different Worlds in Different Sizes

In *Them*, a well-known American science fiction film, giant ants fly hundreds of miles from the desert to the sewers of Los Angeles. The makers of the film, apparently unaware of scaling theory or the influence of gravity, simply scaled the flight ability of small ants to the size of their giants. But since wing surface *area* must supply lift to carry the ant's entire weight, it is quite obvious that such giants could never have got off the ground.

Small animals live in a world utterly different from our own and their form and behavior are adapted to their surroundings. An insect's world is ruled by forces acting upon its relatively enormous surface; gravity is of little consequence. Hence, a fly can walk on ceilings since surface adhesion is so much more powerful than the negligible gravitational force pulling it downward. But a small fly will also drown in a pool of water because the forces of surface tension cannot be overcome. Both the skills and the dangers are different. We, on the other hand, are denizens of a gravitational world. We fear the edges of cliffs (though a falling fly would merely float down uninjured), but we happily take our evening bath.

Nowhere has the theme of different worlds been better treated than in D'Arcy Thompson's classic of scientific prose, *On Growth and Form* [3]. Thompson's own

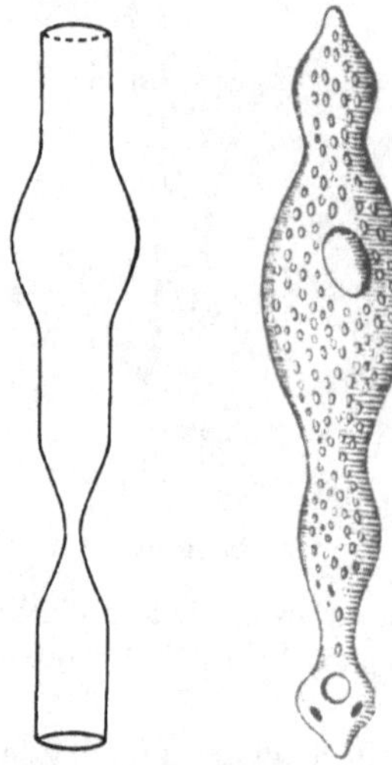

Fig. 4. Comparison between one of Plateau's surfaces of revolution (left, unduloid with positive and negative curvature) and the "flagellate 'monad' *Distigma proteus*". From D'Arcy Thompson [3], Chap. 5, the forms of cells.

theory of form was gloriously idiosyncratic and evidently wrong; yet his insights into the basis of morphology have no equal. He believed, in short, that physical forces *directly* shape the external forms of organisms. And this he attempted to prove by displaying how the impress of *gravity* becomes ever stronger as organisms get larger.

Very small organisms suffer no gravitational influence and are shaped by forces of surface tension alone. Thus Thompson (Fig. 4) compares protozoans with Plateau's surfaces of revolution (surfaces of minimal area radially symmetrical about an axis), and infers from the correspondence that surface tension shapes the simple, single-celled animal. In an intermediate realm of size, surficial and gravitational forces are in balance. Thus, a jellyfish (Fig. 5) is like a drop of viscous material in a less viscous medium: gravity pulls the drop down, but surficial forces retard the descent and spread out the drop in a radially symmetrical

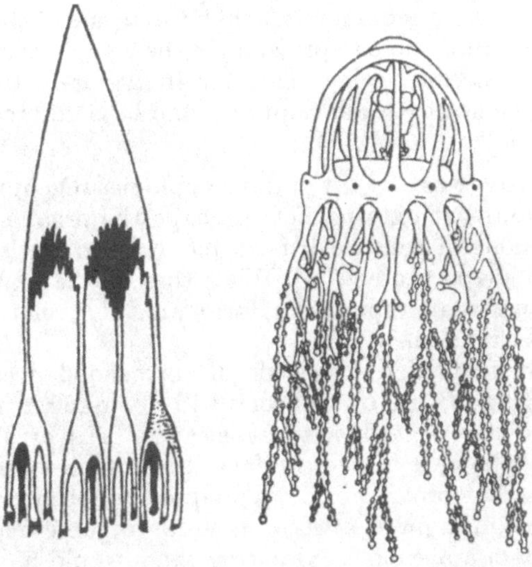

Fig. 5. D'Arcy Thompson's comparison between a falling drop produced by fusel oil in paraffin (left) and the medusoid "jellyfish" *Syncoryme* (right).

pattern. At still larger sizes, surface tension becomes so negligible that rigid hard parts are needed to maintain shape, lest gravity make a world of pancakes. In the internal trabeculae of vertebrate bones, D'Arcy Thompson found patterns that mirror the stresses imposed upon them by the body's weight (Fig. 6). He writes ([3], pp. 976—977):

> A great engineer ... happened (in the year 1866) to come into his colleague Meyer's dissecting room, where the anatomist was contemplating the section of a bone. The engineer, who had been busy designing a new and powerful crane, saw in a moment that the arrangement of the bony

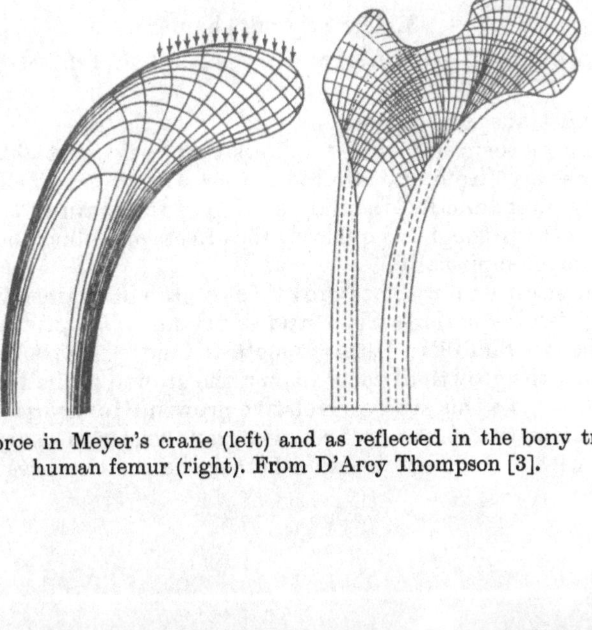

Fig. 6. Lines of force in Meyer's crane (left) and as reflected in the bony trabeculae of the human femur (right). From D'Arcy Thompson [3].

> trabeculae was nothing more nor less than a diagram of the lines of stress, or directions of tension and compression, in the loaded structure: in short, that Nature was strengthening the bone in precisely the manner and direction in which strength was required; and he is said to have cried out, "That's my crane".

Yet, though his prose was elegant and his analogies ingenious, D'Arcy Thompson's simplistic notion that external forces shape the organism cannot be maintained. Surface tensions of cells have been measured directly, for example, and they are too low to shape the cell [4]. What then is the reason for such strong correspondence between an organism's form and the relative strength of the gravitational force at its size?

D'Arcy Thompson identified hundreds of correspondences between physical laws and organic forms. If his own theory will not explain them, then another must be sought. And that other is natural selection. D'Arcy Thompson claimed a direct influence for physical forces; in fact, these forces operate indirectly by specifying the forms that provide optimal adaptation for animals subject to their influence. "Equilibrium figures are common in organic nature", Hutchinson writes, "because any organism not exhibiting them would have to be elaborately protected in other ways against deformation by the stresses and strains imposed by its environment" [5].

Aristotle's great student, in short, mixed up his causes. D'Arcy Thompson's error can be epitomized this way: He viewed physical forces as the efficient cause of form; they are, in fact, formal causes or blueprints of optimum shapes that determine the direction which natural selection (the true efficient cause) must take to produce adaptation. If physical forces are not the kind of cause D'Arcy Thompson thought, they are causes nonetheless and no explanation of form is complete without reference to them. He was right to correlate physical forces with organic forms and to claim that the correspondence was no mere analogy; but he was right for the wrong reason. D'Arcy Thompson thought he had a theory for the efficient cause of good design; he gave us instead the basis for a science of form, an analytic approach to adaptation.

3. The Second Theme: Allometry — Different Shapes for Different Sizes

If animals must adapt their form and habits to the strength of gravitational forces prevailing at their size, it also follows that they must change their shape if they grow or evolve from one size to another. Ever since Snell [6] and Dubois [7] applied the power function to study scaling of the brain in the 1890's, evolutionary biologists have sought to quantify the effects of scaling and to induce some of the laws of form empirically.

Conventional studies of absolute growth plot growth against time using Gompertzian or logistical equations. Such studies say much about size and very little about form. The late Sir Julian Huxley sought to relieve this problem in bypassing time and plotting the growth of parts against the growth of the whole organism — relative growth [8]. To this study of relative growth, Huxley gave the term allometry [9]. By allometry, students of relative growth refer to *changes* in shape that are correlated with differences in size, whether these differences be expressed in growth (ontogeny); in static sequences of individuals at a growth stage (all adults of a population, for example — intraspecific allometry); in static sequences of adults of different, though related, species (interspecific allometry); or in evolutionary sequences (phylogenetic allometry). If shape remains constant as size increases, we have the special condition of isometry. I must emphasize that the standard scaling parameters often differ dramatically for the different categories of allometry; extrapolation must not be made from one to another. For example, the interspecific exponent of brain-body allometry for mouse-to-elephant curves is 0.66; but it runs between 0.2 and 0.4 for intraspecific scaling of adults within a mammalian species. The ontogenetic curve is more complex, with rapid prenatal growth followed by near postnatal cessation [10].

The subject of allometry is chained to no specific form of mathematical expression, but it is one of the striking regularities of relative growth that the simple power function so often fits the data well. One reason for this may be that the predictions for gravitational influence upon form are so often expressed in power functions. Returning again to Galileo's example of relatively thick legs in large animals: in general, weights scale as $(\text{length})^3$ and areas as $(\text{length})^2$; but, in this case, to keep pace with increasing weight, cross-sectional leg area must also increase as $(\text{length})^3$, i.e. the leg must thicken differentially as the animal grows. Now, if we make a conventional plot of leg thickness vs body length, we may predict the form of equation from the previous observation:

$$(\text{cross-sectional area}) \propto (\text{length})^3$$

taking square roots

$$(\text{leg thickness}) \propto (\text{body length})^{1.5}$$

or

$$(\text{leg thickness}) = b\,(\text{body length})^{1.5}$$

which is a specific rendition of the generalized power function or "allometric equation"

$$y = bx^a$$

(where a is the slope and $\log b$ the y-intercept of a plot on logarithmic scales). Such power function plotting has long been the stock-in-trade of empirical studies in allometry. Fig. 7 shows just one example for a phyletic series illustrating increasing size in eight species of the fossil sail-backed reptile *Dimetrodon* [11]. Leg length is isometric with respect to body length, but leg width increases differentially with an exponent near the theoretically predicted value of 1.5.

In recent years, more attention has been paid to the multivariate representation of allometry; the abstration of an integrated morphology in bivariate pairs has always been theoretically unsatisfactory. Techniques are now available for displaying allometric change in any number of variables simultaneously [12—15]. Table 1, for example, shows factor loadings for 14 variables measured in 22 species of fossil pelycosaurian reptiles spanning a large range in size [11]. Bone widths (with the exception of the humerus) join together in a grouping with body weight (estimated). Bone lengths are all together in a second cluster with body length. Despite the very strong correlations of bone lengths with their own widths (r is usually above 0.95), the influence of weight (gravity) over all bones is so general that these correlations are broken to place lengths and widths together. Galileo's principle is confirmed with an entire matrix of data. Some support for this assertion comes from space related research on the growth of animals under chronic acceleration to produce hypergravity in centrifuges. Wunder [16] for mice and Smith and Burton [17] for chickens report that stimulation of skeletal development in such conditions is not restricted to weight bearing bones, but seems to affect the entire skeleton. Again, a general factor in skeletal development is indicated.

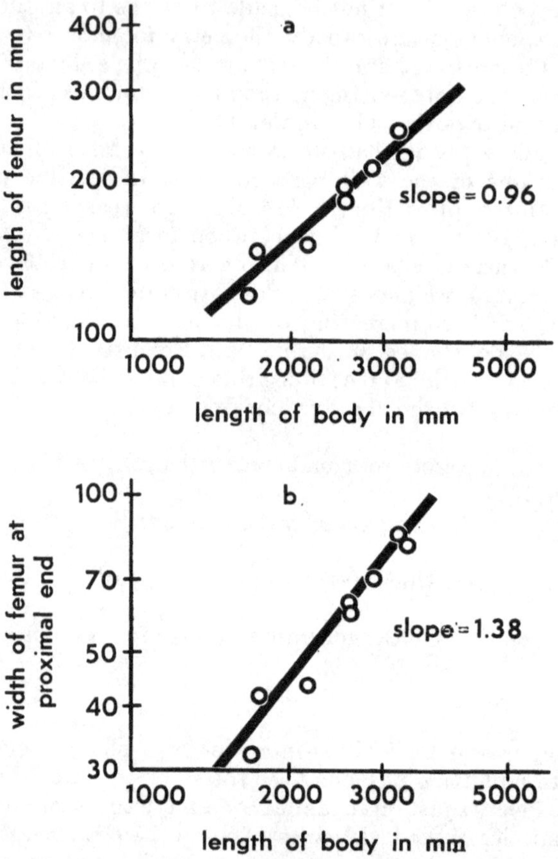

Fig. 7. Positive allometry of femoral thickness (with isometry of length) among eight species of the fossil reptile *Dimetrodon*.

We now know the general laws of scaling for most organs and parts of animals, at least for the higher vertebrate classes. This information is available in several classic sources [18, 19] and recent reviews on the subject of allometry [20—22]. Again, many of the regularities reflect the primary control of gravity as a formal cause in shaping the morphology of organisms.

I have already spoken much of Galileo's principle for leg thickness. But consider some of the other standard scalings that find their explanation in gravity. The height of trees scales to the 2/3 power of their width, as predicted from the theory of elastic buckling under gravitational forces [23]. Even mushrooms are affected in the same way as larger species have relatively thicker stalks [24]. McMahon has

Table 1

Recordered Oblique Projection Matrix for *R*-mode Postcranial Data of Pelycosaurs

Group	Variable	Femur distal width	Lumbar centrum length	Ilium neck width	Ilium base width
I	Femur distal width	1.000	0.000	0.000	0.000
	Body weight	0.854	0.104	−0.160	0.208
	Femur proximal width	0.839	0.060	0.152	−0.044
	Dorsal centrum width	0.775	0.081	−0.161	0.312
	Tibia width	0.587	0.180	0.191	0.053
	Lumbar centrum height	0.480	0.308	0.006	0.221
II	Lumbar centrum length	0.000	1.000	0.000	0.000
	Dorsal centrum length	−0.424	0.918	0.209	0.309
	Humerus length	−0.163	0.917	0.686	−0.422
	Body length	−0.200	0.644	0.453	0.121
	Humerus width	0.348	0.563	0.502	−0.395
	Femur length	0.260	0.518	0.512	−0.270
III	Ilium neck width	0.000	0.000	1.000	0.000
IV	Ilium base width	0.000	0.000	0.000	1.000

recently used the theory of elastic buckling and bending [23] to predict that in terrestrial organisms, general lengths should scale to the 1/4 power of body lengths and widths to the 3/8 power. For more than 3000 Holstein cattle measured by Brody [18], chest girth scales to the 0.36 power of weight while height at the withers scales to the 0.24 power. Stahl and Gummerson [25] for primates and McMahon [26] for ungulates have provided further confirmation. With widths increasing faster than lengths, large mammals have relatively shorter and wider trunks (elephants as long as dachshunds would collapse). Rashevsky [27] used the general theory of uniformly loaded beams (linearized beam bending) to explain this result. He remarked [27, p. 17] that since the shape of supporting extremities and trunk are both conditioned by size itself, gravity "molds" the basic external form of terrestrial quadrupeds.

Even the famous and mysterious 0.75 scaling of metabolism in mammals may prove to be a result of gravitational influences. It has long been appreciated that the 0.75 scaling must have something to do with surfaces (but why isn't the exponent near 0.66 in order to scale as the external surface?). McMahon [23] argues that metabolism should scale as body cross-sectional area (not as external surface) since it depends upon the power output of muscles which scale as their cross-sectional area or d^2. But generalized d scales as weight to the 3/8 power (see above, based on a gravitational argument related to elastic buckling), and d^2 therefore scales to the 0.75 power of body weight, the exponent of metabolism.

Going further into the area of biological frequencies: If organ weights scale as body weight (though they are usually negatively allometric, if slightly [28]), then since metabolism scales to the 0.75 power of weight, frequencies must scale to the −0.25 power (metabolic exponent/weight exponent). It is well known that many important biological frequencies scale to the −0.25 power of body weight (respiratory frequency for example) and this particular value, insofar as it relates to a gravity-conditioned scaling of metabolism, also reflects the pervasive influence of gravity in shaping the form and behavior of organisms.

4. Experimentation and the Control of Scaling through Growth

We do not need direct experimentation to affirm the effects of gravity. We have enough natural experiments to test its influence: Maximum size is smallest where gravitational constraints are strongest (flight) and largest where they are absent (neutrally buoyant organisms in water). Birds are better designed than bumble-bees because they are nearer the gravitational limit for flight (tiny insects can stay aloft with virtually any wing shape). Ciliary locomotion is always replaced by muscular movement above a certain small size because body weight increases so much faster than the number of cilia required to move it (dependent upon surface area).

The great value to evolutionary biology of research in hypo- and hypergravity now being carried out by space scientists lies in what it can teach us about the developmental basis of form. Of course, no one is a naive naturist or nurturist in these days. We dismiss D'Arcy Thompson's theory of pure environmental control as readily as we abandon any notion that genes rigidly determine the form of organisms. We all now speak only of reaction norms — the range (often rather wide) over which environment can channel genetic potential. Still, the empirical determination of reaction norms is of great theoretical interest and of extreme practical importance in any space program. Two recent developments may lead us to suspect that ranges in reaction norms might be characteristically wider than an earlier Mendelian age might have imagined. The development of a more cohesive "functional demand" theory of growth ([29] for example) leads us to suspect that altered conditions of gravity might have a profound effect upon developing form. Moreover, various computer simulations of growth in complex structures [30, 31] have led to the realization that intricate morphologies may be generated with rather few instructions and that even a small alteration in these instructions (perhaps conditioned by novel environments) can have a profound effect upon final form.

Fundamentally, since we have no adequate general theory of development (only the obvious assurance that nature and nurture are both relevant and inexorably intertwined), we must focus gravitational studies on direct experiment to test reaction norms empirically. We may not even extrapolate, as the centrifugationists are well aware, the effects of hypergravity back beyond the $1g$ point into the realm of hypogravity that constitutes the major area of interest for studies in the effects of weightlessness. Allometric theory is not well enough developed (or organisms are just basically too intractable) to permit such extrapolation.

I shall not review the copious literature on experimental alteration of gravity; it is your subject, not the province of this outsider. But I will remark that the literature, as I survey it cursorily, provides the panoply of results that most empirical experimentation conveys in its early stages and in the absence of a general theory — all possible effects seem to arise in studies of long-term exposure (including entire life-times) to hypergravity. Some expected results have not materialized: Oyama and his colleagues could find no consistent thickening of weight-bearing bones in rats exposed to long-term hypergravity [32, 33]. Other positive effects have been attributed to artificial conditions of the experiment: Amtmann and Oyama [33] suspect that alteration in femoral shape of rats is a result of immobilization in the centrifuge. Some gravitational effects are a direct mechanical result of centrifugation: Lim et al. [34] studied the saccules of rats exposed to long-term hypergravity and found that the otoconia redistributed themselves in the direction of the increased gravitational force. Other effects are more in line with expectations of the functional demand theory — positive changes in the direction predicted by Galileo's scaling principle are clearly not a mere mechanical result of spinning, but a growth response of the organism. Thus, Gray and Edwards [35] found that wheat coleoptiles grew straight to 100*g*, bent at higher accelerations and actually broke in two under their own weight in some cases. Thus, the Galilean limits of strength were clearly reached. Compensation was noted in the predicted direction as the coleoptiles increased in relative thickness and altered their shape from elliptical to more circular.

When predicted thickening of weight bearing bones has not been noted, the artificial conditions of experiment may be invoked. Thus, Oyama and his colleagues [32, 33] suggest that the lack of increased femoral thickness in centrifuged rats may simply result from their tendency to lie on their bellies under conditions of spinning. More natural conditions have usually yielded positive results. Tulloh and Romberg [36] draped young lambs with canvas rugs containing lead weights; the metacarpels grew thicker than in unweighted lambs. Since wheat plants were growing in their normal orientation in the centrifuge, positive results on this experiment may reflect the closer simulation of natural conditions.

Allometric theory may have one contribution to make towards these studies. Smith and Burton [17] have emphasized that large organisms generally tolerate increased gravity more poorly than small ones. The general reason for this result may lie in the decreased margins of safety that allometric results indicate for large organisms. It is a largely unnoted regularity in allometric studies that power-function exponents are usually in the right direction but rarely up to the expected magnitude. Leg thickness, for example, almost always undergoes relative increase in larger animals as it should, but the exponential value rarely reaches the predicted 1.5 (1.3—1.4 is more common). Small animals must be considerably "overbuilt" with large margins of safety. As part of the opportunism of evolutionary strategy, phyletic size increase has operated to conserve material and reduce margins of safety. This may present no trouble in natural conditions (behavioral adaptations towards more caution will suffice), but it may be fatal in an unanticipated centrifuge.

5. Epilogue

In 1322, the Norman central tower of Ely Cathedral collapsed. Alan de Walsingham, charged with the rebuilding, could not vault the width of 74 feet in conventional stone. Thus, he vaulted his famous octagon in wood. The same architect built the Lady chapel at Ely, spanning the width of 46 feet in stone, the widest gothic vault in England without intermediary support. Somewhere between 46 and 74 feet lay the gravitational limit for stone vaulting. These large cathedrals were 500 or so feet in length. Given the stringent limitations placed upon their widths by gravitational problems of stone vaults, these churches had to be relatively narrow. But small parish churches were relatively much wider: a fifty-foot church could easily bear a 20 foot vault. I have quantified this relationship for British Romanesque churches and have found a remarkable consistency in relative narrowing with increasing size [37]. The influence of gravity shapes organisms, planets and buildings as well. Goethe once remarked: „Es ist dafür gesorgt, daß die Bäume nicht in den Himmel wachsen." Trees cannot grow to heaven, men cannot be 10 feet tall and healthy; but then, all those upside-down people in Australia don't fall off the earth either. We shall count gravity among our blessings.

References

[1] Galileo Galilei, Discorsi e dimostrazioni matematiche, intorno a due nuove scienze, 1638.
[2] F. W. Went, Am. Scientist **56**, 400 (1968).
[3] D'Arcy Thompson, On Growth and Form, Cambridge Univ. Press, London 1942.
[4] J. T. Bonner, in: Growth and Form, Cambridge Univ. Press, London 1966 (p. 49).
[5] G. E. Hutchinson, Am. Scientist **36**, 581 (1948).
[6] O. Snell, Arch. Psychiat. Nerv Krankh. **23**, 436 (1891).
[7] E. Dubois, Bull. Mem. Soc. Anthrop. Paris **8**, 337 (1897).
[8] J. S. Huxley, Problems of Relative Growth, MacVeagh, London 1932.
[9] J. S. Huxley and G. Teissier, Nature, Lond. **137**, 780 (1936).
[10] H. J. Jerison, Evolution of the Brain and Intelligence, Academic Press, New York 1973.
[11] S. J. Gould, Evolution **21**, 385 (1966).
[12] P. Jolicoeur, Biometrics **19**, 497 (1963).
[13] J. E. Mosimann, J. Am. Statist. Assoc. **65**, 930 (1970).
[14] P. Sprent, Biometrics, 23 (1972).
[15] R. S. Corruccini, Syst. Zool. **21**, 375 (1972).
[16] C. C. Wunder, in: Gravity and the Organism, Univ. of Chicago Press, Chicago 1971 (p. 389).
[17] A. H. Smith and R. R. Burton, in: Gravity and the Organism, Univ. of Chicago Press, Chicago 1971 (p. 371).
[18] S. Brody, Bioenergetics and Growth, New York 1945.
[19] W. R. Stahl, Science **137**, 205 (1962).
[20] S. J. Gould, Biol. Rev. **41**, 587 (1966).
[21] S. J. Gould, in: Approaches to Primate Paleobiology, Karger, Basel 1975 (p. 244).
[22] A. G. Cock, Quart. Rev. Biol. **41**, 131 (1966).
[23] T. McMahon, Science **179**, 1201 (1973).
[24] C. T. Ingold, Trans. Br. Mycol. Soc. **29**, 108 (1946).
[25] W. R. Stahl and J. Y. Gummerson, Growth **31**, 21 (1967).
[26] T. McMahon, Am. Naturalist, in press (1975).
[27] N. Rashevsky, Mathematical Principles in Biology and their Applications, Springfield, Ohio 1961.
[28] L. von Bertalanffy, in: Fundamental Aspects of Normal and Malignant Growth, Amsterdam 1960 (p. 137).
[29] R. J. Goss, Adaptive Growth, Logos Press, London 1964.
[30] D. M. Raup, J. Paleontology **40**, 1178 (1966).
[31] S. J. Gould and M. Katz, Paleobiology **1**, 1 (1975).
[32] J. Oyama and B. Zeitman, Am. J. Physiol. **213**, 1305 (1967).
[33] E. Amtmann and J. Oyama, Z. Anat. Entwicklungsges. **139**, 307 (1973).
[34] D. S. Lim et al., Aerospace Med. **45**, 705 (1974).
[35] S. W. Gray and B. F. Edwards, in: Gravity and the Organism, Univ. of Chicago Press, Chicago 1971 (p. 341).
[36] N. M. Tulloh and B. Romberg, Nature **200**, 438 (1968).
[37] S. J. Gould, Syst. Zool. **22**, 401 (1973).

Life Sciences and Space Research XIV — Akademie-Verlag, Berlin 1976

GRAVITY AND EMBRYONIC DEVELOPMENT

R. S. Young

NASA Headquarters, Washington, D.C., USA

The relationship between the developing embryo (both plant and animal) and a gravitational field has long been contemplated. The difficulty in designing critical experiments on the surface of the earth because of its background of 1 g, has been an obstacle to a resolution of the problem. Biological responses to gravity (particularly in plants) are obvious in many cases; however, the influence of gravity as an environmental input to the developing embryo is not as obvious and has proven to be extremely difficult to define. In spite of this, over the years numerous attempts have been made using a variety of embryonic materials to come to grips with the role of gravity in development. Three research tools are available: the centrifuge, the clinostat, and the orbiting spacecraft. Experimental results are now available from all three sources. Some tenuous conclusions are drawn, and an attempt at a unifying theory on gravitational influence on embryonic development is made.

1. Introduction

Does gravity exert a definitive influence on embryogenesis? The literature on the effects of gravity on animal development has been reviewed recently by Pitts [1]. His review includes clinostat and centrifuge studies on amphibian and a variety of invertebrate eggs, as well as whole animal (turtles, mice, rats and chickens) studies on the centrifuge. It also briefly describes the orbital-weightless studies done to date. Rather than re-review these largely observational papers, it is the purpose of this paper to describe one of the fundamental problems in the field of embryogenesis (induction) and to propose that the use of weightlessness as an experiment research tool can contribute to the formulation of testable theories about the nature of certain cellular processes, such as induction and differentiation. Unfortunately, most early biological experiments flown in orbital spacecraft have been exploratory and observational in nature and have perhaps lacked the sophistication of design and rationale which the rigorous application of the "scientific method" demands. In fact, some have been poorly conceived and performed, to the extent that the "scientific community" has had a negative reaction to the scientific utility of space flight experimentation. The high cost of space flight research demands that a maximum effort be put into experimental design and background study and that rigorously testable hypotheses underlie flight experiments. Purely observational experiments are probably an ill-advised luxury in today's environment.

2. The Problem

I view the egg as a single cell containing sufficient information to develop into a complex multifunctional mass (the organism) upon proper stimulus (fertilization), which does not necessarily include the introduction of new genetic material. The processes by which the egg undergoes this transformation are enormously complex and only partially understood. As cells develop into a multicelled organism, they become morphologically and functionally different from one another even though they are provided with identical genomes. As Jacobson [2] points out, the mechanism of cell differentiation offers to the investigator some of the most provocative problems of modern biology. The mechanism by which many, if not all, animal embryos differentiate is called induction. Induction refers to the interaction between tissues in which one set of cells responds to the presence or juxtaposition of another to proceed on a course of development different from that which it would have followed if the interaction had not occurred. Spemann [3] postulated an organizer tissue in very early embryos which served as a primary inductor. This primary inductor has been traced in transplantation experiments by Curtis [4] to reside in the cortical cytoplasm of the gray crescent region of the undivided egg. Upon fertilization, visible changes occur in the cortical cytoplasm and the inductive properties appear to be quickly spread through the cortex, thus being imparted to the daughter cells in subsequent divisions. The physical-chemical nature of the inductor is not understood. It appears that at least some of the response and inducing capacities are formed during oogenesis, but the unfertilized egg is morphologically dormant until triggered by the act of fertilization. Most embryonic induction studies have been done with amphibian material and it is such work that will be concentrated on in this paper.

The question is, does gravity play a determinative role in the process by which the fertilized egg juxtaposes certain cells or intracellular material in such a way as to determine irrevocably, or at least temporarily, the future role of those cells? The morphogenetic movements which bring inductor and reacting tissues together with such precision must be accounted for in any satisfactory explanation of embryogenesis.

3. The Evidence

3.1. Centrifuge

Gravity has long been suspected of playing a role in development. Most experimental work has been done on the amphibian egg, probably because of its availability and ease of handling in the laboratory and because of its obvious response to gravity. The amphibian egg is telolecithal, that is, the heavier yolk is located in the vegetal hemisphere, which in turn forces the nucleus and much of the cytoplasm into the other (animal) hemisphere. The animal hemisphere then is characterized by its darkly pigmented cytoplasm. The line running between the two poles is the polar axis which is uppermost at the animal pole and lowest at the vegetal pole. During oogenesis in the frog then, the eggs are already oriented with respect to gravity and polarity and metabolic gradients established thereby.

Mature eggs, removed by artificial means from the frog are randomly oriented with respect to gravity and become "frozen" in position rapidly by the imbibition of water by the surrounding jelly mass secreted with the eggs. When fertilized, the first sperm to penetrate the jelly and the vitelline membrane of the egg triggers a number of rapid responses. First the vitelline membrane is elevated and becomes the fertilization membrane, which prevents the entrance of any more sperm. This membrane lifting creates a perivitelline space within which the egg may rotate. At the same time there are metabolic changes, and increase in cytoplasmic viscosity and membrane permeability. Migration of cortical material is also seen at this time. The most striking events, however, are the onset of rotation of the egg so that it becomes reoriented with respect to gravity in a matter of minutes, and the movement of cortical cytoplasm and pigment granules toward the point of sperm penetration with the concurrent appearance of a crescent-shaped area where the pigment is less dense (gray crescent) opposite the point of sperm entry. At this time bilateral symmetry is established.

This rotation of the frog egg was probably the original stimulus for studies on the role of gravity in development. Schultze [5] held fertilized frog eggs between two glass slides in the inverted position and found that a large number of abnormalities developed including a fair proportion of twins. He noted that these results were seen primarily when the experiment was performed at around the two-cell stage. Penners and Schleip [6] extended Schultze's experiments in some detail and with similar results. Tschou-Su [7] reported twinning and other abnormalities in frog eggs contrifuged before cleavage, as did Kas'yanov [8]. At the same time, Young et al. [9] using *Xenopus* eggs showed that the sensitivity of these eggs to gravitational influence was very time dependent (Table 1). In these studies several

Table 1

Xenopus Embryos centrifuged at 40g for 15 min

Time (in min) after Fertilization	Normal	Abnormal	Death	Twins (% live)
15–35	25	60	15	41
30–50	40	25	35	0
45–65	25	15	60	0
60–80	20	15	65	0

experiment protocols were used. Fertilized eggs were placed on the centrifuge within five minutes of fertilization and runs made at 500g for $1^1/_2$ min, and 20 and 40g for 1 hour 15 min respectively. Runs made at 20g for 1 hour or more were lethal regardless of the developmental stage up to the 8–16 cell stage when significant survival was observed. Abnormalities in general decreased in these experiments as the age of the embryos used increased, dropping to very low levels by the early blastula stage. Runs made at 40g for 14 min, showed that these eggs were most sensitive to gravity between fertilization and 35 min after fertilization. This time period was also the only one in which twins were produced. By 40 min after fertilization the eggs were relatively insensitive and a high percentage of normal embryos resulted. At the time of first cleavage a second period of sensitivity was seen with many abnormalities produced, not including twinning. 500g for $1^1/_2$ min produced essentially the same results. After first cleavage the amphibian egg becomes progressively less sensitive to gravitational influence. These experiments were run at 22 °C at which temperature first cell division occurred at 70–75 min.

3.2. Clinostat

In addition to centrifuge studies a considerable effort has been put into clinostat research, particularly with plant material and also with amphibian eggs. The clinostat has the capability of systematically and continuously disorienting the organism with respect to gravitational stimuli by slow rotation. Early studies [10—12] showed no increase in abnormality as a result of slow rotation of frog eggs. On the other hand other workers [13, 14] found increases in abnormalities. Tremor and Souza [15] in detailed clinostat studies found abnormalities in *Rana pipiens* and *Xenopus laevis* eggs and determined that the speed of rotation was a critical factor. The eggs must be rotated at a speed which was properly suited to compensate for the normal rotational speed of the egg. These workers found that rotation at 1/2 rev min^{-1} was optimal for the production of abnormalities in *R. pipiens*, and 1/4, 1 and 2 rev min^{-1} were most effective for *X. laevis* which has a faster rate of rotation. Rates greater than 2 rev min^{-1} were not effective in producing abnormalities. They also found that in *R. pipiens* the period of greatest sensitivity was before the two-cell stage. It was concluded that at higher rotational speeds, the eggs remained in a fixed position with respect to gravity since the directional force of gravity at any point in time was, in effect, counteracted at another. At lower than optimum speeds, the eggs were able to rotate and maintain a normal "yolk-down" position with respect to gravity, at least through first cleavage. The centrifugal force in such experiments is very low ($10^{-2}—10^{-3}g$). Thus, although the mechanism is different, the displacement of cytoplasmic material in eggs maintained in a position abnormal to the earth's field by slow rotation can be compared, in effect, with that obtained by high speed centrifugation. The key to such experiments is in rotational timing. in the case of the clinostat, to speed and duration of centrifugation and in both cases to the stage of egg development at which treatment begins. This latter observation is consistent with the centrifuge studies of Banta and Gortner [16].

3.3. Weightlessness

Yet another type of experiment has been performed in orbital space flight by Young and Tremor [17, 18]. In these experiments, near weightlessness was experienced for 4 days in the Gemini 8 and 12 flights and for over 2 days in Biosatellite 2. In all cases, eggs of *R. pipiens* were used and the effects of orbital flight on cell division and subsequent embryogenesis were studied. From such experiments it was concluded that a gravitational field is not necessary for the egg to divide normally, nor is it necessary for differentiation and morphogenesis. In the Gemini 12 flight normal swimming tadpoles were recovered. Unfortunately, in all these cases, the eggs were fertilized on the ground and did not experience the weightless condition until after the 2-cell stage, which virtually all ground-based studies show to be critical. Such results are consistent with the studies of Vinnikov et al. (19) on the vestibular development of *R. temporaria* in weightlessness. Here too, normal development was observed, but only later stages of development were exposed to orbital flight. To date, technical difficulties in experimental implementation have made it impossible to do the crucial experiments, namely, getting embryonic material into the weightless condition before fertilization so that all phases of development are devoid of gravitational input.

4. Discussion

The experimental observations briefly described above seem to suggest rather strongly that embryonic material is especially sensitive to the influence of gravity during the earliest stages of development, including oogenesis, see, for example, the work of Kochav and Eyal-Gilali [20]. By the time of first cleavage, the embryo has become much less responsive to gravitational input although, of course, physical damage can be done to the developing embryo with very high *g* loading. The centrifuge, clinostat, and weightless results seem quite consistent with the concept of induction. Developmental programming begins during oogenesis when the cytoplasm of the egg apparently becomes regionally differentiated to some degree, depending on the egg. At fertilization, dynamic processes are triggered which further program the fate of the various regions of the cytoplasm, so that by the time of first cleavage the two daughter cells are already differentiated to a degree and a great deal of the future history of the embryo is imprinted. Organs arise as a result of such early determination and the process of induction. Steinberg [21] feels that the mechanism underlying morphogenetic movements which bring inductor and reactive tissues together may be a simple hierarchy of adhesives in a number of randomly distributed adhesive sites on cell surfaces. Certainly, the first visible suggestion of induction is seen in a behavioral change in the reacting cells, that is, shaping or clumping or a variety of movements. In most systems in which the chemical basis of induction was studied, it has been found that most determination has already occurred before the study began, usually before cleavage. As development proceeds with time, the responding cells become more and more firmly committed to a particular course of differentiation, perhaps ultimately with the production of cell-specific proteins.

There would seem to be sufficient experimental evidence to suggest very strongly that an embryo, after a few cell divisions, has already been sufficiently differentiated so that external influences will have little, if any, effect on subsequent morphogenesis. Gravity seems to play no determinative role after a few cleavages as illustrated in centrifuge, clinostat and space flight experiments. In fact, if determination occurs as early as most data indicate, one should not expect a gravitational influence on later developmental processes and continued attempts to fly such experiments in spacecraft may indeed be wasteful.

On the other hand, there is clearly a gravitational influence on morphogenesis before first cleavage, with perhaps the most sensitive period being shortly after fertilization, at least in the amphibian egg. It is also clear that there are considerable differences between eggs of different organisms. These differences may well be due to differences in the structure of the egg as it is formed in the ovary or later in the oviduct. The amphibian egg has a great deal of yolk in the cytoplasm. This yolk is of greater density and responds accordingly to gravity. Thus this type of egg is gravity oriented from the beginning. This is also true of the avian egg where bilateral symmetry is established by the gravitational forcing of the egg (and blastodisc) into an oblique position in the uterus [20].

It seems safe to assume that differences in density of cytoplasmic constituents is responsible, during oogenesis, for the earliest events related to subsequent differentiation. A localization of certain materials in the cortical cytoplasm of the egg as a response to gravity should be expected. That such localizations exist can be seen from early studies [22—24]. In some cases (e.g. extremely teleolecithal eggs) such localizations are already sufficient to serve as determinants, in a limited sense, for future development (e.g. bilateral symmetry in the chick egg), however such early determination can be altered in many cases by external means. At the time of fertilization, a mechanical stimulus (e.g. sperm entry) is the event which actuates the system now poised for a rapid sequence of physical-chemical changes (Fig. 1). By inducing changes in solubility, permeability and viscosity of egg components, fertilization causes the egg to respond very quickly to external input (e.g. gravity) again. This response is very complex, but among other things, cytoplasmic material undergoes chemical reaction and change and is relocated in the egg, very likely in response to gravitational input. At this time the egg is most susceptible to external input (centrifuge, clinostat, weightlessness, etc.) and is very labile. Once the reaction to the stimulus of fertilization is complete the situation tends toward stability once more, and considerable stability is evident by the time of first cell division. By this time presumptive material has been relocated and has begun its job of induction on receptor material and the process of determination becomes less and less reversible or labile.

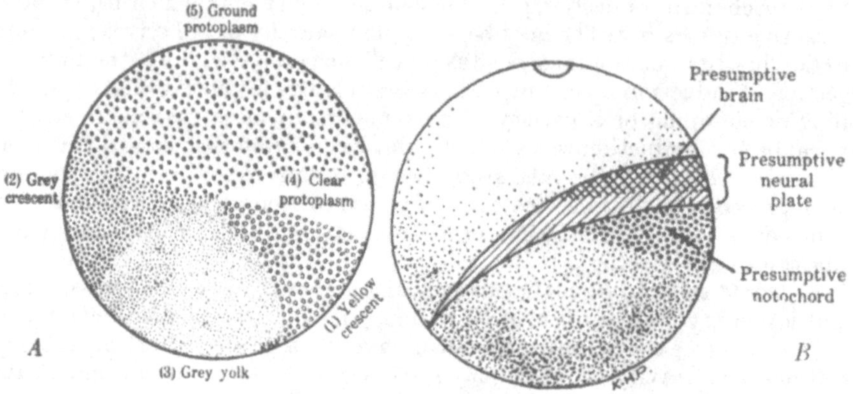

Fig. 1. Organ-forming areas in tunicate eggs. A, fertilized egg of *Styela*, from left side (*after* Canklin); B, unfertilized egg of *Ascidiella*, from right side (*after* Vandebroek, in Dalcq). Approx ×250.

It should be pointed out that the time schedule for transition from lability to stability in the process of induction will vary from organism to organism and will very likely be found to be related to density differences within the egg. In this context, it would be interesting to re-study the literature on echinoderm development, since these eggs have relatively little yolk. Experiments on *Ascaris*, such as those of Schatz and Teuchert [25], should also be reconciled with these concepts.

I would suggest that weightless flight offers a unique opportunity to test the hypothesis that gravity indeed plays a determinative role in embryogenesis, but that experiments to date have not had the proper design to answer the question, in that flight experiments to date have not placed material into the weightless condition well before first cell division. I would further suggest that the whole animal (frog, etc.) should be used so that the process of oogenesis can also be experienced in the virtual absence of gravitational stimuli. A method should be devised to go through the entire gamut of development from oogenesis, ovulation, fertilization, to subsequent embryogenesis. It is not clear that centrifuge and more particularly clinostat studies can be considered adequate substitutes for weightlessness, but in any case, both centrifuge and clinostat studies should be extended backward to include the period of oogenesis to complete the picture, if such experiments are feasible. The feasibility of flight of the whole animal (*Xenopus*) has been explored by the author. The flight of the whole animal, with the induction of ovulation and fertilization appears to be possible, but very difficult. On the other hand, such an experiment has the potential of resolving one of the fundamental problems of developmental biology.

References

[1] G. C. Pitts, Life Science and Space Research XI, 171 (1973).
[2] A. G. Jacobson, Science **152**, 25 (1966).
[3] H. Spemann, Zool. Jahrb. Abt. Allgem. Zool. Physiol. Tiere **32**, 1 (1912).
[4] A. S. G. Curtis, Endeavour **22**, 134 (1963).
[5] O. Schultze, Arch. Entwicklungsmech. Organismen **1**, 269 (1894).
[6] A. Penners and W. Schleip, Z. Wiss-Zool. **130**, 305 and **131**, 1 (1928).
[7] Tchou-Su, Comptes Rend. Soc. Biol. **122**, 1043 (1936).
[8] V. L. Kas'yanov, Dokl. Akad. Nauk. SSSR **180**, 1008 (1968).
[9] R. S. Young, P. Deal, K. Souza and O. Whitfield, Exper. Cell. Res. **59**, 267 (1970).
[10] W. Roux, Bresl. Aertzt. Z. **6**, 57 (1884).
[11] C. Kathariner, Archiv. Entwick.-Mech. **12**, 597 (1901).
[12] T. H. Morgan, Anat. Anz. **21**, 313 (1902).
[13] O. Schultze, Phys. Med. Ges. **4**, 41 (1897).
[14] S. Gordon, personal communication (1969).
[15] J. W. Tremor and K. A. Souza, Space Life Sci. **3**, 179 (1972).
[16] A. M. Banta and R. A. Gortner, J. Exp. Zool. **17**, 433 (1915).
[17] R. S. Young and J. W. Tremor, Life Sciences and Space Research VI, 87 (1968).
[18] R. S. Young and J. W. Tremor, Bioscience **18**, 609 (1968).
[10] Y. A. Vinnikov et al., Z. Evol. Biok. i Fiziol. **8**, 343 (1972).
[20] S. Kochav and H. Eyal-Gilali, Science **171**, 1027 (1971).
[21] M. Steinberg, in: Cellular Membranes in Development, Academic Press, New York 1964 (p. 321).
[22] P. Vintemberger and J. Clavert, C.R. Soc. Biol. Paris **154**, 1072 (1960).
[23] P. Ancel and P. Vintemberger, Bull Biol. Suppl. **31**, 177 (1948).
[24] J. Pasteels, Arch. Biol. **52**, 321 (1941).
[25] A. Schatz and G. Teuchert, Aerospace Med. **43**, 614 (1972).

Life Sciences and Space Research XIV — Akademie-Verlag, Berlin 1976

PHYSIOLOGICAL EFFECTS OF SUSTAINED ACCELERATION

L. H. Vogt

DFVLR Institut für Flugmedizin, Bonn-Bad Godesberg, FRG

A review is given of the literature concerning the physiological and pathophysiological changes in humans, caused by sustained acceleration. After definition of the acting forces and an introduction into terminology, circulatory and respiratory mechanisms are described which are active under sustained acceleration. The origin of visual disturbances associated with acceleration is discussed. Acceleration tolerance is influenced by magnitude, duration, direction, and rate of application of G-forces together with environmental conditions and the condition of the subject. Acceleration protection may be achieved by technical devices (anti-g suits), voluntary manoeuvres and change of posture. Recent work dealing with high sustained $+G_z$ (HSG) is reviewed and the associated problems are discussed.

1. Introduction

Sustained acceleration has been part of the environment of aviators and astronauts for many years. Research in this field has been stimulated by new technologies and challenges like space flight, and recently by the development of aircraft with low wing-loading and high thrust-to-weight ratios. Since the start of acceleration research in the early thirties several thousand articles have been published on this and it is the objective of this paper to review some of them.

1.1. Definitions

Accelerations are vectors which have to be referenced to the axes of the human body. The unit for acceleration in aerospace physiology is G. ($1G = 0.9807$ m s^{-2}.) It represents the total reactive force divided by the body mass. The axes therefore display the reactive displacements of organs within the body with respect to the skeleton. According to this the Z axis is down the spine, with $+G_z$ designations for accelerations causing the internal organs to displace footwards. In the same way the X axis runs from front to back and the Y axis from right to left. In a similar manner angular accelerations are defined by the three body axes. The term sustained acceleration is not clearly defined. In this presentation it will describe an acceleration pulse with a plateau duration of 15 sec or longer. This figure is rather arbitrarily chosen but is in good agreement with procedures used in various laboratories.

This presentation will only cover the effects of accelerations acting along the positive Z axis and the positive X axis and combinations of these directions. They produce the major problems in aviation and space flight.

2. Tolerance Criteria

If man is subjected to increasing $+G_z$ a sequence of symptoms develops in a predetermined order. Some of them are of visual origin involving various stages of light loss. They are known as peripheral light loss, central light dim and central light loss or "black-out". Together with other criteria they have been used as end-points to establish human tolerance to sustained acceleration [1].

3. The Effects of Sustained Acceleration on Circulation

The mechanism causing black-out has been intensively investigated since Andina [2] in 1937 suggested an explanation when he found that similar symptoms could be produced by applying positive pressure to the eyes with a tonometer. Later (1946—1954) other authors [3—5] could confirm that failure of retinal blood flow is the cause of central light loss. They showed that pulsations occurred in the arterioles of the retina shortly before black-out. Loss of vision was accompanied by emptying and collapse of these vessels. For these reasons the first decades of acceleration research were devoted to circulatory physiology.

The changes in arterial blood pressure under $+G_z$ were investigated by Ranke [6] in Europe and Lambert and Wood [7] in the United States, to name just a few. Their findings may be summarized by way of a simple circulation model [8]: If the blood pressure at heart level is kept constant at 115 mmHg, the pressure 30 cm up (that is the vertical distance between the heart and the head in a seated human) is reduced by the hydrostatic pressure gradient to 92 mmHg at $1G$, and to 23 mmHg at $4G$. In this case the blood pressure at eye level will equal the intraocular tension provided that the original pressure of 115 mmHg remains constant. Retinal blood flow will stop and central light loss occurs. If acceleration increases to $+5G_z$ blood pressure at head level will drop to zero which would theoretically stop blood flow in the brain, causing unconsciousness. This pressure distribution in the arterial tree can only be seen during accelerations with a rapid rate of onset during the first seconds of exposure. After that time it is changed by compensatory reflexes from the circulatory system.

Associated with the arterial pressure drop is a change in the distribution of circulating blood volume [10]. By the increasing hydrostatic pressure in the lower parts of the body, the circulating blood volume is reduced by pooling into dependent limbs, the splanchnic area and into the lower parts of the lung. There is still some argument about the exact amount of blood pooling into these areas. Since the available literature dealing with this question quotes various experimental conditions like rate of onset, maximum acceleration, and time spent at maximum G it is impossible to compare findings. However the results indicate that two stages of volume change of the legs can be distinguished as shown by Fig. 1 [5]. The first stage is a rapid increase together with the applied acceleration and is due to pooling, while a second, slower stage, is attributable to filtration of evasculated fluid. The falling blood pressure is to a certain degree compensated by reflex

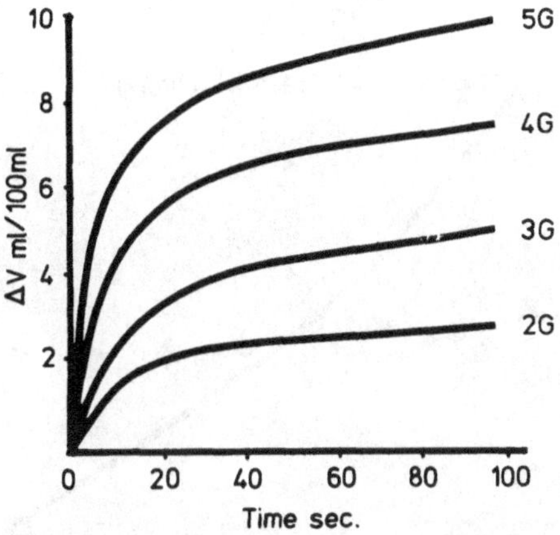

Fig. 1. Changes in calf volume with acceleration (redrawn after [5]).

activity originating from the receptors located in the aortic arch and the carotid artery. They cause an increase in heart rate and peripheral resistance.

Fig. 2 [9] shows arterial blood pressure, venous pressure of the jugular bulb, and acceleration plotted against time. Arterial pressure drops with increasing acceleration and recovers to a certain degree after plateau acceleration is reached.

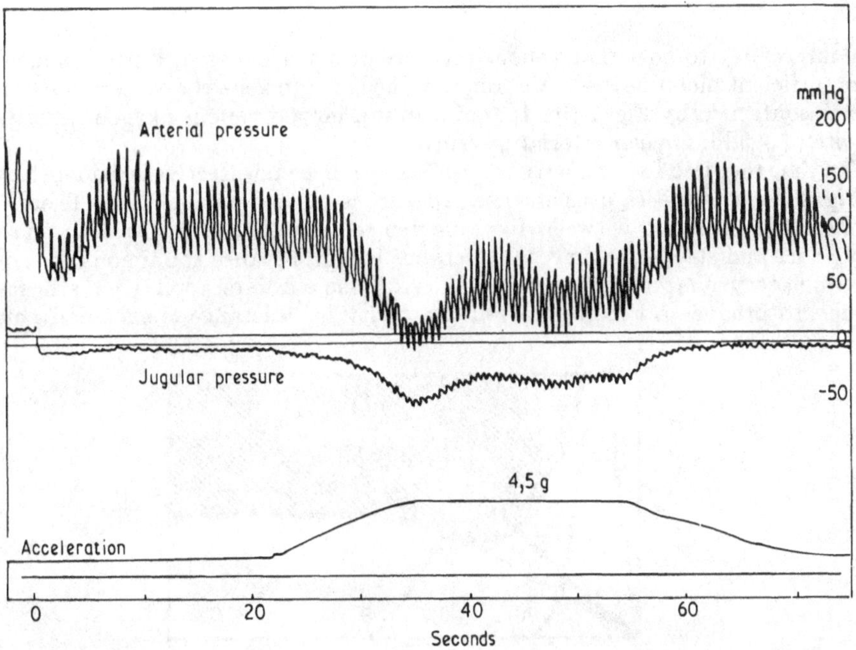

Fig. 2. Arterial pressure at head level and pressure in the jugular venous bulb under $+G_z$ acceleration exposure [9].

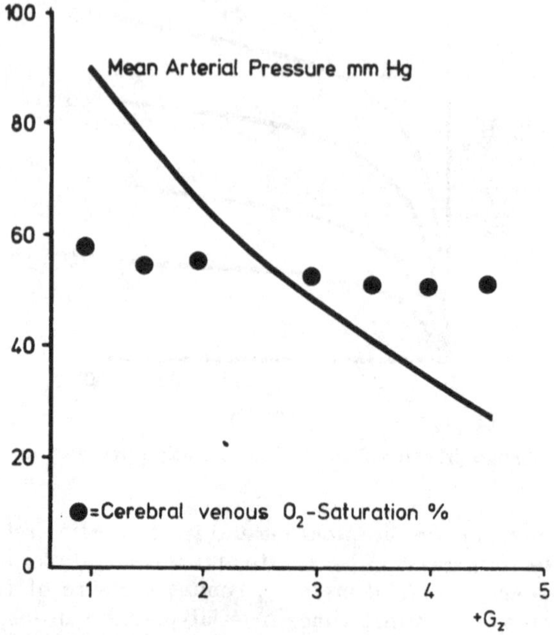

Fig. 3. Mean arterial blood pressure and cerebral venous oxygen saturation under $+G_z$ exposure (redrawn after [9]).

It is interesting to note that venous pressure drops below zero. With this mechanism sufficient blood flow is maintained in the brain to keep the subject conscious. This is confirmed by Fig. 3 [9]. It shows an unchanged venous oxygen saturation despite of a falling mean arterial pressure.

The time required for compensatory reflexes to become effective is demonstrated in Fig. 4 [9] where $+G_z$ acceleration exposure is plotted against time. There is a minimum in tolerance between five and ten seconds. If the acceleration pulse is very short and steep like curve A, tolerance is high because the time is too short for circulatory symptoms to develop. Curve B has a rate of application still steep enough to produce a time lag in reflex activity, so tolerance is exceeded before

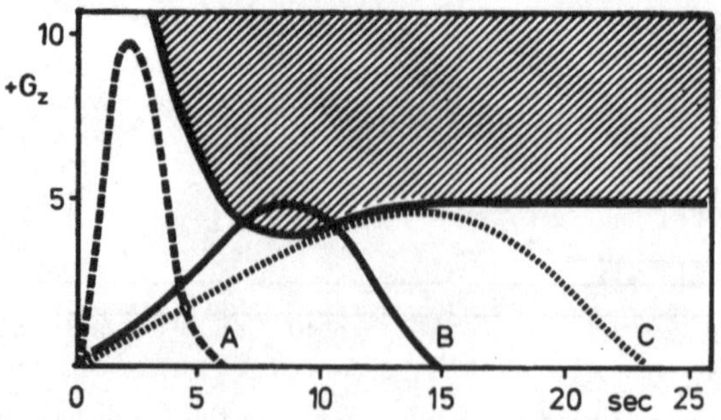

Fig. 4. G-time tolerance for upright posture (redrawn after [9]).

compensatory action can be developed by the body. Curve C rises slowly to allow full circulatory compensation at every moment during the exposure. This is in very short terms and very simplified form the circulatory mechanism that determines man's tolerance to sustained acceleration. According to Meehan [11] the mean value for unprotected relaxed $+G_z$ acceleration tolerance for black-out

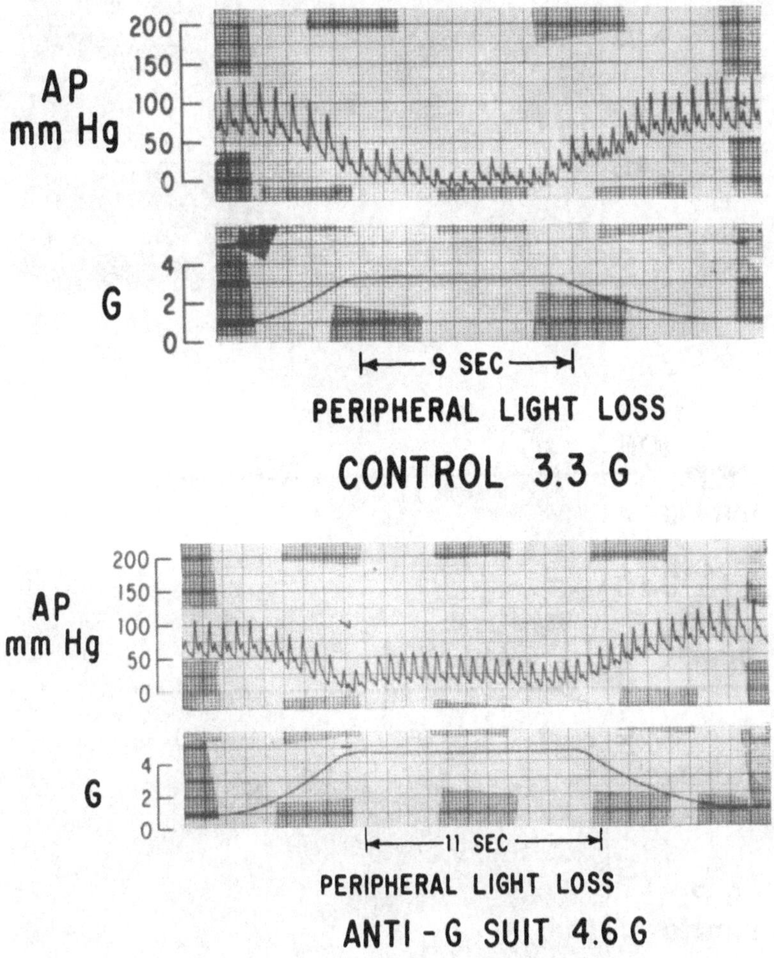

Fig. 5. The effect of an anti-*g* suit on arterial eye level pressure [13] (by permission of Aerospace Medicine).

is $+4.7G_z$. However, in 1972 Parkhurst and coworkers published a paper entitled: "Human tolerance to high sustained $+G_z$ acceleration (HSG)" [12] which stated that the majority of their subjects were able to tolerate $+9G_z$ for 45 seconds. He defined HSG as a positive acceleration force exposure of $+6G_z$ and greater for periods longer than 15 seconds. This extreme environment can only be tolerated by man if he takes certain protective measures; they are (a) the use of mechanical devices like the anti-*g* suit and positive pressure breathing, (b) voluntary acts

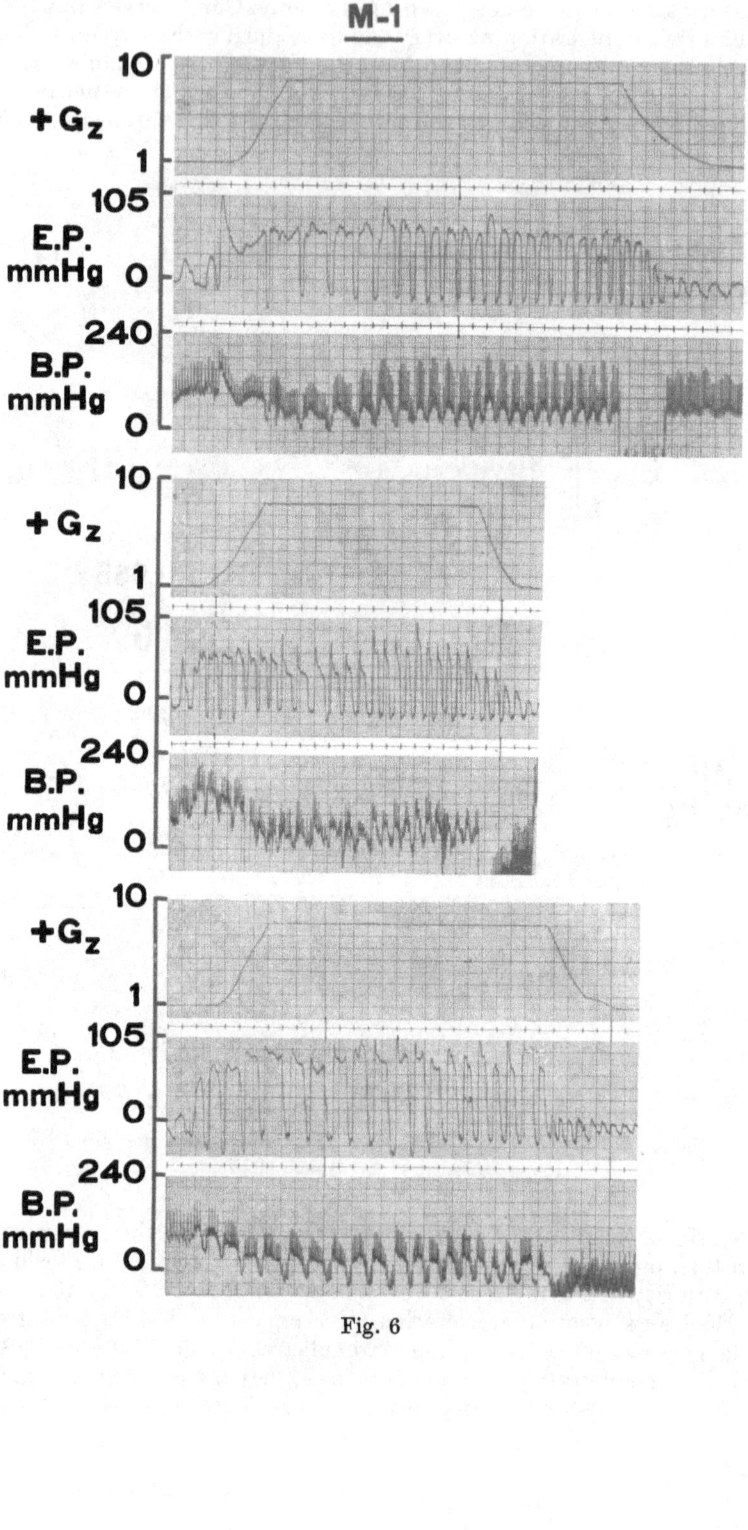
M-1
10
+Gz
1
105
E.P.
mmHg
0
240
B.P.
mmHg
0
10
+Gz
1
105
E.P.
mmHg
0
240
B.P.
mmHg
0
10
+Gz
1
105
E.P.
mmHg
0
240
B.P.
mmHg
0

Fig. 6

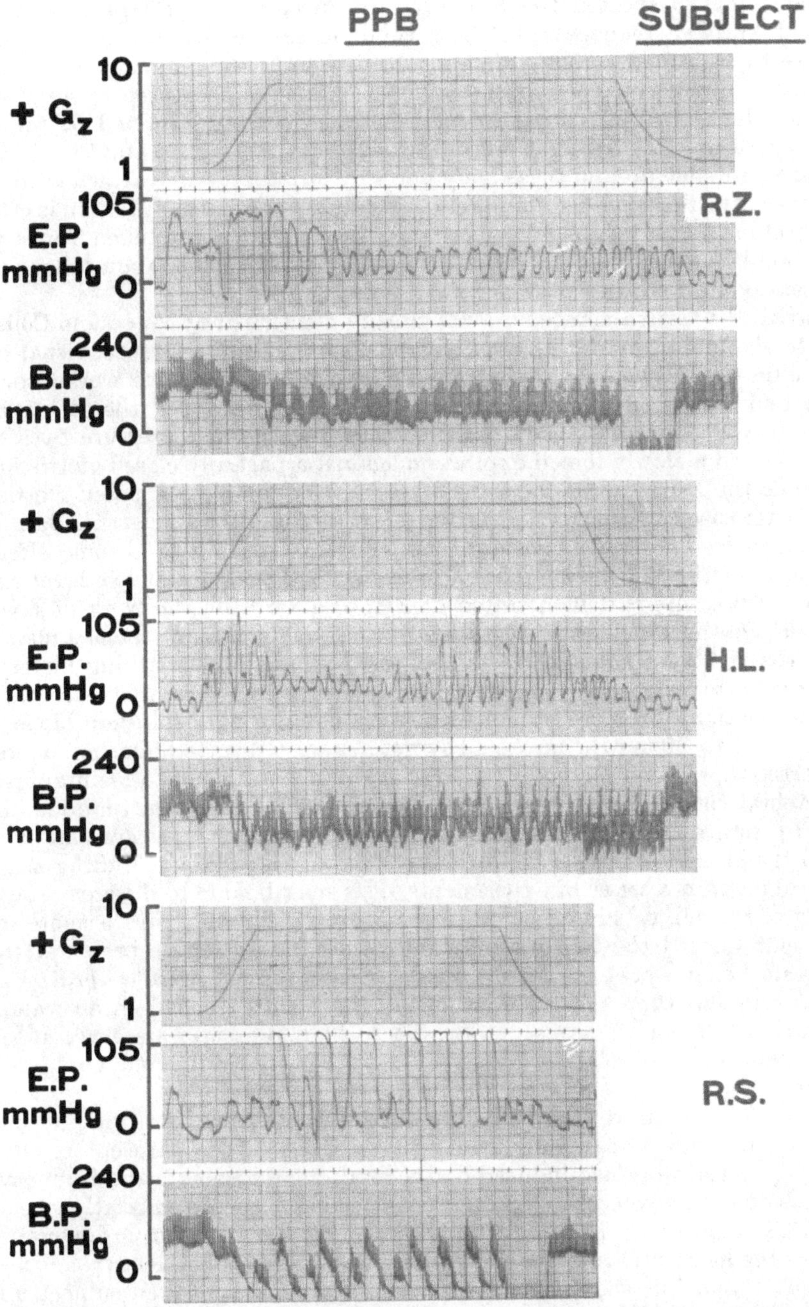

Fig. 6. Comparison of eye level arterial pressure and esophageal pressure from 3 subjects exposed to $+8G_z$, while using either the M 1 or positive pressure breathing [15] (by permission of Aerospace Medicine).

like crouching to shorten the heart-eye distance, muscular straining and the modified Valsalva manoeuvres. These countermeasures change the physiologic response to sustained acceleration considerably from the sequence described for relaxed and unprotected G-exposure. Anti-g suits have been used by aviators for some decades as a countermeasure against the effects of sustained G. An anti-g suit is a garment made of tight fabric enclosing the lower limbs and the abdomen. They are tightened around these body parts by inflatable tubes attached to them, thus preventing the pooling of fluid into dependent body parts. Inflation is effected by an acceleration sensitive valve located in the aircraft. Fig. 5 taken from a paper by Shubrooks and Leverett [13] indicates that by use of an anti-g suit tolerance can be raised by approximately $2G$ units.

A modified Valsalva manoeuvre was already mentioned by Wood and Code [14] in 1947, who found that by its use before and during $+G_z$ exposure, visual symptoms could be delayed. Since that time many researchers have worked on this matter and suggested modifications. Today two manoeuvres, the M 1 and the L 1 are used by aviators. They both increase intrathoracic pressure by a rapid inhalation and a slowly forced expiration against a partially closed glottis for the M 1, while the L 1 requires the same act against a completely closed glottis. So, in the latter case the gasp period includes exhalation plus inhalation. Both manoeuvres require the crouch position and muscular tensing to become effective. The effects of the M 1 manoeuvre on arterial blood pressure at eye level and on esophageal pressure is demonstrated in Fig. 6 taken from the work of Leverett et al. [15]. On the left side are recordings of three subjects using the M 1 manoeuvre; on the right the same subjects use positive pressure breathing instead. In both cases arterial pressure fluctuates with esophageal pressure. With the M 1 these fluctuations are more pronounced. Although during inspiration blood pressure dropped to 20 mmHg no visual symptoms were reported. During expiration mean arterial pressure stayed at 90 mmHg which is sufficient to maintain retinal and cerebral circulation. Using positive pressure breathing, fluctuations in esophageal pressure are less pronounced. Arterial pressure changes are half as great as with the M 1. However, as indicated in Fig. 6, great individual differences are not uncommon in a series of experiments. This may be due to different stages of training of the subjects in the pressure breathing technique. Furthermore in the experiment quoted the less trained subjects used less positive pressure. In the same tests venous pressure at the ankle of the subject were recorded. They reveal a similar time relationship as in Fig. 1 but reached mean values of 380 mmHg after one minute at $+8G_z$. These high pressures were accompanied by leg pain but no relation could be found between pain and ankle venous pressure.

During high sustained G heart rates up to 200 beats per minute are not uncommon. They do not originate only from reflex response of the baroceptors but also from the physical workload induced by the straining manoeuvre and from psychic tension before the exposure. Parkhurst [12] found a direct correlation between pre-run heart rate and the anticipated G-load. Fig. 7 derived from Leverett [15] indicates the heart rate increase due to performing the M 1 manoeuvre and acceleration. The solid line shows the heart rate response of a relaxed subject undergoing $+3G_z$. The dotted line represents the heart rate of a subject performing only the M 1 manoeuvre with the centrifuge at rest. There is no difference between maximum frequency for this condition compared with a $+6G_z$ run using the M 1 as represented by filled circles. These high heart rates and unusual arterial pressures are often associated with changes in heart rhythm and in the electrocardiogram (ECG).

Premature ventricular contractions are nearly always present in the late stages of centrifuge runs exceeding $+6G_z$. They showed abnormal QRS complexes, no P waves and frequently a compensatory pause [16] but were never of a parasystolic origin. Discussing these results the author suggests the following items to cause these changes: high heart rate, changes in heart position and shape and a certain degree of myocardial ischeamia and/or possible changes in cardiac filling and blood volumes. Some degree of myocardial ischeamia is pointed out by S-T segment changes in the ECG. Leverett [15] proved that the M 1 manoeuvre alone could not produce any of these symptoms.

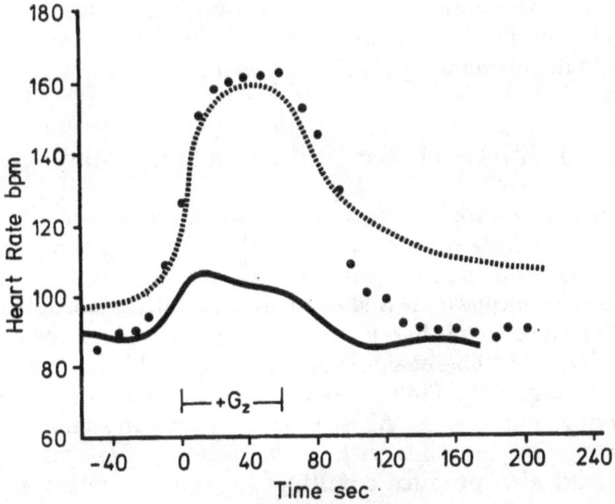

Fig. 7. The effect of $+G_z$ and the M 1 on heart rate (modified from [15]). Dotted line, $1G_z$; solid line, $3G_z$; circles, $6G_z$.

A sudden decrease in heart rate under sustained acceleration is not uncommon. Even under moderate G-loads sudden slowing of heart rate and unconsciousness of the subject which lasts longer than the normal unconsciousness under G of a few seconds has been described earlier and is explained by the activities of a depressor reflex from the left ventricle of the heart by Gauer and Zuidema [9]. Similar symptoms were described by Shubrooks [16] who termed them "high G bradycardia". He also explains them by the depressor reflex but adds that "increased vasovagal activity leading to syncope, predisposed to by the decrease in cardiac filling" may also have been a factor. Up to the present time no other abnormalities from the cardiovascular system have been reported from high sustained G experiments in man. However several animal studies with miniature swine and other animals show serious pathological damage of the heart. Especially the miniature pig has been used as an analog for human subjects. Burton and McKenzie [17] claim that they are good models for acceleration patho-physiology because they have a vascular hydrostatic column length similar to that in man, perform an M 1 like manoeuvre, and have similar acceleration tolerance levels. These animals after exposure to 5, 7 and $9G_z$ sustained acceleration for 45 seconds, showed haemorrhage of the left ventricle and subendocardial haemorrhage. Furthermore histologic examination of the heart muscle revealed myocardial rhexis and degeneration, intracellular and extracellular oedema and haemorrhage, the same picture that can be seen after haemorrhagic shock. Another probable cause that the authors discuss, had already been suggested by Gauer [18] in the late forties. He had strong evidence that the heart might empty during acceleration stress and the walls of the ventricles slammed together during violent contractions. The latter might also be supported by the fact that heart lesions in animals can be avoided by suppressing tachycardia through proper medication during the centrifuge run. But as Burton et al. [19] state in their review: "These animal experiments are relevant to human exposure to high sustained G_z only to the extent that high sustained acceleration has been demonstrated to produce pathology in animals which exhibit several human acceleration-induced physiologic similarities". Further research and better methods are required to determine whether high acceleration forces also produce pathologic findings in man.

4. Effects of Acceleration on Respiration

The lung is an organ which is particulary sensitive to sustained acceleration. Here the mechanical effects on increased gravity mix with the ruling laws of lung ventilation and perfusion making this field of acceleration physiology very complicated, and hard to explain in a short time. I will try to outline some of the most important mechanisms inherent in the lung that determine human tolerance to sustained acceleration. The basic research in this field was carried out by Barr, Bjurstedt and Coleridge [20]. They measured the oxygen saturation and pH of the arterial blood in dogs under $+G_z$. Although they found in some cases, as expected, a decrease in saturation together with an increase in pH, hyperventilation with 100% oxygen could also produce a falling oxygen saturation or a rise in pH. Henry [21] found similar results under negative acceleration and attributed them to a disturbance in pulmonary circulation.

Pulmonary perfusion and ventilation depends, together with the distribution of the ventilation/perfusion ratio within the lung, on the direction of gravity. It is a well-known fact that, in the upright position, even under normal gravity, the apical parts of the lungs are less perfused than the basal parts. The inequality of the ventilation/perfusion ratio, caused mainly by change of perfusion, influences the respiratory gas exchange, resulting in an increasing pressure gradient for CO_2 and oxygen. In 1944 Gauer [22] showed by X-ray-cinematography how intrathoracic blood volume shifts under accelerative forces. His findings were proven later by Bryan et al. [23] and Glaister [24] who found with more sophisticated techniques that a progressive decrease of perfusion in the apical parts of the lung occurs, associated with an increase in the basal regions depending on the magnitude of G. As a result of the changed ventilation/perfusion ratio during acceleration a decrease in arterial oxygen saturation takes place independent from the acting G-vector. This is a cause for the respiratory gas exchange to fail progressively with G.

In our laboratory v. Nieding et al. [25] used a mass spectrometer for continuous recording of respiratory gas exchange from human subjects on the centrifuge. Distribution of lung perfusion was measured by means of double nuclide perfusion scintigraphy. Arterial O_2 and CO_2 partial pressures and pH were determined from blood samples. The findings can be summarized that with increasing acceleration those parts of the lungs that experience a normal perfusion are gradually diminished and that only small areas remain for an uninhibited gas exchange. Together with this, arterial oxygen partial pressure decreases linearly with G. Arterial CO_2 did not change up to $+3G_z$ and pH stayed constant up to $3G_z$.

In high sustained acceleration experiments this mechanism is changed by the M 1 manoeuvre and positive pressure breathing. Yet the result is still an exponentially declining curve for arterial partial oxygen pressure as shown in Fig. 8 taken from Leverett [15]. He also calculated arterial oxygen saturation from

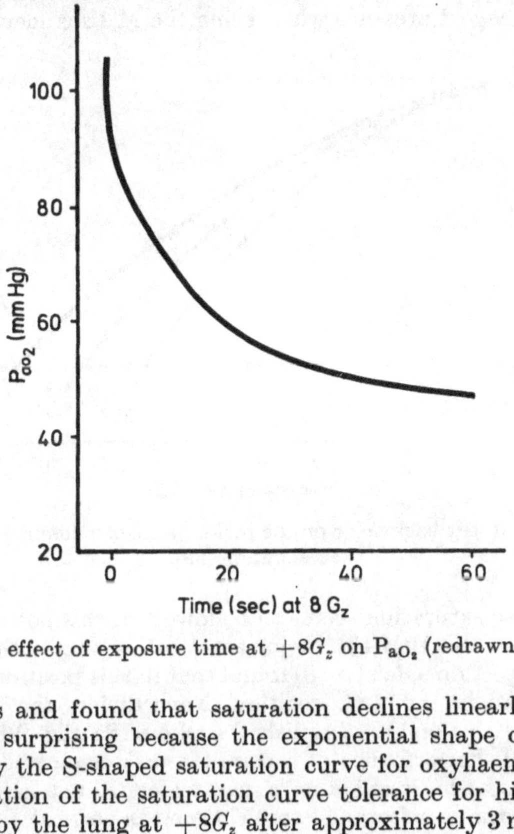

Fig. 8. The effect of exposure time at $+8G_z$ on P_{aO_2} (redrawn after [15]).

blood gas analysis and found that saturation declines linearly with increasing $+G_z$. This is not surprising because the exponential shape of the O_2-pressure curve is altered by the S-shaped saturation curve for oxyhaemoglobin. Furthermore by extrapolation of the saturation curve tolerance for high sustained $+G_z$ would be limited by the lung at $+8G_z$ after approximately 3 minutes. After that time 60% saturation would be reached. The author however suggests that earlier fatigue would probably be the limiting factor, in which case the subject is no longer able to strain actively against the forces of gravity.

5. Other Means of $+G_z$ Protection

Because the protective measures against the effects of sustained acceleration are in general very stressful and pathologic reactions cannot be completely avoided up to this time, other means of G-protection have been considered. It is well known that a subject's G-tolerance is increased by changing his position to the vector of gravity. By reclining the seat from the upright position the subject is exposed to various combinations of $+G_z$ and $+G_x$. The result of reclining is a shortening of the effective distance between heart and brain. This effect has been applied in aviation since 1940 when the first reclining seats were flown in Germany [26]. Because of the visual requirements in operational aviation they have not been used in standard aircraft. Recently reclining seats have gained new interest [27]. The investigations concerning tilt-back and pelvis and legs elevating seats (PALE) can be summarized that a back angle of 65 degrees from the vertical seems to be the most favourable position. High sustained acceleration can be tolerated with less physical effort as indicated by heart rate, eye-level arterial pressure and esophageal pressure when using the M 1 manoeuvre. In contrast to

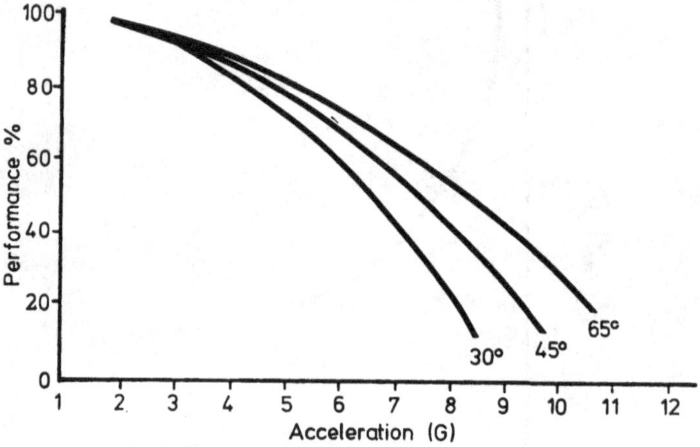

Fig. 9. The influence of seat-back angle on the performance of a complex task under sustained acceleration [29].

this, arterial oxygen saturation seems to be lowest in this position [28]. This may be explained by the fact that the major acceleration vector through the body is now in the $+G_x$ direction. Glaister [8] found that in this position oxygen desaturation is greater than in the $+G_z$ position. Nevertheless the symptoms of high sustained $+G_z$ acceleration are diminished as shown in Fig. 9 taken from Rogers and coworkers [29]. Performance of a complex task is plotted against sustained acceleration with seat-back angle as a parameter. A net gain of almost $2G$ is gained by rotating the seat from 30° to 65° from the vertical.

6. Concluding Remarks

In conclusion it can be stated that there is only limited protection against the effects of sustained acceleration. Tolerance levels have been raised considerably with the onset of research into high sustained acceleration. There is still an open question what mechanism really determines the final limits of tolerance. Is it a failure of the circulatory system, ventilation/perfusion inadequacies of the lung or overall fatigue? Since high sustained acceleration is potentially harmful to man a cautious approach is recommended.

References

[1] K. R. COBURN, Aerospace Med. **41**, 5 (1970).
[2] F. ANDINA, Schweiz. Med. Wschr. **67**, 753 (1937).
[3] G. B. BIETTI and A. SCANO, Riv. Med. Aero **9**, 138 (1946).
[4] T. D. DUANE, Arch. Ophthal. NY **51**, 343 (1954).
[5] P. HOWARD, in J. A. Gillies, A Textbook of Aviation Physiology Pergamon Press, Oxford (1965).
[6] O. F. RANKE, Luftfahrtmedizin **2**, 243 (1938).
[7] E. H. LAMBERT and E. H. WOOD, Fed. Proc. Fed. Am. Soc. Exp. Biol. **5**, 59 (1946).
[8] D. H. GLAISTER, AGARDograph 1933, Technivision Services, England (1970).
[9] O. H. GAUER and G. D. ZUIDEMA, Gravitational Stress in Aerospace Medicine, Little, Brown & Co., Boston 1961.
[10] W. E. CLARK, I. GARDINER, A. MCINTYRE and H. GORGENSEN, Rep. No. 468, Committee on Aviation Medicine, Nat. Res. Co. USA.
[11] J. P. MEEHAN, in: O. H. Gauer and G. D. Zuidema, Gravitational Stress in Aerospace Medicine, Little, Brown & Co., Boston 1961.
[12] M. J. PARKHURST and S. D. LEVERETT, JR, and S. J. SHUBROOKS, JR, Aerospace Med. **43**, 708 (1972).
[13] S. J. SHUBROOKS and S. D. LEVERETT, J. Appl. Physiol. **34**, 460 (1973).
[14] E. H. WOOD and C. F. CODE, Fed. Proc. **4**, 78 (1947).
[15] S. D. LEVERETT, R. R. BURTON, R. J. CROSSLEY, E. D. MICHAELSON and S. J. SHUBROOKS JR, USAFSAM-TR-73-21 (1973).
[16] S. J. SHUBROOKS, JR, Aerospace Med. **43**, 1200 (1972).
[17] R. R. BURTON and W. F. MCKENZIE, Aerospace Med. Ass. Preprints, Ann. Scient. Meeting, Washington 1974 (p. 40).
[18] O. H. GAUER, in: World War II., **1**, 544, US Government Pointing Office, Washington D.C. (1950).
[19] R. R. BURTON, S. D. LEVERETT, JR, and E. D. MICHAELSON, Aerospace Med. **45**, 1115 (1974).
[20] P. O. BARR, H. BJURSTEDT and J. C. G. COLERIDGE, J. Physiol. **142**, 60 (1958).
[21] J. P. HENRY, USAF Tech. Rep. 5953, Wright Patterson AFB, Ohio (1950).
[22] O. H. GAUER, Luftfahrtmed. **9**, 109 (1944).
[23] A. C. BRYAN, W. D. MACNAMARA, J. SIMPSON and H. N. WAGNER, J. Appl. Physiol. **20**, 1129 (1965).
[24] D. H. GLAISTER, Proc. Roy. Soc. B **168**, 311 (1967).
[25] G. VON NIEDING, H. KREKELER, K. KOPPENHAGEN and S. RUFF, Pflügers Arch. **342**, 159 (1973).
[26] H. WIESEHÖFER, Luftfahrtmed. **4**, 145 (1940).
[27] H. J. VAN BECKH, Techn. Rep. NADL-72262-CS (1972).
[28] J. W. BURNS, Aerospace Med. Ass. Ann. Scient. Meeting, Washington D.C. 1975 (p. 45).
[29] D. A. ROGERS, A. B. ASHARE, J. W. SMILES, J. W. FRAZIER, V. D. SKOWRONSKI and F. M. HOLDEN, USAF-AMRL-TR-73-5 (1973).

Life Sciences and Space Research XIV — Akademie-Verlag, Berlin 1976

PHYSIOLOGICAL CHANGES ASSOCIATED WITH LONG-TERM INCREASES IN ACCELERATION

A. H. Smith

Department of Animal Physiology, University of California, Davis, Calif., USA

The long-term retention of organisms in increased acceleration fields is an experimental approach towards understanding the biological consequences of earth gravity. Such treatment, called chronic acceleration, simulates a change in gravity and requires exposed organisms to adapt physiologically to the new environment. Information from chronic acceleration supplements that from space physiology in understanding gravitational physiology.

Many of the responses in long-term exposure to increased acceleration are those which would be anticipated from the imposed symmetrical loading. For example, increased requirements for posture and locomotion induce appropriate changes in musculo-skeletal organs. Displacement of body fluids and increased hydrostatic pressures lead to greater blood volumes and increased tissue hydration. However, there are also specific acceleration effects which cannot be so directly interpreted. Among these are decreases of mature body size and of depot fat, which are proportional in degree to field strength.

The role of chronic acceleration research in the development of gravitational biology and its relationship to earth-orbital experiments are considered. The applicability of chronic acceleration studies with human subjects towards planning of deep space exploration is also discussed.

1. Introduction

Simulations of altered gravity states, by summation of gravitational and inertial fields, are the best current means for research in gravitational biology. Since the effects of gravitational and inertial forces are indistinguishable, as proposed by Einstein in the principle of equivalence [1, 2], it is reasonable to interpret the results from such investigation in general terms, as equivalent to those which would be obtained with a change in gravity. In this way a gravity-free state has been simulated in orbiting vehicles, by producing an inertial field equal and opposite to the ambient gravity, such that the algebraic sum is zero. Increased gravity is readily simulated on the earth's surface as the resultant of gravity and inertial fields produced by motion. These simulations have the desirable property of symmetrically loading and unloading exposed organisms, which is not the case with other more conveniently arranged simulations (e.g. chronic recumbency or immersion).

2. Acceleration Stress and Physiological Adaptation

The response of an organism to an altered acceleration field is greatly dependent upon duration of the exposure, as well as field intensity. The immediate responses are largely mechanical in nature, and generally reflect a limitation of physiological function and homeostatic compensations. With sufficient exposure to fields greater than some threshold intensity, patho-physiological states develop, and a condition of stress is recognized by a frank acceleration sickness. This has been investigated so far only in chickens [3], but presumably the principles apply generally. A mortality roughly proportional to field strength accompanies acceleration sickness. Since postmortem examination does not reveal any systematic or limiting organic lesion, either grossly or microscopically, the responsible mechanisms are presumed to be sub-microscopic, and perhaps metabolic.

Animals surviving the period of acceleration sickness become physiologically adapted, much like the responses to other, more familiar environmental stressors. A transient lymphopenia indicates the participation of adreno-cortical mechanisms in the process [4, 5]. The greatly enhanced susceptibility of reserpene-treated animals to centrifugation also indicates a continuing sympathico-adrenal participation in the tolerance of acceleration. The endocrinology of acceleration stress and physiological adaptation to it has not yet been sufficiently investigated to permit a satisfactory description of hormonal participation. A particularly revealing aspect of the physiological adaptation to acceleration is its great heritability [6]. Reproduction of survivors (i.e. acceleration selection) for five generations reduces the mortality from acceleration sickness during centrifugation to approximately 10% of that in unselected populations. Such rapid selection progress is interpreted as indicating a metabolic basis, because of the direct relationship between enzymes and genes. Consequently the toleration of acceleration exposure, as well as the susceptibility of it, appears to have important metabolic elements which are not currently identified. The physiologically adapted state also appears to require an active maintenance. Individuals that have adapted physiologically and tolerated an increased acceleration environment for a long period (e.g. for a year) may become stressed and die from acceleration sickness. This exhaustion phenomenon [7] is comparable to Monge's disease, the abrupt appearance of altitude sickness in Andean natives of pre-Columbian stock [8].

The period of stress and development of physiological adaptation in accelerated animals, which may require weeks or months for completion, is characterized by a changing biology which is difficult to rationalize. However, with the establishment of physiological adaptation, a steady state is achieved and the accelerated animals exhibit a stable physiology indefinitely — chronically. This condition, distinguished as *chronic acceleration*, is comparable to the natural earthly existence. The *acute* and *chronic* acceleration designations correspond to the terms *soujourner* and *resident* that are used in high altitude physiology.

3. Anatomic Responses to Gravitational Loading

The most obvious changes in chronically accelerated animals are those related to load, the artificially increased weight induced by centrifugation. Since load will be a function of body size, there is a general proportionality of severity of acceleration effects and body size. For example, 30 g mice can become physiologically adapted to a $7g$ field [9], and 250 g rats to $5g$ [10], but 2 kg chickens can tolerate only $3g$ [3]. The degree of physiological change encountered likewise tends to be proportional to body size, as well as field strength.

The responsiveness of bone morphology to load, Wolff's law, was described in 1892 [11, 12]. Under terrestrial conditions, this phenomenon requires a relatively greater skeletal size in larger animals [13]:

$$\text{kg skeleton} = 230\ (\text{metric tons body mass})^{1.15}.$$

That this hypergonic relationship is in response to load is indicated by the relative indifference of skeletal size to body size in aquatic mammals [14]:

$$\text{kg skeleton} = 137\ (\text{metric tons body mass})^{1.02}.$$

The artificial loading of bones produces changes in skeletal growth similar to those of body size. This has been demonstrated in dogs [15] and sheep [16] under earth gravity through the application of weighted saddles. In the lamb studies [16], animals were maintained on two nutritive levels, optimal and sub-optimal, and bone growth as well as body growth were proportional to nutritional intake. However, when the load of the sub-optimally nourished animals was increased by weighted saddles to that exerted by the body weight of the well-nourished, the growth of selected bones became comparable for the two groups.

Increases in relative skeletal size resulting from centrifugation during the period of skeletal development have been reported for mice [17], rats [10] and chickens [18]. Evaluation of the effect is complicated by accompanying changes in the soft tissues (a decreased body mass) and uncertainties of the similitude relationships. This treatment also upsets the normal chronological relationship between bone development and load. Ordinarily skeletal maturity precedes the body weight maturity which is a biological necessity in large animals since part of the developing bone, cartilage, has poor mechanical properties. Consequently in animals centrifuged prior to skeletal maturity, rather gross bone deformations are encountered, and in the centrifuging chick these were considered sufficient to affect the normal function of the viscera [3]. To avoid such complications, subsequent acceleration experiments have been performed on skeletally mature animals ($>$ 90 days age).

Loaded bone also undergoes a change in shape, becoming shorter and thicker, a phenomenon described by Galileo in 1638 [19] and formalized by Wolff [11]. In terrestrial animals, bones support the load of the soft tissues, and this places a compressive force on the bone. As first noted by Galileo, this mechanical situation is reversed in aquatic animals with the less dense soft tissues buoyantly supporting the load of the denser hard tissue and placing the bone under tension. As a result, increasing the gravitational loading in aquatic animals causes their bones to become longer and thinner. Such a response has been observed in centrifuging chick embryos [20], which can be quite properly considered as aquatic animals until a few days before hatch.

In terrestrial situations postural maintenance requires a loading of skeletal muscles as well as the skeleton. As a result, there is also an increase in relative muscle size with increasing body size [14]:

$$\text{kg muscle} = 357\ (\text{metric tons body mass})^{1.05}.$$

However, in aquatic mammals, the role of blubber as the load bearing tissue results in muscle being hypogonically related to body size [14]:

$$\text{kg muscle} = 752\ (\text{metric tons body mass})^{0.83}.$$

The influence of artificial loading, by centrifugation, upon muscular development has been reported in hamsters [21, 22] and rats [23]. However, as with the skeletal investigations, the alteration of body size and uncertainties of similitude relationships, limit the interpretation of the results. The comparison of size changes in an antagonistic pair, a flexor and an extensor, circumvents much of the uncertainty since both share a somatic influence equally. Such a comparison has been made on a pair of hip muscles of comparable size in the chicken — the adductor, an extensor, and the sartorius, a flexor [24]. Under chronic acceleration the extensor increased, and the flexor decreased in size. The rate of change was slow, and a year's treatment at $2g$ was required to obtain 85% of the mathematically predicted maximum change. A comparison of the equilibrium mass ratio ($E{:}F$) with field strength (g) indicated an arithmetic relationship:

$$E{:}F = 0.59 + 0.28g.$$

There also are functional changes in muscles of chronically centrifuged animals. The first chronic acceleration experiments were undertaken by Matthews [25] to determine the influence of symmetrical loading upon muscle function. He selected rats as subjects since they are small quadrupeds, and loading, rather than circulatory impairment, would be the principal effect. After a year at $3g$, he found that the centrifuged animals exhibited a marked decerebrate extensor tonus. Canonica [22] found that prepared skeletal muscles from animals exposed to $4g$ were much more resistant to fatigue than were those from their gravity controls. In chickens [26] the maximum work capacity at earth gravity, measured by exhaustion time while running at 37 m min^{-1}, was three-fold greater in animals centrifuged six months at $1.75g$ as compared with non-centrifuged controls.

The circulatory problems resulting from acceleration exposure are rather familiar because of their importance to aviation medicine. With chronic acceleration, the continued inferior displacement of the blood leads to an increased plasma volume as a result of a diminished activation of the Gauer-Henry reflex [27] and a relatively greater production of anti-diaretic hormone (ADH). An increased relative plasma volume ($\text{PV} = \text{ml} \times 100 \times \text{g body mass}^{-1}$) has been reported in rats [28] and in chickens [29], the latter with an exponential relation to field strength (g):

$$\text{PV} = 2.9e^{0.16g}.$$

Appropriate changes in vascular resistance (increased) and baroreceptor responsiveness (enhanced sensitivity) have been reported in rats at earth gravity after four weeks exposure to $3g$ [30].

4. Metabolic Response to Gravitational Loading

The load induced by chronic acceleration also imposes additional metabolic requirements, since the mechanical work requirements for posture and locomotion are determined by weight, rather than mass. After a transient anorexia, accelerated animals exhibit increased feed intake rates, and this has been reported for rats [31, 32] and chickens [33]. The maintenance feed requirement [34, 35] has been reported for chronically accelerated chickens [33], and it has been interpreted as being arithmetically related to field strength:

maintenance feed: $F_g = 26.6 + 9.6g$;
maintenance metabolizable feed: $F_g = 17.3 + 6.5g$;

where: F_g is feed intake (gm feed × kg body mass^{-1} day^{-1}); metabolizable feed is the feed intake less excreta. These observations indicate that earth gravity determines approximately 27% of the natural maintenance requirement, or, in absolute terms, 16 kcal × kg mass^{-1} g^{-1}. As a physically determined requirement, this increase should apply equally to each kg of body mass in all animals, irrespective of size. But, since the metabolic rate is proportional only to body mass$^{3/4}$, gravity is responsible for a much greater part of the maintenance requirement of large animals. Kleiber [36] has estimated from such data that 40% of the human (75 kg) maintenance requirement would be gravity determined.

5. Influence of the Ambient Acceleration upon Body Size and Composition

Many other anatomic and physiological changes have been reported in chronically accelerated animals, for which the relationships to loading are not obvious. One of the more general phenomena is a repressed growth rate and diminished mature body size which has been reported in mice [9, 37], hamsters [38], rats [10, 37], rabbits [39], and domestic birds [40, 41]. The effect appears to be proportional to the mature body size, and in similar series (domestic birds from 0.1 to 5 kg) it follows the relationship [41]

$$M_g = M_0 - kg$$

where M is the body mass; M_0 the intercept (the indicated value of M when $g = 0$) and M_g the mass in a field of strength g; k is the proportionality coefficient, the change in M per unit increase in g. Then for various birds:

Quail:	$M_g = 127\text{ g} - 6.93g$
Young chickens:	$M_g = 2.10\text{ kg} - 0.36g$
Old chickens:	$M_g = 2.46\text{ kg} - 0.26g$
Large chickens:	$M_g = 6.76\text{ kg} - 1.37g$.

The influence of body size on the loss of mass can be estimated by comparing the ratios of the proportionality constants k with the potential size M_0:

$$I(\%) = (k/M_0) \times 100$$

where I is the inhibitory effect of acceleration upon growth, literally, the effect of earth gravity upon body size. In a similar series (domestic birds 0.1—5 kg body mass) this inhibitory effect of earth gravity I upon body size is arithmetically related to potential body size M_0:

$$I = 7.5 + 2.0M_0.$$

When mature chronically centrifuged animals are fasted an additional increment of body mass is lost — perhaps 10% in a three-day fast. However, upon re-alimentation, the lost body substance is restored at least as rapidly as with earth gravity controls [40]. Obviously, the decreased body size in the hypergravic environment is a regulated phenomenon, and not merely the result of an inability to acquire food or an incapacity for synthesis. In terms of current theories of growth regulation [42, 43] this indicates that gravity (as well as genetic factors) is a determinant in the *set-point* towards which growth proceeds. This is not the case with other environmental stressors which limit growth only through some form of metabolic restriction.

The limitation of body size in chronically accelerated animals usually involves a selective loss of fat. This visually obvious depletion of fat depots at autopsy has been reported for mice [9, 44], hamsters [22, 47], rats [10, 32, 45, 46] and domestic birds [3, 18, 40, 48]. A single exception, based on *in vivo* measurements, has been reported for monkeys, which actually gained body fat while on the centrifuge [49]. A prominent abdominal fat pad (AFP) is found in most animals and it is conveniently and accurately dissected. The size of this fat pad provides a reliable index of body fat content. The influence of chronic acceleration on the size of fat pads (AFP, % body mass) has been measured in rats [50] and chickens:

$$\text{Chicken:}\quad \text{AFP}_{(\%\ \text{carcass})} = 11.5e^{-1.7g}$$

$$\text{Rat:}\quad \text{AFP}_{(\%\ \text{body})} = 10.2e^{-0.9g}.$$

From these equations, it appears that the influence of acceleration on abdominal fat pad size is greater in the larger animal — about double that for rats. However, at the limit of acceleration tolerance ($3g$ for chickens and $5g$ for rats) the decrease in fat pad size is equivalent for both species.

The influence of the ambient acceleration field on body fat content has been reported only for rats [46], chickens [18, 48], and some other domestic birds [41]. In all cases the relationship is arithmetic, each increment of increasing field strength inducing equal reductions in fat:

Rats [46]: $\text{Body fat}_{(g)} = 53.3 - 9.5g$
$[M_0 = 367\ \text{g};\ I_f = 17.8\%]$

Quail [48]: $\text{Carcass fat conc.}_{(\%)} = 15.35 - 3.38g$
$[M_0 = 127\ \text{g};\ I_f = 22.0\%]$

Young chickens [48]: $\text{Carcass fat conc.}_{(\%)} = 11.07 - 3.21g$
$[M_0 = 2.10\ \text{kg};\ I_f = 29.0\%]$

Old chickens [48]: $\text{Carcass fat conc.}_{(\%)} = 35.40 - 14.90g$
$[M_0 = 2.46\ \text{kg};\ I_f = 42.1\%]$

Large chickens [48]: $\text{Carcass fat conc.}_{(\%)} = 27.22 - 12.28g$
$[M_0 = 6.76\ \text{kg};\ I_f = 45.1\%]$

As with the other acceleration responses, there is a direct relationship between body size and the inhibiting effect of acceleration on body fat (I_f, the % reduction of potential fat concentration, F_0, by earth gravity). Among a series of domestic birds [49], the relationship between body size (M, kg) and gravitational inhibition of fat concentration (I_f) is arithmetic:

$$I_f = 25.5 + 3.19M_0.$$

The acceleration-induced reduction in body fat content is accompanied by changes in the intermediate metabolism of fat and carbohydrate, which have been reported in rats [45, 51, 52] and in chickens [48, 53]. Generally these indicate a transient decrease in fat utilization during the period of acceleration stress and adaptation, but returning to normal utilization levels in the chronic acceleration stage. Simultaneously there is an increased carbohydrate utilization which remains elevated in chronic acceleration. However, the principal mechanism for the reduction of body fat in chronic acceleration appears to involve a specific humoral agent, similar to that involved in lipodystrophy [54, 55], a human syndrome that is characterized by a loss of depot fat. This material, the fat mobilizing substance (FMS), is a polypeptide of 5000 daltons, which appears to be produced in the pituitary of mammals [56] and the hypothalamus of birds [57]. It is readily recovered from the urine of appropriately treated animals [58] and induces anorexia and fat mobilization when administered to untreated animals [59—61]. Some experiments still in progress indicate that extracts of urine from centrifuging chickens and rabbits exhibit both an anorexic and lipotropic activity. An attractive hypothesis is that in chronic acceleration the brain, which has a small negative buoyancy, is mechanically stimulated as it is loaded and forced against the floor of the cranium, releasing FMS. The arithmetic relationship between acceleration field strength and fat loss would support such a scheme — equal increments of acceleration producing equal increments of mechanical stimulation and FMS production.

A difficulty with such a model is the lack of fat loss in centrifuging monkeys, even after treatments which would markedly de-fat smaller animals [49]. However, a comparison of brain geometries, particularly the region and angle of entry of the cord into the brain, indicated that primates may be systematically different from other animals. In most, the cord enters the brain posteriorly and obliquely, which would offer little support for any load developing in the hypothalamic region. In man, as exemplifying primates, the cord appears to enter the brain vertically and near the brain's center of mass [49], an arrangement which might attenuate load and prevent mechanical stimulation of the hypothalamic region in moderate acceleration fields. Unfortunately, information needed to test such hypotheses (e.g. the negative buoyancy of brains and the identity of anatomic structures attenuating the forces involved) is not readily apparent in the literature. It also is possible that loading affects the functional properties of neural tissue. Winget et al. [62, 63] have reported a maintained labyrinthine sensitivity of centrifuging chickens, whereas ordinarily there is a habituation, with decreasing responses to repeated rotatory stimulations. These observations are consistent with those from Skylab [64], that the orbiting astronauts appeared to be insensitive to rotatory stimulation.

6. Outlook for Chronic Acceleration Research

Chronic acceleration and space research share a common philosophic goal, an understanding of the biological consequences of earth gravity. This constitutes a very comprehensive task since gravity is the common factor in all terrestrial phenomena. Both activities also share a common technique — the artificial alteration of the ambient acceleration field, and a comparison of the responses with those obtained under earth gravity. How mutually supportive these approaches may be is dependent upon the continuity of acceleration effects above and below natural gravity. There is no basis at present for deciding the nature of this relationship, although arguments can be offered for the concept of continuity [65]. In either event, both procedures will be necessary for a satisfactory understanding of gravitational biology. With only two points, earth gravity and orbital weightlessness, no generalizations will be possible. But, without satellite experiments, the purely mass-determined biological processes, those which persist in weightlessness, will not be known, and acceleration biology will lack a foundation.

Greater use should be made of chronic acceleration procedures, to develop a background of information with which to anticipate and interpret findings in hypogravity. It would also be advantageous if these included chronic studies with humans. For such investigations, new devices would be required capable of developing fields of $1.5-2g$ while limiting rotation rates to $20°-30°\ s^{-1}$, rates which have been found to be tolerable to humans [66]. For such performance a device of 100—200 m diameter capable of circumferential velocities in the order of $100\ km\ hr^{-1}$ would be required. This could be arranged with a circular track on which self-propelled vehicles operate. With a suitable switching system, vehicles could be periodically exchanged while maintaining subjects under hypergravic conditions, perhaps for a year or more.

With such a device it should be possible to estimate the maximal rate of physiological adaptation ($\Delta g\ day^{-1}$) for a $1g$ increase in the ambient field, i.e. to $2g$. Alternatively the optimum daily exposure to $2g$ ($hrs\ day^{-1}$) to induce physiological adaptation [67] could be examined to determine the preferable procedure. Such information would be of great value in planning a re-adaptation program for gravity de-adapted astronauts following a deep space voyage of many months duration. Of particular importance would be an identification of the earliest signs of *acceleration sickness* in humans, to provide a rational basis for physiological monitoring of re-adapting astronauts.

The rate of loss of a hypergravic adaptation, from $2g$ to earth gravity, would also be quite useful. This can be determined by periodically returning previously adapted individuals abruptly to the hypergravic environment, and observing the degree of tolerance retained. The gravity adapted state appears to be lost quite slowly, compared with the de-adaptation to other environmental extremes. This has been examined in chickens which were returned to earth gravity after having been adapted to a $2g$ field. At monthly intervals, they were returned abruptly to a $2g$ field, and their response was compared with that of some accompanying, but previously unexposed chickens over a 24-hour period. Largely on the basis of lymphocyte response, it was estimated that they retained 70% of their acceleration-adaptation for six months. This retained physiological adaptation permitted an indefinite return to the higher-g environment without untoward effects. Similar information with humans would provide some basis for estimating the duration of a space flight compatible with direct earth return. The feasibility of extending the duration of such deep space flights by previous adaptation of astronauts to a hypergravic environment could also be examined.

References

[1] W. G. V. Rosser, An Introduction to the Theory of Relativity, Butterworth, London 1964.

[2] L. Witten, Gravitation: An Introduction to Current Research, Wiley, New York 1962.

[3] R. R. BURTON and A. H. SMITH, Aerospace Med. **36**, 39 (1965).
[4] R. R. BURTON, S. J. SLUKA, E. L. BESCH and A. H. SMITH, Aerospace Med. **38**, 1240 (1967).
[5] R. R. BURTON and A. H. SMITH, Proc. Soc. Exp. Biol. Med. **128**, 608 (1968).
[6] A. H. SMITH and C. F. KELLY, Physiologist **4**, 111 (1961).
[7] H. SELYE, Stress, Acta Inc., Montreal 1950.
[8] C. MONGE M., and C. MONGE C., High Altitude Diseases, Thomas, Spingfield, Ill. 1966.
[9] C. C. WUNDER, Aerospace Med. **33**, 866 (1962).
[10] J. OYAMA and W. T. PLATT, Am. J. Physiol. **209**, 611 (1965).
[11] J. WOLFF, Das Gesetz der Transformation der Knochen, Hirschwald, Berlin 1892.
[12] A. W. HAM, Histology, 5th ed., Lippincott, Philadelphia 1965.
[13] CH. KEYSER and A. HEUSNER, J. Physiol. (Paris) **56**, 489 (1964).
[14] A. H. SMITH and N. PACE, Environ. Physiol. **1**, 122 (1972).
[15] YE. I. DENILOVA and A. I. SVIRIDOV, Zool. Zh. **32**, 4 (1953).
[16] N. M. TULLOH and B. ROMBERG, Nature **200**, 438 (1963).
[17] C. C. WUNDER, S. R. BRINEY, M. KRAL and C. SKAUGSTAD, Nature **188**, 151 (1960).
[18] A. H. SMITH and C. F. KELLY, Ann. NY Acad. Sci. **110**, 410 (1963).
[19] Galileo GALILEI, Dialogues Concerning Two Sciences, 1638 (Engl. Trans. H. CREW and A. DESALVIO, Macmillan, New York 1914).
[20] D. R. REDDEN, Am. J. Physiol. **218**, 310 (1970).
[21] S. R. BRINEY and C. C. WUNDER, Am. J. Physiol. **202**, 461 (1962).
[22] P. C. CANONICA, Master's dissertation, Univ. South Carolina (1966).
[23] J. OYAMA and W. T. PLATT, Am. J. Physiol. **212**, 164 (1967).
[24] R. R. BURTON, E. L. BESCH, S. J. SLUKA and A. H. SMITH, J. Appl. Physiol. **23**, 80 (1967).
[25] B. H. C. MATTHEWS, J. Physiol. **122**, 31P (1953).
[26] R. R. BURTON and A. H. SMITH, Proc. XVI Intl. Congr. Aviat. Space Med., Lisbon (1967).
[27] O. H. GAUER, J. P. HENRY and C. BEHN, Ann. Rev. Physiol. **32**, 547 (1970).
[28] B. R. DULING, Doctoral dissertation, Univ. Iowa (1967).
[29] R. R. BURTON and A. H. SMITH, Space Life Sci. **1**, 503 (1969).
[30] B. R. DULING, Am. J. Physiol. **213**, 466 (1967).
[31] H. W. CASEY, D. R. CORDY, M. GOLDMAN and A. H. SMITH, Aerospace Med. **38**, 451 (1967).
[32] F. L. D. STEEL, Nature **193**, 583 (1962).
[33] A. H. SMITH, R. R. BURTON and C. F. KELLY, J. Nutr. **101**, 13 (1971).
[34] S. BRODY, Bioenergetics and Growth, Reinhold, New York 1945.
[35] M. KLEIBER, The Fire of Life, Wiley, New York 1961.
[36] M. KLEIBER, in: Energy Metabolism of Farm Animals, Oriel Press, Newcastle on Tyne 1969.
[37] J. OYAMA and W. T. PLATT, Nature **203**, 766 (1964).
[38] S. R. BRINEY and C. C. WUNDER, Am. J. Physiol. **202**, 461 (1962).
[39] A. H. SMITH, Life Sciences and Space Research XI, 201 (1973).
[40] A. H. SMITH and R. R. BURTON, Growth **31**, 317 (1967).
[41] A. H. SMITH, O. SANCHEZ P., and R. R. BURTON, Life Sciences and Space Research XIII, 21 (1975).
[42] R. J. GOSS, Regulation of Organ and Tissue Growth, Academic Press, New York 1972.
[43] A. E. NEEDHAM, The Growth Process in Animals, Van Nostrand, Princeton 1964.
[44] L. C. KEIL, Growth **33**, 83 (1969).
[45] D. D. FELLER, E. D. NEVILLE, J. OYAMA and E. G. AVERKIN, Proc. Soc. Exp. Biol. Med. **19**, 522 (1965).
[46] G. C. PITTS, L. S. BULL and J. OYAMA, Am. J. Physiol. **223**, 1044 (1972).
[47] S. R. BRINEY and C. C. WUNDER, Proc. Iowa Acad. Sci. **67**, 495 (1960).
[48] J. W. EVANS, A. H. SMITH and J. M. BODA, Am. J. Physiol. **216**, 1468 (1969).
[49] A. H. SMITH, D. F. RAHLMANN, A. M. KODAMA and N. PACE, Life Sciences and Space Research XII, 129 (1974).

[50] J. OYAMA and B. ZEITMAN, Am. J. Physiol. **213**, 1305 (1967).
[51] B. C. DALIGCON and J. OYAMA, Physiologist **13**, 174 (1970).
[52] D. D. FELLER and E. D. NEVILLE, Am. J. Physiol. **208**, 892 (1965).
[53] J. W. EVANS and J. M. BODA, Am. J. Physiol. **219**, 893 (1970).
[54] T. M. CHALMERS, in: Handbook of Physiology, Section 5, Adipose Tissue, Waverly Press, Baltimore 1965 (p. 542).
[55] B. SENIOR, in: Handbook of Physiology, Section 5, Adipose Tissue, Waverly Press, Baltimore 1965 (p. 662).
[56] G. L. S. PAWAN, in: Polypeptide Hormones, Akad. Kiado, Budapest 1971.
[57] I. NIR, M. K. DIMMICK and S. LEPKOVSKY, Can. J. Physiol. Pharmacol. **47**, 435 (1969).
[58] J. A. F. STEVENSON and J. R. BEATON, Ann. NY Acad. Sci. **131**, 189 (1965).
[59] T. M. CHALMERS, A. KECKWICK and G. L. S. PAWAN, Am. J. Clin. Nutr. **8**, 728 (1960).
[60] A. KECKWICK and G. L. S. PAWAN, Lancet **2**, 198 (1968).
[61] G. L. S. PAWAN, Biochem. J. **110**, 52P (1968).
[62] C. M. WINGET and A. H. SMITH, J. Appl. Physiol. **17**, 721 (1962).
[63] C. M. WINGET, A. H. SMITH and C. F. KELLY, J. Appl. Physiol. **17**, 709 (1962).
[64] A. GRAYBIEL, E. F. MILLER and J. L. HOMICK, in: Proc. Skylab Life Sciences Symp., NASA TMX-58154, 1974 (p. 169).
[65] A. H. SMITH and R. R. BURTON, in: Gravity and the Organism, Univ. Chicago Press 1971.
[66] F. E. GUEDRY, JR., R. S. KENNEDY, C. S. HARRIS and A. GRAYBIEL, Aerospace Med. **35**, 1071 (1964).
[67] R. R. BURTON and A. H. SMITH, Am. J. Physiol. **222**, 1505 (1972).

Life Sciences and Space Research XIV — Akademie-Verlag, Berlin 1976

PHYSIOLOGICAL EFFECTS INDUCED BY ANTIORTHOSTATIC HYPOKINESIA

L. I. Kakurin, M. P. Kuzmin, E. I. Matsnev and V. M. Mikhailov

USSR

The main purpose of the present investigation was to reproduce some physiological reactions in the sensory system which are similar to those observed during the first day of the adaptation to weightlessness. This was achieved by a 5-day bed rest experiment during which the test subjects were kept in the antiorthostatic position at anlges of 0°, −4°, −8°, −12°.

Our investigations allowed simulation of the acute stage of adaptation to weightlessness and assessment of gravity-induced blood redistribution in the development of the above physiological reactions.

1. Introduction

Simulation of physiological reactions typical of the weightless state is very important for space medicine. Up to the present time disorders occurring in simulated weightlessness have been fully studied, and experimental evaluations of the efficiency of different countermeasures have been developed. Recently the extensive investigations carried out in the USSR to simulate physiological effects of weightlessness have been supplemented by a new method, antiorthostatic position of the subject during bed rest. The first antiorthostatic test at an angle of −4° was carried out in 1970 to evaluate the efficiency of countermeasures that were recommended for the Salyut 1 orbital flight [5].

The accumulated theoretical and factual data show convincingly that many physiological effects found in simulation experiments have much in common with those noted in real flight. The most important thing is that the same physiological systems show the highest reactivity in real and simulated weightlessness. In other words, problems involved in cardiovascular deconditioning, nervous skeletomuscular disturbances and changes in fluid electrolyte balance are common ones for real and simulated flights and play an outstanding role in space medicine.

Our experience gained in real space flights and in simulated weightlessness experiments suggests that the major pathogenic factor which causes functional disorders in flight is gravity-induced redistribution of fluids in the body. Assuming this, we postulated that short-term experiments in which adverse effects of bed-rest are minimized may simulate responses of the vestibular, optic and other analysers which are typical of the acute period of adaptation to weightlessness. In this connection a special investigation was carried out in order to demonstrate that antiorthostatic hypokinesia is an adequate model of zero gravity.

2. Experimental Conditions

Four experimental runs were carried out in which eight healthy male test subjects (aged 27 to 37) participated. The details of the experiment are given in Table 1. The subjects were kept in bed for five days; those of the first group were kept in the horizontal position (0°), those of the second, third and fourth groups were kept in bed at an angle of −4°, −8° and −12°, respectively.

Table 1

Characteristics of the Test Subjects

Subject	Age (years)	Height (cm)	Weight (kg)
1	26	173	81.0
2	36	176	76.7
3	27	175	73.0
4	29	178	68.5
5	33	178	70.0
6	27	158	54.0
7	28	168	60
8	37	165.5	60.5

3. Methods

During the experiment the test subjects underwent regular physical examinations (measurement of heart rate, arterial pressure, body temperature) and special examinations by doctors, ENT specialists and eye specialists. Special investigations included: 1, ENT examination; 2, eye examination; 3, postural equilibrium using two ataxic tests (standing and walking) [4]; 4, passive tilt test at an angle of 75° for 10 min; 5, bicycle ergometry at a work load of 600 kg min^{-1} for 7 minutes in the sitting position.

4. Results

Throughout the experiments the condition of health of the test subjects remained satisfactory. In the antiorthostatic position they reported backache and fever in the legs, whereas at the end of the experiment they complained of chilled legs. Some subjects showed changes in taste and olfactory sensitivity. Similar changes were observed in the Skylab crewmembers.

In comparison with the degree of recumbency the antiorthostatic subjects reported a more marked rush of blood to the head, heaviness in the head, nasal congestion and impaired nasal breathing. They also showed facial puffiness, and reddening of the sclera and conjunctiva. Some test subjects felt overfilling and heaviness in the nasal sinuses, "clogging" of the ears, uncomfortable feelings in the nose and pharynx, and a hoarse voice. In three subjects these symptoms caused headache and disturbed sleep. The subjects who were exposed to −8° and −12° bed rest showed greater nasal congestion during prolonged stay on one side. The congestion was so great that it simulated rhinitis.

Rhinoscopic examination revealed noticeable swelling and enlargement of nasal conchae and edema of the nasal mucosa; the latter was bright red in some test subjects and cyanotic in the others. The rhinopneumometric studies indicated an increase of the intranasal resistance, the degree of which depended on the angle of the antiorthostatic position. For instance, the subjects who were exposed to $-12°$ showed a rise of the intranasal resistance to 40—50 mm H_2O, whereas recumbent subjects displayed 10—12 mm H_2O.

Swelling of the mucous membrane of the posterior wall of the pharynx which was well seen in the soft palate, lateral vallum, uvula and epiglottis and engorgement of the vessels was noted in the larynx and pharynx.

Otoscopic examination showed moderate skin hyperemia, especially of the osseous compartments of the acoustic ducts and engorged vessels of the tympanic membrane.

Some test subjects who were in the antiorthostatic position ($-4°$, $-8°$ and $-12°$) reported mild dizziness and nausea upon abrupt head movements and nystagmus during the first day of bed rest. They also showed nystagmoid tremor of the eye which was recorded electrographically, during the first and second days of bed rest. Most subjects noted various illusory sensations such as slight tilting, a head-down position, a sensation of falling, especially with closed eyes. In other cases when the subjects fixed their eyes on a single stationary point a few seconds later they felt as if the point began to move; in other words, they developed the so-called autokinetic illusion. One test subject exposed to $-12°$ experienced with his eyes closed optic hallucinations which involved different geometric figures that moved and superimposed.

Similar disturbances were noted in cosmonauts during their first day of adaptation to zero gravity [1—3, 8—10]. Most Soviet cosmonauts and American astronauts reported a rush of blood to the head, different illusory sensations and symptoms of motion sickness. For instance, out of 27 US astronauts who participated in Apollo 7—15 missions several displayed illusions of rotation (3 cases), stomach awareness (9 cases), nausea (4 cases) and vomiting (3 cases). Symptoms of motion sickness were found in the three Skylab 3 crew members and in one Skylab 4 astronaut.

The test subjects kept in the antiorthostatic position complained of heaviness in the eyes, fatigue in reading and reduction of acuity of central vision. The diastolic pressure in the retinal central artery and the retinobrachial index gradually increased, reaching a maximum on the third day of bed rest at 0° and $-4°$ ($P < 0.05$) and at the 5th—6th hour at $-8°$ and $-12°$. In the latter case the parameters remained elevated till the end of the experiment (Fig. 1). The intraocular pressure varied widely in all test subjects, rising to the upper limit of normality or exceeding it at $-8°$ and $-12°$. Some subjects showed deterioration of visual acuity.

During an exposure to $-8°$ and $-12°$ the retinal arteries dilated moderately, and the retinal veins enlarged markedly in proportion to the angle of inclination. The $-8°$ test subjects showed a mild pericapillary edema of the retina and the $-12°$ subjects displayed noticeable edema and congestion of the discs of the optic nerves. A day after termination of the experiment these parameters returned to normal. A comparison of the sensation of eye heaviness felt by test subjects and cosmonauts leads to the conclusion that antiorthostatic hypokinesia at an angle of $-8°$ is the most suitable model of zero gravity.

The assessment of postural equilibrium immediately after the experiment

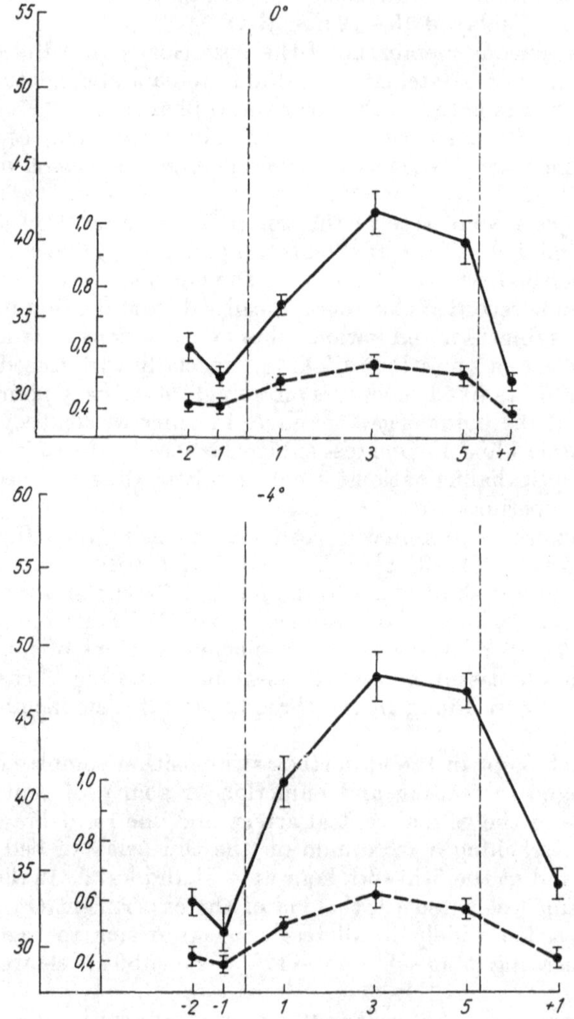

0°
55
50
45
40
35
30
1,0
0,8
0,6
0,4
-2 -1 1 3 5 +1
-4°
60
55
50
45
40
35
30
1,0
0,8
0,6
0,4
-2 -1 1 3 5 +1

Fig. 1

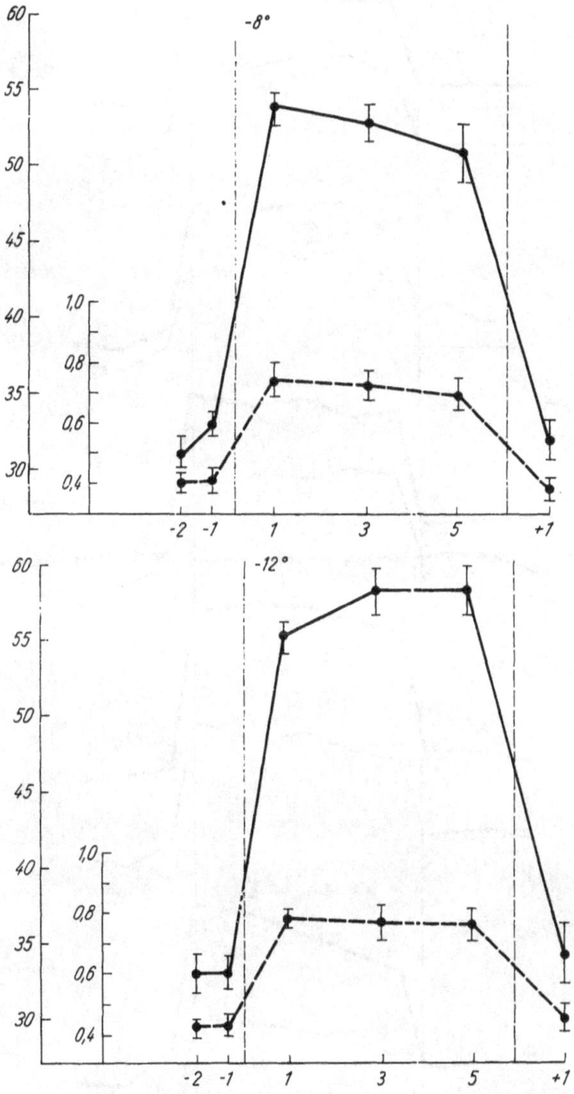

Fig. 1. Changes in the diastolic pressure in the central retinal artery (solid line) and retinobrachial index (dashed line) during the experiment.

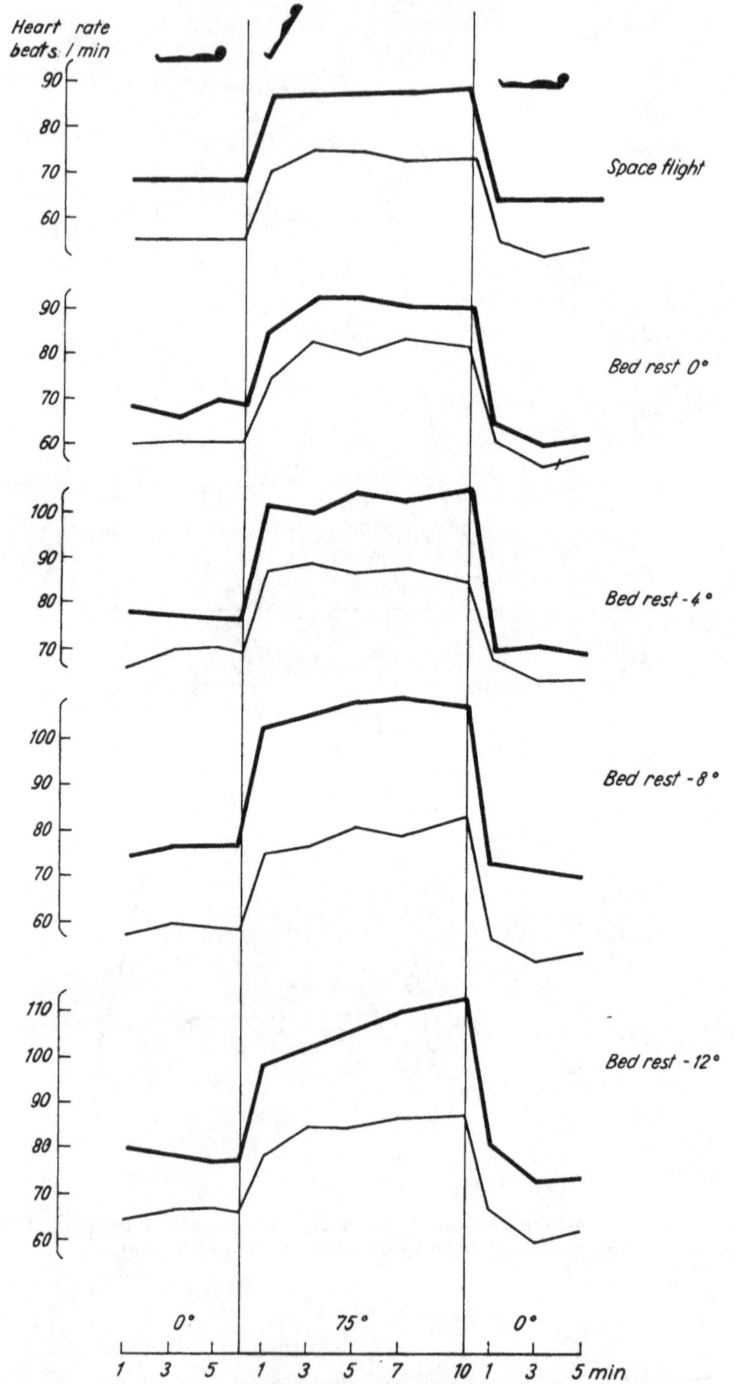

Fig. 2. Heart rate during the passive orthostatic test (tilt at 75° for 10 min) (before test, thin line; after test, thick line).

demonstrated significant impairment of standing and walking. The subjects could hardly keep their balance when standing, they tried to balance while walking, stepping back and sideways from the required direction, and did not pace rhythmically. Changes in the postural equilibrium after antiorthostatic hypokinesia were more marked than after horizontal recumbency; the changes were most significant after exposure to −12°. It is important to note another characteristic feature: the test subjects leaned or fell down in the direction of the position of the body during bed rest.

Similar results were obtained as a result of 3- and 5-day space flights: the crew members displayed certain statokinetic disturbances, e.g. instability in the Romberg test and deteriorated coordination of movements when walking [6]. It should be mentioned that Skylab 2 crew members could stand on the 75 cm bar with the eyes closed for a few seconds only [7]. After bed rest orthostatic tolerance decreased in all test subjects, being correlated with the angle of inclination (Fig. 2).

Comparison of the heart rate of seven cosmonauts before and after the 5-day space flights of Soyuz 6, 7 and 8 during the orthostatic test shows that the changes were similar to those in the subjects exposed to −4° and −8°.

During bicycle ergometry the heart rate remained essentially unaltered in the recumbent test subjects (decreased by 3 beats/min) and increased by 17, 8 and 8 beats/min in the antiorthostatic subjects. Oxygen consumption varied differently: it decreased by 3% in 0° test subjects, did not change in −4° test subjects and increased by 2 and 3% in −8° and −12° test subjects. Oxygen pulse rate did not alter in the recumbent test subjects and decreased by 11, 4 and 3% respectively in the other groups.

The comparison of the test with the physical exercise test performed before and after the 5-day space flights of Soyuz 6, 7 and 8 indicates that the heart rate, oxygen consumption and oxygen pulse rate of −4° and −12° test subjects are closer to those of the cosmonauts. It should be also noted that cardiorespiratory responses to the physical exercises were minimal both after the space flight and after the bed rest experiment. This makes a reliable comparison of the results difficult.

In conclusion it should be stated that the above experiments allow simulation of the early stage of human adaptation to zero gravity and, particularly, assessment of the contribution of gravity-induced blood redistribution to the development of the physiological reactions noted in test subjects and cosmonauts. Our findings provide material for a new interpretation of the functional disturbances in the sensory system of man (vestibular, optic, motor, interoceptive and taste) which occur during adaptation to zero gravity. They also indicate the major role of haemodynamic changes in the above disorders at an early stage of adaptation.

References

[1] C. A. Berry, Aerospace Med. **40**, 245 (1969).
[2] C. A. Berry, Aerospace Med. **41**, 500 (1970).
[3] C. A. Berry, 4th Int. Symp. on Basic Environmental Problems of Man in Space, Yerevan, USSR 1972.
[4] A. Fregly, A. Graybiel and M. Smith, Aerospace Med. **43**, 395 (1972).
[5] A. M. Genin and L. I. Kakurin, Kosm. Biol. Med. **6**, No. 4, 26 (1972).

[6] YU. N. PURAKHIN, L. I. KAKURIN and V. S. GEORGIEVSKY et al., Kosm. Biol. Med. 6, 47 (1972).
[7] C. E. ROSS and J. R. HORDINSKY, Skylab 1/2 Preliminary Biomedical Report, Dec. 1973 (p. 31).
[8] E. I. VOROBIEV, A. D. EGOROV, L. I. KAKURIN et al., Kosm. Biol. Med. 4, 26 (1970).
[9] E. I. VOROBIEV, YU. G. NEFEDOV, L. I. KAKURIN, et al., Kosm. Biol. Med. 3, 46 (1969).
[10] E. I. VOROBIEV, Kosm. Biol. Med. 4, 65 (1970).

Life Sciences and Space Research XIV — Akademie-Verlag, Berlin 1976

THE PREVENTION OF MOTION SICKNESS IN ORBITAL FLIGHT

A. Graybiel

Naval Aerospace Medical Research Laboratory, Pensacola, Florida, USA

A question has arisen whether zero gravity qualifies as just another "motion environment" in which motion sickness may be elicited, or if one or more undetermined etiological factors are present which render such terms as "motion environment" and "motion sickness" inappropriate. The question is of more than academic interest for the reason that in the forthcoming Shuttle Program we have either one problem or two: one if we are concerned solely with prevention, but two if we must tackle the problem of covert etiological factors as well. This problem points up the need to define motion sickness if investigators, world-wide, hope to resolve the matter in the most economical manner.

Meanwhile, substantial reliance must be placed on the use of antimotion sickness drugs and some recent findings are presented based on tests carried out in a rotating room using a new procedure. Whereas previous findings in our laboratory dealt with group responses, the new findings are not only valid for a group, but also valid for each subject tested. It was demonstrated that the effects of a drug may be efficacious for a group yet may be detrimental for one or more individuals in that group.

1. Introduction

This report falls into three parts. The first is an attempt briefly to define motion sickness. The second raises the question whether zero gravity qualifies as a motion environment, or whether we are dealing simply with "symptoms characteristic of motion sickness". The third part summarizes some recent findings dealing with the effectiveness of representative antimotion sickness drugs.

2. Motion Sickness: A Brief Definition

2.1. Terminology

The term motion sickness, proposed by Irwin [1] in 1881, has slowly gained wide acceptance because it met the test of convenience by its etiologic and symptomatic connotations. Objections have been raised to the use of "motion sickness" in scientific discourse but it is satisfactory for our use here in treating motion sickness as a clinical syndrome that may be experienced in conveyances (motion environments) of different types. *Experimental motion sickness* broadens the use of the term, especially if we include responses elicited in conducting a wide variety of laboratory tests which involve the vestibular system. The term *amotion motion sickness* identifies the category when the vestibular and organs are not stimulated such as may occur in simulators [2], when viewing movies projected on a hemispherical screen and during optokinetic stimulation [3]. In the clinic patients suffering from a variety of afflictions may experience *symptoms characteristic of motion sickness* either spontaneously or as a consequence of head and body movements. All of these distinctions are usually clear from the context.

2.2. Etiology

The side effects elicited as a result of a too rapid transition into certain stressful motion environments tend to fall into two categories [4]. One comprises the immediate reflex disturbances (IRD) such as nystagmus and postural and visual illusions. The other, i.e. motion sickness, is a delayed epiphenomenon comprising a constellation of symptoms and syndromes resulting from a failure in homeostatic mechanisms that permits irradiation of impulses or influences to sites outside the vestibular and visual systems where first-order responses have their origin. First-order responses, acting as stimuli, may elicit second and higher-order responses and so on, until the organism is generally involved. Although the vestibular system is essential for the elicitation of motion sickness (persons who have lost vestibular function are immune), an array of secondary etiological factors among which visual influences are prominent, are always involved even in healthy persons.

These secondary influences constitute an important yet poorly understood aspect of motion sickness and account for much of the lack of agreement (even among students of motion sickness) concerning its "cause". It may be assumed that any factor tending either to evoke or inhibit a response characteristic of motion sickness will affect susceptibility; "cold" sweating for example is elicited quickly in a warm environment but much later if the environment is cold [5]. Vision is ambivalent in this regard; sea sickness is more readily elicited below deck (or even with eyes closed) than while fixating on the horizon. The great importance of secondary influences is best demonstrated by circumstances in which they play a greater etiologic role than does stimulation of the vestibular end organs, i.e. psychological influences, notably conditioned responses. It is also important to distinguish between persons who are healthy and persons who are either in ill health or experiencing a concomitant functional disturbance that is not secondary or of a higher order of response or even a complication induced by such responses. For example, alterations such as changes in electrolyte balance and distribution of body fluids may precede the elicitation of motion sickness or be one of its consequences. The distinction is important from the standpoint of causation of symptoms.

2.3. Incidence

All normal persons are probably susceptible to motion sickness unless countermeasures are taken. All persons with bilateral labyrinthine defects [6] are probably immune. Persons with loss of vision remain susceptible including those who have never perceived light [7]. Variations in susceptibility are so great, however, that some persons do not experience motion sickness in conveyances.

2.4. Symptomatology and derived Phenomena

It is convenient to distinguish not only between acute brief episodes and chronic or prolonged manifestations of motion sickness but also between the responses *per se* and what might be termed derived phenomena.

Acute Motion Sickness. Symptoms of acute motion sickness useful in making a clinical diagnosis include pallor, sweating, salivation, drowsiness, and most important from a practical standpoint, the nausea syndrome. Release of the antidiuretic hormone [8] and urinary excretion of 17-hydroxycorticosteroids and catechol amines [9] are among the many biochemical responses that may be manifested within a short time. Acute motion sickness may be experienced under operational conditions and brief exposures are commonly used to test for susceptibility under laboratory conditions.

Derived Phenomena. From an analysis of the symptomatology in a typical instance of acute motion sickness the following characteristics emerge: (i) delay in appearance of symptoms, (ii) temporal summation, and either (iii) preservation of symptoms after cessation of stressful stimuli or (iv) acquisition of adaptation effects. From an analysis of the symptomatology obtained from repeated exposure of one person either to the same level of stress or to the same incremental increases in levels of stressful stimuli, variations in responses are predictable only within a certain range. From analysis of data on groups of persons, derived phenomena include individual differences in (i) susceptibility to motion sickness, (ii) rate of acquisition and decay of adaptation effects, (iii) transfer of adaptation from one motion environment to another, (iv) amount of performance decrement for comparable levels of severity of symptoms, and (v) responses to countermeasures.

Chronic Motion Sickness. A sharp distinction cannot be made between acute and chronic motion sickness which includes the recovery period or convalescence. A person severely ill for no longer than 1 to 2 hours may require "days" to recover. Moreover, it may become difficult or impossible to distinguish between the period of susceptibility (with some response being made to eliciting stimuli) and the recovery period unless there is a cessation of eliciting stimuli.

Symptoms. Symptoms of chronic motion sickness [10, 11] may or may not have an acute onset. With a gradual onset the symptoms do not appear, ramify and intensify in the order characteristic of acute motion sickness. One extreme is the so-called "avalanche phenomenon" and the other extreme is a syndrome characterized subjectively by apathy, drowsiness and fatigue, and objectively by sluggishness and performance decrement. It has been termed the sopite syndrome [12] and, under experimental conditions, it may be the only definite overt evidence of motion sickness. Although systematic observations are few, there is a clear indication that the time course for different symptoms may be different.

The recovery from motion sickness during continual exposure to the stressful accelerations is complicated. First, the sites of origin of symptoms must either be freed of eliciting influences or the intensity of restoring mechanisms must be greater than the eliciting influences. Thereafter restoration takes place continuously but not necessarily rapidly through homeostatic events and processes. Adaptation in the vestibular system is essential to prevent irradiation of vestibular activity beyond its confines. If a chemical etiological factor is also involved, considerable time may be required for its spontaneous disappearance. Thus, although it appears that adaptation of symptoms of motion sickness directly involves their underlying systems, this is not the case.

Derived Phenomena. For self evident reasons, in chronic motion sickness, the identification of such important derivatives as temporal summation, perseveration of symptoms and acquisition of adaptation effects may be difficult to determine.

2.5. Diagnosis

The diagnosis of acute frank motion sickness is so easy that it is not a matter of professional concern, but two sorts of diagnostic problems do exist. One involves criteria to indicate levels of severity ranging from "frank sickness" to the point where motion sickness is no longer present. Unfortunately, there has not been a systematic attempt to categorize these manifestations as first, second or higher order effects, and agreement is lacking with regard to diagnostic criteria. The other problem involves chronic motion sickness when there is not a clear relation between adequate eliciting stimuli and responses. The sopite syndrome is readily mistaken for fatigue or boredom unless specifically kept in mind.

3. Etiology of Motion Sickness in Orbital Flight

There are important differences in the opinions expressed regarding the etiology of motion sickness in orbital flight. At one extreme, zero gravity is regarded as a typical motion environment. At the other extreme, motion sickness is likened to a "vestibular storm", i.e. symptoms appear spontaneously as in disease states. Between these extreme positions a combination of pathological and functional etiological factors are thought to play a role. These differences in opinion have practical as well as theoretical implications and will be discussed briefly in the light of findings obtained in orbital [13—15] and parabolic flight [16, 17].

3.1. Orbital Flight

Pre-Skylab Findings dealing with motion sickness aloft are summarized in Table 1. The US data clearly rule out the vestibular storm theory as the sole etiological factor. The USSR data suggest, but do not prove, that zero gravity should be

Table 1

Manned Space Flight Programs

United States			Russia		
Program	Number of space pilots	Incidence of motion sickness	Program	Number of space pilots	Incidence of motion sickness
Mercury	6	0	Vostok	6	1
Gemini	10	0	Voskhod	5	3
Apollo Command Module	25	9	Soyuz	13	0
Apollo Lunar Landing	12	0			

regarded as just another motion environment. This conclusion is inferred from the absence of motion sickness in Soyuz astronauts, suggesting that ground-based training was effective.

Skylab Findings. Fig. 1 shows [15] that under operational conditions motion sickness was experienced aloft by five of the nine astronauts. In the Command Module symptoms appeared after approximately 30 minutes in one astronaut and after several hours in two additional astronauts. Symptoms appeared in two astro-

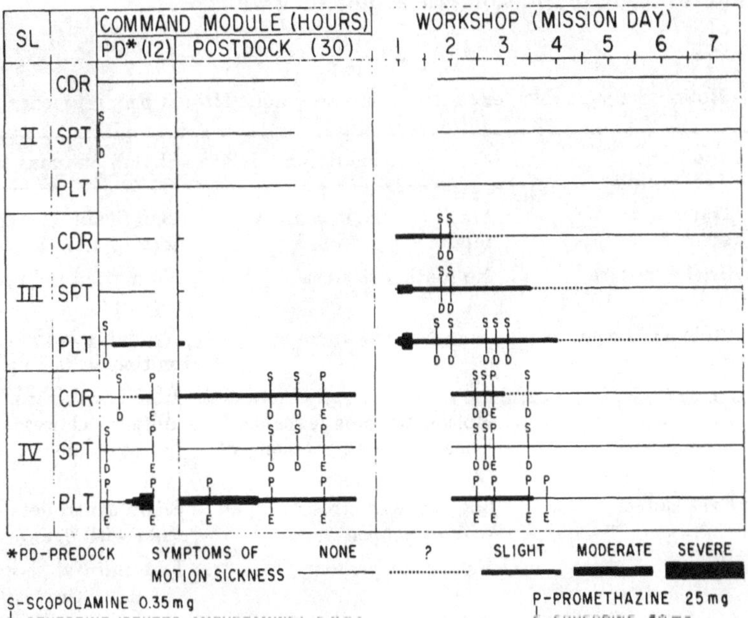

Fig. 1. Time course of motion sickness in five of nine astronauts during the orbital phase of Skylab Missions. The horizontal lines reflect two things. First, where the astronauts were based during the first week in orbit; the thickness and continuity of the lines indicate the onset and probable disappearance of symptoms of motion sickness. The vertical lines indicate when an antimotion sickness drug was taken and its composition.

nauts soon after their transition into the workshop where stimulus conditions were more stressful. In all five astronauts we are dealing with what might be termed chronic motion sickness.

Table 2 compares the findings in a slow rotation room (SRR) and in the orbital phase of Skylab missions. There is much resemblance between stimulus conditions in the two environments and, of course, substantial differences. Resemblances include the relation between activities that involve head movements or free floating activities and the elicitation of motion sickness. The second similarity involves the characteristic time course of the illness in both environments. Two differences deserve emphasis. First, the two vestibular organs are affected very differently during natural activities; the canals are stimulated normally while the constant stimulus to the otolith organs due to gravity is abolished. In the

SRR the canals are stimulated abnormally (except when turning in the plane of rotation) and the linear accelerations combine with the gravitational vector in stimulating the macular receptors in the otolith organs. The second difference concerns the potentialities aloft, at once limiting a person's natural movements and encouraging unnatural movements, that may result in unusual vestibular and visual sensory inputs, whereas, in the SRR, visual cues are familiar. In astronauts experiencing motion sickness it was demonstrated that (i) vestibular and visual inputs played important etiological roles, (ii) that symptoms were rarely severe, and (iii) that adaptation was acquired slowly.

Table 2
Motion Sickness (MS) experienced in two unique Motion Environments

Stimulus Conditions		Slow Rotation Room (SRR)	Skylab Missions (aloft)
Self-restraint; No head movements (HM)	Transition	Small change in sensory inputs	Significant change in sensory inputs
Self-restraint; No head movements (HM)	Brief exposure	No motion sickness	No motion sickness reported
Self-restraint; No head movements (HM)	Prolonged exposure	No motion sickness	Available evidence: no motion sickness (systematic studies not conducted)
Self-restraint; Head movements	Eyes open	Motion sickness (except when executed in plane of rotation)	Motion sickness (all rotations stressful)
Self-restraint; Head movements	Eyes closed	Slightly less stressful than with eyes open	Significantly less stressful than with eyes open
General activities:		Natural movements	Few natural movements possible. Unnatural movements generate unusual stimuli
All stressful stimuli "on-off". Alternate "disturbance-recovery"		Duration of motion sickness: "days"	Much the same as in SRR (Self-diagnosis difficult)
SRR: Predetermined angular velocity Skylab: Command Module to workshop		Worst case situation	Worst case situation
Programmed adaptation		Highly successful	More difficult than in SRR but possible

Parabolic Flight. Findings in parabolic flight must be included here although their simulation of orbital flight is obviously imperfect. A systematic series of tests using a procedure similar to that used in the workshop revealed (Fig. 2) that a small majority of the subjects were less susceptible than they were on the ground and a large minority of the subjects were less susceptible aloft than in preflight [17]. Moreover it is well known that persons not experiencing symptoms while seated may experience motion sickness under free floating stimulus conditions.

In summary, there are good reasons to regard stimulus conditions in zero gravity as a motion environment with unique features in which persons not only adapt but retain their adaptation, in many instances, for prolonged periods. If there is a concomitant etiological factor such as that caused by the headward shift of body fluids this factor must be relatively unimportant. Systematic studies, however, will be required to establish the presence and significance of any such additional etiological factors.

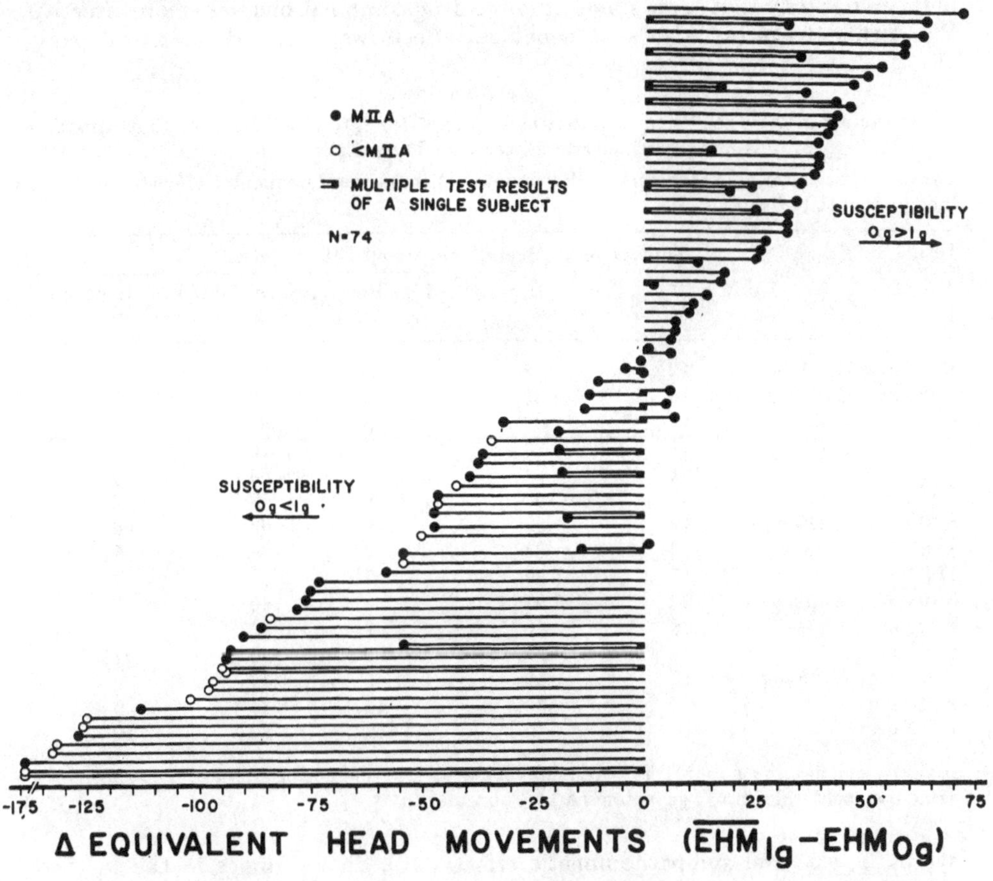

Fig. 2. Range of altered susceptibility among 74 subjects tested in zero *g* as generated by parabolic flight. Their susceptibility change is expressed as the difference between the number of equivalent head movements executed under zero- and earth-gravity conditions, reaching the endpoint of malaise IIA, or a limit of head movements imposed by the test conditions.

4. Antimotion Sickness Drugs

The efficacy of antimotion sickness drugs may be evaluated in any motion environment but it must be kept in mind that persons may differ in their susceptibility to motion sickness in different motion environments.

Most of our bioassays for testing the influence of drugs on motion sickness have been carried out using a slow rotation room (SRR) in a laboratory setting.

In the past, the procedure (described in more detail in [18]) involve (i) the selection of subjects, which included not only a comprehensive medical evaluation but also assessment of canalicular and otolith function, (ii) familiarization, (iii) generation of stressful stimuli by the execution of standardized head movements during rotation, (iv) the use of a motion sickness endpoint (or the execution of a given number of head movements), (v) administration of drugs and placebos using a 10-unit Latin-square design, and a double blind technique. The findings indicate not only that some single drugs or drug combinations were more effective than other drugs but also that beneficial effects were related to central para-

Table 3

Drugs ranked in Terms of Percentage Response (in 225 tests) to Usual Doses of 15 Antimotion Sickness Drugs administered in Three Experiments

Weighted response take account of detrimental effects; one detrimental effect equals two inconsequential effects.

Drug	Number of Subjects	Overall Beneficial Effectiveness			
		Unweighted %	Response Rank	Weighted %	Response Rank
P (25 mg) **E** (25 mg)	12	92	1	92	1
P (25 mg) **E** (50 mg)	18	83	2	79	3
S (0.3 mg) **E** (25 mg)	11	82	3	82	2
S (0.6 mg) A (5 mg)	11	73	4	73	4
S (0.3 mg)	11	64	5	64	5
S (0.6 mg)	30	63	6	63	6
S (0.6 mg) A (10 mg)	19	63	6	63	6
D (50 mg) **E** (50 mg)	19	63	6	63	6
D (50 mg)	17	59	7	50	7
S (0.3 mg) A (5 mg)	22	55	8	50	7
P (25 mg)	8	50	9	50	7
E (50 mg)	8	50	9	50	7
D (50 mg) **E** (25 mg)	12	42	10	42	8
A (10 mg)	17	35	11	35	9
E (25 mg)	10	10	12	10	10

A = d-amphetamine sulfate; D = dimenhydrinate; **E** = ephedrine sulfate; **P** = promethazine hydrochloride; S = *l*-scopolamine hydrobromide.

sympatholytic and sympathomimetic effects [19]. Shortcomings in the method were also revealed and it was evident that while the results had validity for groups of subjects, this did not apply to the individuals within these groups.

The changes in the procedure involved substitution of an incremental stress profile [20] for a predetermined constant level of stress, modification of the Latin-square design to provide a better placebo baseline, and distinguishing three categories of response: inconsequential, substantially beneficial (hereafter beneficial) and substantially detrimental.

Some of the findings in three experiments [21] are summarized in Table 3. The 15 drugs administered are ranked in terms of the percentage of doses (225 evaluations in 31 subjects) that were substantially beneficial. The weighted response takes into account detrimental effects; one detrimental effect is made the equivalent of two inconsequential effects. The efficacy of scopolamine alone or in combination with amphetamine was an expected finding but results involving promethazine and ephedrine were somewhat surprising. The combination promethazine and ephedrine, 25 mg each, would have reached 100% effectiveness if one subject had executed a few more head movements. This resulted despite the fact that, administered as single doses, ephedrine was notably ineffective. In combination they may be regarded as homergic drugs manifesting suprasummation.

There were seven substantially detrimental responses (bold D's in Table 4) involving four subjects and four drugs. Only eight beneficial responses were elicited in these four subjects and subject 3 accounted for half of the total. Indeed, subjects 4, 11 and 19 manifested the worst responses among the entire group; the average number of beneficial responses was 4. It is interesting that among subject 3's four best responses, promethazine (25 mg) plus ephedrine (50 mg) ranked best.

Table 4

Drugs Involved in Seven Detrimental Responses among 225 Evaluations in 31 Subjects

Subject No.	B	I	D*	Dimen-hydrinate (50 mg)	Promethazine (25 mg) Ephedrine (50 mg)	Scopolamine (0.3 mg) Amphetamine (5 mg)	Scopolamine (0.6 mg) Amphetamine (10 mg)
3	4	1	2	**D** – 23%	(B – best of 4 + 54%)	**D** – 17%	(I 0% change)
4	1	5	1	(I + 3%)	(only B response +58%)	**D** – 25%	(I + 9%)
11	1	3	3	**D** – 27%	**D** – 18%	(I 0% change)	**D** – 18%
19	2	5	1	**D** – 25%	(I + 13%)	Not adm.	(I – 13%)

*B = beneficial I = inconsequential D = detrimental

Moreover, the fact that subject 3 manifested his only inconsequential responses when scopolamine (0.6 mg) plus amphetamine (10 mg) was administered, indicated that the detrimental response with half the dose represented a valid test. The responses of subject 4 closely resembled those of subject 3. The responses of subject 19 resembled those of subject 11; his two beneficial responses (not shown in the table) followed the administration of promethazine (25 mg) and the combination dimenhydrinate (50 mg) plus ephedrine (50 mg).

The chief conclusion to be drawn from the findings in these experiments is that there are substantial individual differences in response to some of the best anti-motion sickness remedies available, implying that careful assessment should be carried out on an individual basis.

5. Discussion

An attempt has been made to demonstrate the resemblance between motion sickness in a rotating environment and in a weightless spacecraft. In the rotating environment not only can motion sickness be prevented by an incremental introduction to the stressful stimuli but also this method of acquiring adaptation results with little or no penalty. In other words, the CNS patterning does not render a person susceptible to symptoms on return to a stationary environment; indeed the subject is rendered less susceptible to motion sickness in some but not all conveyances provided there has not been too great a decay in the adaptation acquired in the rotating room.

On transition into orbital flight there is much evidence that symptoms of motion sickness may be elicited by activities that stimulate mechano-receptors in the vestibular organs and that visual inputs may also play an important etiological role. The vestibular inputs are unique for, even without moving, the constant stimulus due to gravity is absent thus ensuring an unusual input from the macular receptor system. After one week in orbit, when the motionsick Skylab astronauts had recovered, they were less susceptible to the stressful type of canalicular stimuli than pre-flight. If the findings using the same rotation test in parabolic flight are valid, then it would appear that some persons are more susceptible and some less susceptible to motion sickness in zero gravity than on the ground.

Yet the exposures are too brief to invoke substantial charges in endolymph and perilymph. Moreover, the fact that adaptation is acquired and retained in zero gravity suggests the involvement of CNS patterning rather than temporary changes in circulation or changes in electrolyte balance. Admittedly, none of these arguments completely rule out a concomitant unknown secondary etiological factor but it would seem to rule out a primary factor.

The prevention and treatment of motion sickness by means of drugs leaves much to be desired. Involved here is not only the selection of a drug used as a single dose on an individual basis but the assessment of drug therapy covering periods measured in days.

References

[1] J. A. IRWIN, Lancet, 907 (1881).
[2] J. W. MILLER and J. E. GOODSON, Aerospace Med. **31**, 204 (1960).
[3] TH. BRANDT, R. WIST and J. DICHGANS, Archiv. Psychiatry and Neurology **214**, 365 (1971).
[4] A. GRAYBIEL, Astronautica Acta **17**, 5 (1972).
[5] J. A. MCCLURE and A. R. FREGLY, Aerospace Med. **43**, 959 (1972).
[6] A. SJOBERG, Acta oto-laryng., Stockh., Suppl. **14**, 1 (1931).
[7] A. GRAYBIEL, Aerospace Med. **44**, 593 (1970).
[8] N. B. G. TAYLOR, J. HUNTER and W. H. JOHNSON, Canad. J. Biochem. Physiol. **35**, 1017 (1957).
[9] J. K. COLEHOUR and A. GRAYBIEL, Aerospace Med. **35**, 370 (1964).
[10] K. E. MONEY, Physiol. Rev. **50**, 1 (1970).
[11] W. H. JOHNSON and N. B. G. TAYLOR, Acta Otolaryngol. **53**, (1961).
[12] A. GRAYBIEL and J. KNEPTON, Aerospace Med., in press.
[13] G. L. KOMENDANTOV and V. I. KOPANEV, in: Problems of Space Biology **2**, 80 (1962).
[14] N. N. GUROVSKIY, I. I. BRYANOV and A. D. YEGOROV, Fifth Man in Space Symp., Acta Astronautica, **2**, No. 3–4 (1975).
[15] Proc. Skylab Life Sciences Symp. NASA TMX-5815, **1**, 169.
[16] YE. M. YUGANOV, I. A. SIDEL'NIKOV, A. I. GORSHKOV and I. I. KAS'YAN, Izv. Akad. Nauk SSSR **3**, 369 (1964).
[17] E. F. MILLER II and A. GRAYBIEL, Space Life Sciences **4**, 295 (1973).
[18] C. D. WOOD and A. GRAYBIEL, Aerospace Med. **39**, 1341 (1968).
[19] C. D. WOOD and A. GRAYBIEL, Clin. Pharm. Therapeutics, **11**, 621 (1970).
[20] W. J. OOSTERVELD, A. GRAYBIEL and D. B. CRAMER, Aerospace Med. **43**, 1005 (1972).
[21] A. GRAYBIEL, C. D. WOOD, J. KNEPTON, J. P. HOCHE and G. F. PERKINS, Aerospace Med. **46**, 1107 (1975).

Life Sciences and Space Research XIV — Akademie-Verlag, Berlin 1976

MINERAL AND NITROGEN METABOLIC STUDIES ON SKYLAB FLIGHTS AND COMPARISON WITH EFFECTS OF EARTH LONG-TERM RECUMBENCY

G. D. Whedon[a], L. Lutwak[b], P. Rambaut[c], M. Whittle[c], C. Leach[c], J. Reid[a] and M. Smith[c]

[a] National Institute of Arthritis, Metabolism, and Digestive Diseases, National Institutes of Health, Bethesda, Md, USA
[b] University of California at Los Angeles School of Medicine, Veterans Administration Hospital, Sepulveda, Calif., USA
[c] Biomedical Research Division, Johnson Space Center, NASA, Houston, Texas, USA

A metabolic study of the effects of space flight on various chemical elements, particularly those with special relevance to the musculo-skeletal system, was carried out on the nine astronauts who participated in the three Skylab flights of 28, 59 and 84 days in 1973—1974. The study required of the cooperating crewmen constant dietary intake, continuous 24-hour urine collections and total fecal collections for 21—31 days before each flight, throughout each flight and for 17—18 days post-flight.

Increases in urinary calcium and negative calcium balances during space flight were generally similar to those found in previous immobilization and bedrest studies. The persistence of these alterations in calcium metabolism throughout the flights suggested that calcium losses would contine in weightlessness for a very long time. Significant losses of nitrogen and phosphorus occurred, associated with observed reduction in muscle tissue. Both mineral and muscle losses occurred despite vigorous exercise regimens in flight. It was concluded that unless protective measures can be developed, capable musculo-skeletal function is likely to be impaired in space flights, ultimately to be conducted to Mars, of $1^1/_2$ to 3 years duration.

1. Introduction

After eight years of preparation, which took place before and during the Apollo-Moon space flight series, the National Aeronautics and Space Administration in May 1973 launched into orbit a cylinder 22 feet in diameter and 48 feet in length called the Skylab Orbital Workshop. The living quarters and storage area of this cylinder contained nearly all of the life support and scientific experimentation equipment and supplies for three separately launched three-man astronaut crews. The first crew remained in orbit for 28 days, the second for 59 days, and the last crew for 84 days.

Of the more than 100 scientific experiments or observations in the Skylab Program, 16 were devoted to the medical or physiological performance of man in the weightless state in space. This report provides the principal results of one of these studies, that on mineral and nitrogen balance. This "experiment" required of the cooperating astronauts fairly constant dietary intake, continuous 24-hour urine collections and total fecal collections for 21—31 days before each flight, throughout each flight and for 17—18 days post-flight for a total of 909 man days of metabolic study.

Prediction that the various stresses of space flight, particularly weightlessness, would bring about significant derangements in the metabolism of the musculoskeletal system has been based on various mineral and nitrogen balance study observations of long immobilized or inactive bedrest. The earliest was that of Deitrick, Shorr and Whedon [1] in 1948. Immobilization of four healthy young men in body casts for 6—7 weeks (Fig. 1) led to marked increases in urinary cal-

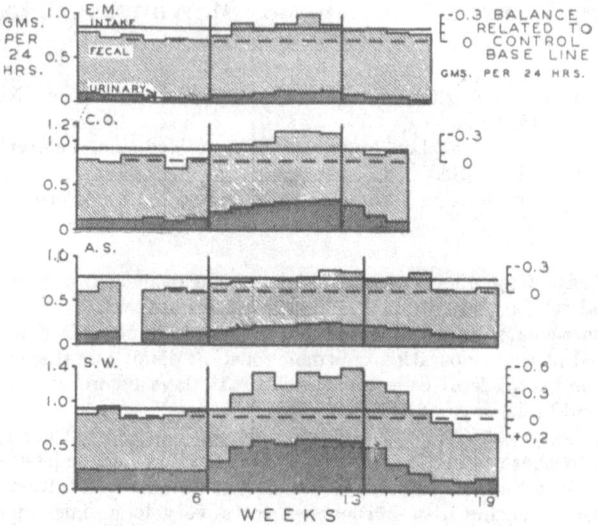

Fig. 1. Effect of immobilization on the calcium metabolism of four normal male subjects. In each subject the daily calcium intake was kept constant throughout all periods of the experiment. For each subject the control base-line (broken horizontal line) is an average of the total outputs of the last four control weeks. In this graph the intake and output are plotted upward from the zero base-line. (Reproduced by permission of Medical Clinics of North America, **35**, No. 2: 545, March 1951.)

cium and significantly negative calcium balances, and there were related losses of nitrogen and phosphorus. Several subsequent bedrest studies of normal subjects confirmed these substantial metabolic derangements [2]. The only attempt — not fully successful — at controlled metabolic observations in space flight prior to Skylab was performed by us [3] in conjunction with the 14-day Gemini 7 flight in 1965. That relatively short study revealed quite modest losses of calcium and phosphorus and varied changes in the metabolism of other elements.

In preparing for the opportunity in the Skylab flights to answer the scientific question: "in weightlessness would there occur disuse of bone and muscle that would lead to significant atrophy of these tissues?", the following methodologic question was extremely important: how best could one determine whether these changes would occur and, as accurately as possible, to what extent? Despite the sophistication of radioisotopic studies, by that method one only obtains rates over a period of time (and one still needs dietary control and collections for reliable isotopic studies), and bone densitometry, though improving, would still give only a total change from *before* to *after* a flight, with no greater precision. Hence, we felt that only by the ancient metabolic balance technique could we hope to get quantitative results which, if well done, could indicate changes in calcium and nitrogen in something close to milligram quantities and could describe *patterns* of changes, if not day-to-day at least week-to-week. No metabolic balance study was ever perfect, and this extraordinary venture under bizarre circumstances certainly lived up to that tradition; but the studies seem to have been done well enough to have provided some quite clear answers.

2. Procedure

A cardinal principle of metabolic study is that changes in the excretion of key nutrient elements, such as calcium or nitrogen, can only be interpreted as due to the influence or agent under test if environmental factors are kept as constant as possible from phase to phase and from day to day. One of the most important of these factors in metabolic study is the dietary intake. We had relatively little control over the selection by NASA, for various reasons of stability and acceptability, of the 70-odd space food items, the main types of which were solid foods in plastic packages, sticky semi-liquid foods in metallic tins, magnetically held in flight to a warming tray, and drinks in closed plastic cylinders. Selection was constrained for most items by the requirement of stability at room temperature in space for more than a year; only seven frozen food items could be used. In addition, in an effort to achieve best acceptability, many items were mixtures of foods and thus not conducive to exactness of composition in their production. Although these foods were far from ideal for balance studies, nevertheless by skillful, lengthy consultations with the astronaut crew members, our dietitians developed for each crewman sequences of six menus of similar elemental composition which were rotated steadily throughout the pre-flight, in flight and post-flight study phases. Whenever a particular food could not be consumed, a system of rapid calculation and provision of supplement tablets for pertinent elements helped to maintain dietary element constancy. During the flight phase, the crew's evening report to Ground Control included the relatively infrequent dietary omissions; rapid calculations were then made for deficits in key elements; the correct number of supplement tablets or capsules were prescribed up to the flight; then from previously stowed supplies, the prescribed supplement tablets (or capsules) were taken next morning.

Our relative success in dietary control is indicated in Table 1 which shows for one crewman, as representative of the group, the Commander of the 60-day flight, the phase-by-phase means of *actual consumption* of a few key elements and shows also the standard deviations from these means of day-by-day actual consumption.

Table 1

Mean (± s.d.) Daily Dietary Intake, Commander of Skylab 3

	Pre flight		In flight		Post flight	
kcal	2732	(±113)	2781	(±259)	2940	(±149)
Protein (g)	95	(±5)	85	(±11)	96	(±6)
Nitrogen (g)	15.2	(±0.9)	13.6	(±1.7)	15.4	(±1.0)
Phosphorus (mg)	1517	(±57)	1431	(±116)	1537	(±68)
Calcium (mg)	725	(±31)	729	(±72)	742	(±40)

Twenty-four-hour urine collections were made throughout the studies, but in flight, because of limitations in return weight and volume, approximately 120 ml aliquots of each days' urine collection were taken, frozen and returned to earth, using a very complex system because of the absence of gravity. In weightlessness there are difficult technical problems of collecting urine, separating liquid from air and taking a well-mixed measured aliquot, all without the aid of gravity which we take so much for granted. In addition, because volume cannot be measured in the weightless state in the same way as on earth, in-flight 24-hour urine volumes were determined by a tracer dilution technique, using lithium chloride, pre-injected into the 24-hour collection bags. In flight stool samples were dried in the workshop in flight and returned to earth *in toto*.

3. Results

Urinary creatinine excretion revealed considerably more fluctuation than is found under ideal urine collection conditions, but the values were consistent enough to indicate that average 24-hour urinary creatinine excretion was not changed by space flight.

Fig. 2 shows the urinary calcium excretion on the 28-day flight. [It bears the notation SL-2 because SL-1 is the NASA label for the first unit launched, the

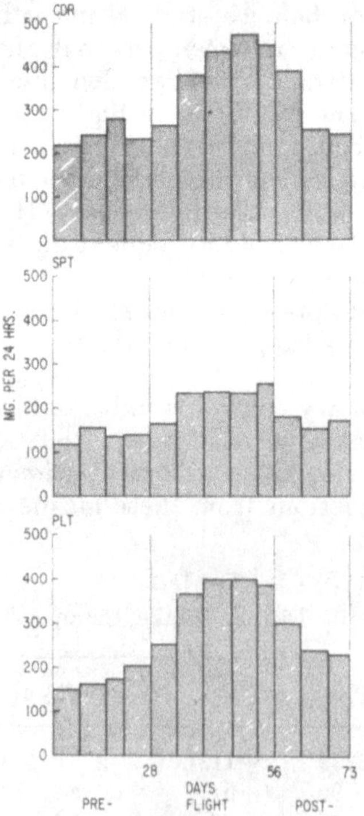

Fig. 2. Effect of space flight on urinary calcium excretion in the astronauts of the 28-day flight (SL-2).

Workshop, initially unmanned.] Urinary calcium in-flight increased steadily to a more or less plateau in virtually the same pattern and degree as in bedrest studies. As seen in bedrest, inter-individual variation occurred in degree of loss, but the peak reached during the latter part of flight was from 80% greater to more than double the control, pre-flight levels. During recovery, post-flight, urinary calcium excretion subsided promptly toward control levels. Fig. 3, for the 59-day flight,

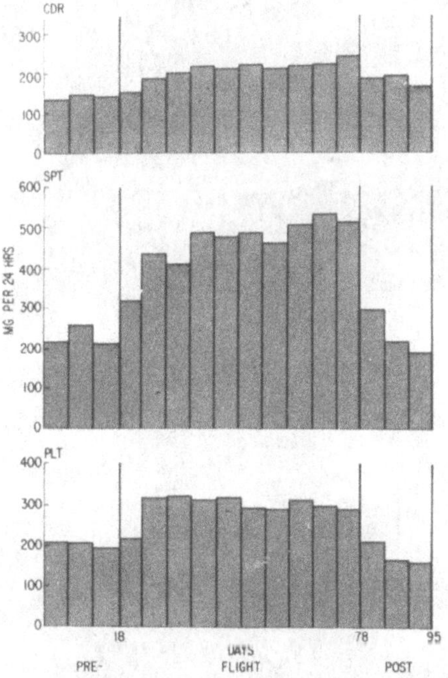

Fig. 3. Effect of space flight on urinary calcium excretion in the astronauts of the 59-day flight (SL-3).

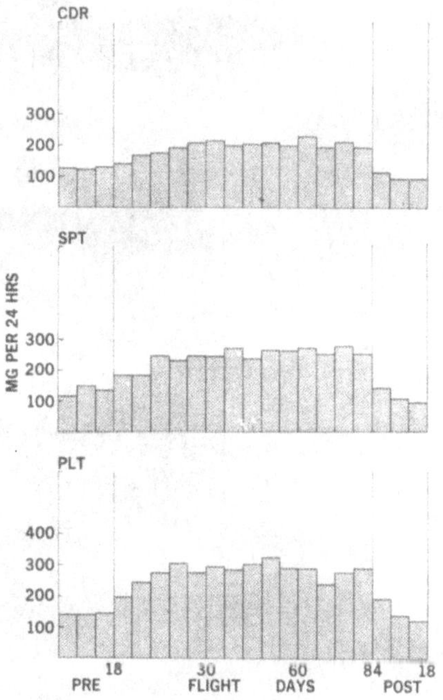

Fig. 4. Effect of space flight on urinary calcium excretion in the astronauts of the 84-day flight (SL-4).

shows the same pattern of gradual rise in urinary calcium in two crewmen and a rather abrupt rise in the third, and also shows inter-individual variation in degree of loss, which in one was to much more than double control levels. Urinary calcium data in the 84-day flight (Fig. 4) showed the same characteristics, plus an added point of interest: there was no suggestion of decline in the high level of excretion toward the end of the flight.

Urinary hydroxyproline increased in flight with considerable inter-individual differences; the mean increase for the six crewmen of the first two flights was 33%.

With regard to calcium balance, in the 28-day flight fecal calcium increased during flight in one crewman (CDR) and decreased slightly in the other two, and the balance became negative in two crewmen and changed imperceptibly in the third (SPT). In the 59-day flight, the negative shift in calcium balance was more marked (Fig. 5), resulting from increases in both urinary and fecal calcium. Analyses of fecal calcium for the 84-day flight are undergoing checking and re-analysis. In this longest flight significant decreases in bone density as measured in the *os calcis* (J. M. Vogel, using the Cameron-Sorenson technique) occurred in the two astronauts with the greatest increases in urinary calcium excretion.

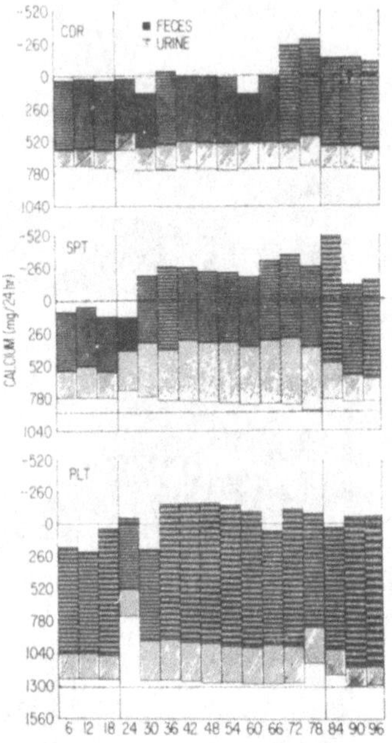

Fig. 5

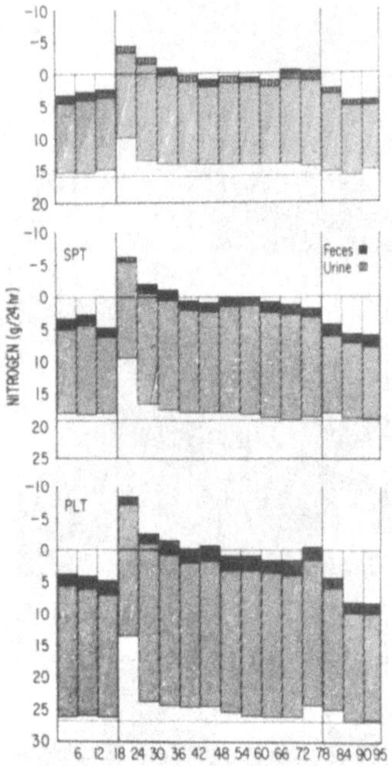

Fig. 6

Fig. 5. Calcium balances before, during and after space flight in the astronauts of the 59-day flight (SL-3). In this and subsequent balance graphs the data are plotted in conventional Albright-Reifenstein style, the intake downward from the zero base-line, then urinary (light shading) and fecal (heavy shading) excretion upward from the intake lines; shaded area above the zero base-line indicate negative balance or loss.

Fig. 6. Nitrogen balances before, during and after space flight in the astronauts of the 59-day flight (SL-3).

Phosphorus data (completed thus far for SL-2 and SL-3) showed a distinct increase in flight in urinary phosphorus, a small increase in fecal phosphorus, and negative balance in all six astronauts.

Nitrogen balance data revealed during flight a pronounced increase in urinary nitrogen excretion, while fecal nitrogen remained characteristically unchanged. In the 59-day flight (Fig. 6), the highly negative balance of the first 6-day period was due to the lowered intake resulting from anorexia during the first two to three days in the new weightless environment; nitrogen balance continued negative for a few weeks and then was only slightly positive despite high protein and calorie intake levels. The mean *shift* in nitrogen balance (for the six crewmen of the first two flights) from pre-flight phase to flight was 4.5 g/day. In the 84-day flight (Fig. 7) increases in nitrogen excretion of similar magnitude were observed.

With respect to other elements, significant, though variable and generally small losses occurred in flight in magnesium, potassium and sodium.

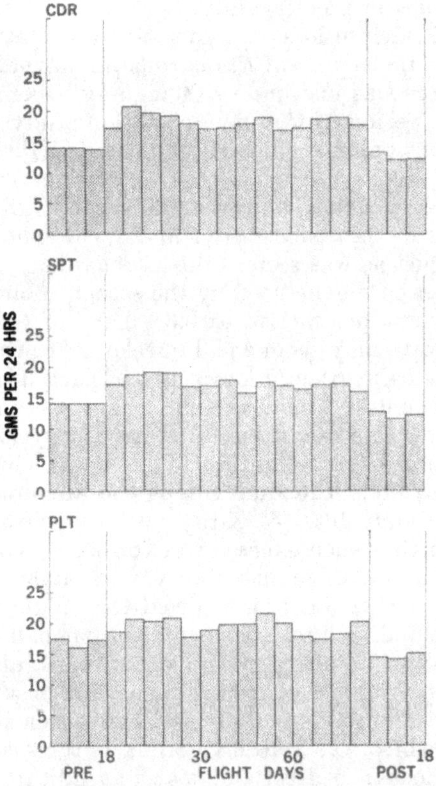

Fig. 7. Urinary nitrogen excretion before, during and after space flight in the astronauts of the 84-day flight (SL-4).

4. Discussion

Comparison of these results of calcium and nitrogen metabolic study of astronauts on the Skylab flights with prior studies of recumbency or bedrest on earth in general shows important and rather striking overall similarities, but there were some modest, interesting differences in pattern and degree.

Urinary calcium in both situations, space and bedrest, usually increased gradually over the first 3—4 weeks. In space the levels reached were from 50% to a little more than 100% greater than in the ambulatory control, whereas in immobilized bedrest the urinary calcium levels reached were always at least double the control level. The elevated levels in both situations were maintained in a plateau profile. In the longest bedrest observation [4], in which the normal subjects were *not* immobilized, the elevated urinary calcium subsided partially during the third and fourth months of bedrest to a level about the same extent above control as was seen in the astronauts. The important point to note, however, is that in this and other bedrest studies and in space flight, urinary calcium remained elevated and did not fall to normal until after the subjects began to bear weight on their feet again.

As regards *degree* of calcium loss, as indicated by the calcium balance results, the mean negative calcium balance of the astronauts on the first two flights after the first 12 days in space was just under 100 mg/day, whereas the mean negative balance for the similar period in the immobilization study was 154 mg/day. The negative *shift* in calcium balance similarly indicated slightly greater loss during immobilization, —280 mg/day versus —184 mg/day for the six astronauts. In *un*immobilized or *ordinary bedrest*, however, the negative shift in calcium balance was about the same as in the space study. Again, calcium loss persisted as long as bedrest continued, just as was seen in the astronauts.

The total calcium loss rate generated by the second month in space amounted to approximately 4 grams per month or 0.3—0.4% of total body calcium per month. Although this rate may seem small in relation to the whole skeleton, the general similarity to bedrest calcium losses and in particular the failure to show any tendency to abatement in 3 months' time makes it necessary to deal with an assumption that mineral loss might continue for a very long time, presumably many months. By 6 months of weightless flight, the calcium loss would amount to about 2.0% of total body calcium. This is the amount of loss at which in studies of paralytic poliomyelitis [5] X-ray visible rarefaction appeared in the distal tibiae. Assuming that such excess excretion would continue, then, the calcium loss rate of 0.3—0.4% per month observed in Skylab takes on clearer and more ominous significance when it is realized that flights to Mars and return, when ultimately conducted, will take from one and onehalf to three years.

The increased excretion of nitrogen and phosphorus, also similar to that in bedrest, reflected substantial loss of muscle tissue, which was clearly observed in the astronauts' legs. Both muscle and mineral loss occurred despite an exercise regimen on all flights, which was indeed vigorous on the second and third flights.

In degree of muscle loss in bedrest and space, as indicated by nitrogen balance measurements, there was a curious contrast. In bedrest studies, the majority of the nitrogen loss occurred during the early weeks, and the shift in nitrogen balance even in immobilized bedrest averaged only a little more than 1 gram/day. On the other hand, the pattern of increased nitrogen excretion in space flight indicated continuing loss throughout flight, and the mean shift in balance was —4.5 g/day. The significance of such large nitrogen balance changes for alterations in muscle protoplasm and protein stores must be explored by examination of excretory ratios of nitrogen to phosphorus, to sulfur and to potassium. Nitrogen metabolism studies obviously merit high priority in future space flight research opportunities because of the importance of maintenance of muscle function in long space flights.

We concluded, then, that it seems reasonable to predict musculo-skeletal "safety" in space flight for at least 6 months and probably for 9 months, but that musculo-skeletal function might be seriously impaired in crews on space flights of extreme duration *unless protective measures can be developed*. The likelihood of need for protective measures in space flight is accentuated by the following consideration: although the bone losses thus far observed have been reversible upon return to normal gravity (or to ambulation after bedrest), no observations are available to permit estimation of a magnitude of loss that would represent a "point of no return". Thin trabeculae in bone can be returned to normal thickness but, from our present understanding of the adult skeleton, completely lost trabeculae cannot be restored. In anticipation of these results, NASA about 1967 began sponsoring studies, using bedrest as the analogue for weightlessness, designed to test a variety of hormonal, dietary and physical measures which might protect the skeleton in long space flight. So far, there have been no convincingly positive results in these time-consuming studies, but several leads for further investigation have been developed.

In a final comment, these observations in space may have significance for earth medicine. In reminding us of the deleterious effects of disuse on bone mass, they re-emphasize the importance of direct physical longitudinal stress (weight bearing) to the integrity of bone. In continuing search for effective protective or even therapeutic measures in osteoporosis, the most prevalent clinical disorder of bone, more attention might well be given to determining just how weight bearing affects the balance between bone formation and loss. Once we know the specific characteristics of that mechanism, perhaps we can take practical advantage of that knowledge in dealing with osteoporosis more effectively.

References

[1] J. E. Deitrick, E. Shorr and G. D. Whedon, Am. J. Med. **4**, 3 (1948).
[2] S. J. Birge, Jr., and G. D. Whedon, in: Hypodynamics and Hypogravics, Academic Press, New York 1968 (p. 213).
[3] L. Lutwak, G. D. Whedon, P. A. LaChance, J. M. Reid and H. S. Lipscomb, J. Clin. Endocr. Metab. **29**, 1140 (1969).
[4] C. L. Donaldson, S. B. Hulley, J. M. Vogel, R. S. Hattner, J. H. Bayers and D. McMillan, Metab. **19**, 1071 (1970).
[5] G. D. Whedon and E. Shorr, J. Clin. Invest. **36**, 966, Part II (1957).

Life Sciences and Space Research XIV — Akademie-Verlag, Berlin 1976

HUMAN TOLERANCE TO ACCELERATION AFTER EXPOSURE TO WEIGHTLESSNESS

A. R. KOTOVSKAYA

Institute of Medical and Biological Problems, Moscow, USSR

The major rôle in the genesis of varying human tolerance to decelerations that follow weightlessness is evidently played by hypodynamic and hydrostatic factors.

Long disuse of compensatory antigravity mechanisms in weightlessness may bring about their deconditioning and reduction of their functional capabilities, and may finally affect general tolerance of crewmembers to decelerations.

Laboratory experiments demonstrated changes in the human tolerance to G_x accelerations of varying duration (from 3 to 100 days) and tested the efficacy of different countermeasures. A decrease in the human tolerance to $+G_x$ is on the average $-2.0g$. It should be noted that an elongation of simulated weightlessness (from 7 to 100 days) caused no further decrease in the $+G_x$ tolerance.

Our investigations helped to assess the threshold of human tolerance to accelerations after an exposure to simulated weightlessness and to delineate the value of real risk. The tolerance limit to $+G_x$ accelerations which followed simulated weightlessness of the above duration ranged from 9.5 to $13.0g$, averaging $11.6 \pm 1.6g$.

The information on the tolerance of Soviet and American astronauts to decelerations shown during re-entry in real space flights give support to the laboratory results and predictions.

When carrying out space programmes it is important to understand clearly what influence zero gravity may exert on man and to what extent he can maintain his mental and physical performance during re-entry and recovery. This problem is very important because, on the one hand, asthenization of the organism may develop by the end of mission and, on the other hand, deceleration during re-entry may reach high values. The success of the entire mission depends essentially on the human tolerance to deceleration during re-entry.

Human tolerance to accelerations depends on many factors, including the value, direction, duration and other parameters of the physical effect as well as on the functional state of the cosmonaut before re-entry. At present information on the human tolerance to accelerations has been obtained from real space flights of different duration (up to three months in Skylab 4) and ground-based bed rest experiments of comparable duration (up to 100 days). Therefore, it is very important to review and analyse the available data.

It has been demonstrated in many experiments that decrease of gravity effects and diminished motor activity reduce significantly human tolerance to orthostatic tests which may result in precollapse and collapse states. Beckman et al. [1], Miller and Leverett [2] and the present author [3, 4] have shown a reduction of human tolerance to $+G_x$ after simulated weightlessness.

It seems appropriate to discuss first of all the changes which typically appear in men exposed to maximally permissible accelerations $+G_x$ after bed rest. In order to plan prolonged space missions it is important to identify the level of the decrease of human tolerance after an exposure to weightlessness, to determine whether human tolerance will continue to deteriorate with flight time or will stabilize at a certain level, and to assess the role of mechanisms responsible for the decrease of tolerance to acceleration after simulated weightlessness. This problem is of particular importance because its solution may help develop countermeasures against unfavourable effects of weightlessness.

Our investigations have shown that, regardless of the length of simulated weightlessness (from 7 to 100 days), during and immediately after an exposure to acceleration typical changes in the physiological reactions of the body were recorded. Subjectively, all the test subjects were much less tolerant of exposure to acceleration after simulated weightlessness. Muscular weakness upon static tension to maintain the position, impaired respiration and earlier development of visual disturbances up to complete loss of vision were noted. In all cases an exposure to identical accelerations required higher functional tension in the physiological systems of the body than in the control. The most common cardiovascular changes were tachycardiac increase with subsequent development of bradycardia in some cases. The most frequent after effect was prolonged and cyclic recovery of arterial pressure and heart rate. During this period the pulse pressure decreased, sometimes significantly, due to an increase of the diastolic pressure which remained higher than in the norm. Arterial pressure returned to normal within 30—60 min and sometimes 1.5—2.5 hours after the exposure. Under usual conditions the time for cardiovascular parameters, including arterial pressure, to return to normal is not more than 10—15 min.

The development of autonomic disorders which limited further increase of acceleration in most experiments was followed by appreciable ECG changes suggesting cardiac disorders. These changes included significant and progressive decrease of heart rate and other rhythmic disorders (extra-systoles, migration of the source of myocardial excitation, blockage). The $+G_x$ acceleration at which bradycardia developed was on the average $11.2 \pm 1.05 G_z$ (at $14.0 \pm 0.86 g$ in the control study). After bed rest visual disturbances became more pronounced and usually developed earlier than decrease in the heart rate; sometimes they coincided. The visual disorders appeared at accelerations of $11.6 \pm 0.45 G_z$ and before bed-rest at $13.6 \pm 0.35\ g$.

As to the respiratory function, the breathing depth showed the greatest changes. After acceleration of 6—8g it became significantly lower than the control and than normal values established for healthy people under normal conditions. This decline of the respiratory volume was at first made up for by increase in the respiration rate and, therefore, the minute respiration volume did not differ from the baselines. Later, in spite of greater increase in the respiration rate, the minute volume of respiration decreased. Superficial breathing led to its reduced effectiveness and to decrease in amount of oxygen consumed. This affected the value of oxygen deficiency which, as a rule, was greater after bed rest than in pre-test centrifugation.

It was very important from the practical point of view to find out whether the degree of tolerance reduction grew with an increase in length of bed rest. No strict correlation between the decline of tolerance to acceleration $+G_x$ and time of simulated weightlessness was established. After 3 days of bed rest acceleration tolerance remained essentially unaltered and after 7—20 days of bed rest it was reduced by 2.2g on the average. Prolongation of simulated weightlessness to 100 days brought about no further decrease of acceleration tolerance. This observation was also confirmed by the study of physiological reactions of man. There are grounds for believing that during prolonged bed rest the human body develops a peculiar adaptation to simulated weightlessness and that its asthenization measured with respect to G_x acceleration tolerance does not increase with an increase of bed rest.

It has been demonstrated that unfavourable effects of simulated weightlessness can be alleviated by means of physical exercises and drugs. The combination of these countermeasures is more effective in increasing acceleration tolerance than each of them used separately. The tolerance limit after simulated weightlessness when these countermeasures were used was the same as in control studies. However, this tolerance was only achieved with great strain on the physiological systems of the body. The cardiovascular system showed the largest changes. This indicated that the countermeasures did not eliminate a state of deconditioning.

The experience gained in the Soviet and American space missions shows no noticeable deterioration of work capacity or acceleration tolerance. Physiological changes of nervous-emotional origin developed during weightlessness before re-entry. By the end of flight all cosmonauts showed increased heart rate and respiration rate. This is very likely to occur in anticipation of the re-entry as this is a very demanding terminal stage of the flight. Essentially all cosmonauts displayed increased sensitivity to acceleration. They had the sensation of being exposed to acceleration of several G, although it was only the beginning of the re-entry. According to most cosmonauts, during an exposure to maximum accelerations their feelings were similar to those during simulated descent in the centrifuge. However, many cosmonauts reported that subjectively during actual flight they tolerated decelerations worse than during pre-flight centrifugation.

Physiological reactions during re-entry and recovery were more pronounced in all the cosmonauts than during pre-flight exposures. Heart rate and respiration rate were higher than during centrifugation. In some Soviet and American cosmonauts heart rate was 160—190 beats/min. The highest heart rate values were recorded during descent immediately after attainment of maximum deceleration at the stage of re-entry and braking after parachute deployment. In some cosmonauts ectopic beats were recorded electrocardiographically, their number being larger than in similar exposures on the ground. In some cases cosmonauts reported visual disturbances during re-entry. Such disturbances did not occur during centrifugation. Greyout during re-entry was suggestive of the deconditioning effect of the orbital flight on the cardiovascular system. Nevertheless, the work capacity of cosmonauts during recovery remained adequately high.

The study of the results of the Soviet and American space missions has given an answer to a very important question, from the practical point of view, on human tolerance to accelerations in flights of varying length. It has been established that with an increase of the flight duration no deterioration of acceleration tolerance occurs during recovery. These results of actual space flights are in good agreement with the data of ground-based experiments and were predicted long before man went into space. In ground-based experiments more marked changes were observed than after real weightlessness. This difference can be attributed to the fact that in ground-based experiments the test subjects were exposed to maximum accelerations whereas in flight they were exposed less extreme values. In other words, the acceleration profile in the centrifuge was much more demanding than that in real flight.

However, one should take into consideration the possibility of emergency situations in flight when accelerations will be high and asthenization of the body will be pronounced. The laboratory experiments suggest that cosmonauts can well endure accelerations of 9.5—13.0g (average, 11.6 $\pm$ 1.6g) with an onset rate of 0.2g s^{-1} after a 100-day exposure to weightlessness. The average value of acceleration which can be tolerated during re-entry can be increased to 12.5g by the use of different countermeasures (physical exercises, drugs). The major role in the changes of human tolerance to acceleration turns upon hypodynamic and hydrostatic factors. Long-term disuse of compensatory, antigravity mechanisms in weightlessness may lead to deconditioning and decline of functional capabilities and finally affect the human tolerance to acceleration in general. It is important to investigate the mechanisms responsible for the decrease of human tolerance to transverse acceleration after an exposure to weightlessness.

We shall discuss certain changes in physiological reactions which develop as a result of space flight and which may adversely influence human tolerance to acceleration. At present it has been established that diminished muscular activity induces polymorphic disorders far beyond locomotor functions. The experimental findings give evidence for the development of muscular atrophies, demineralization of bone tissue, decrease of the breadth of the heart, orthostatic hypotension, etc. According to Kakurin [5], long-term restraint of sensory analysers, circulation receptors and the neuromuscular system may bring about functional disturbances. Their origin seems to be associated with deconditioning of nervous and humoral mechanisms of the regulation of circulatory homeostasis which have been developed in the course of evolution. They largely depend on the state of the receptor field of the skeletal muscles.

It is known that the muscular system plays a major role in the development of compensatory reactions of the human body to acceleration. The level of the muscular development, especially of brachial and abdominal muscles, determines to a certain extent individual tolerance of men to acceleration. The tension of brachial and abdominal muscles increases tolerance to $+G_z$ acceleration by 1.8 $\pm$ 0.3g on the average. The contribution of muscular activity to the tolerance of transverse accelerations is very important because in these circumstances man actively fights against accelerations which cause deformation of the chest and abdominal cavity and disturb the act of respiration. The increased load on the chest impairs inhalation. In order to breathe in, great physical efforts are required to overcome the acceleration force which squeezes the chest and abdomen and forces them in the exhalation position. As is known, upon high accelerations the act of respiration involves not only the muscles of the chest but also auxiliary muscles of the torso.

Other muscular groups of the torso and the limbs are also in a strained position or under static load typical of acceleration exposure, in order to maintain man's posture. Also, inclination of the couch forms the longitudinal head-to-feet component $+G_z$ which is the larger the closer the man's position is to the sitting position. This longitudinal component gives rise to general haemodynamic disorders. Therefore, the contribution of the muscular system to the development of compensatory reactions in response to this component naturally increases.

The reduction of the breathing level in our post-bed rest experiments was associated, first of all, with decrease in the capacity for tonic tension of the musculature involved in respiration during an exposure to acceleration. In addition, gas exchange takes place in the alveoli; therefore alveolar ventilation depends on the air volume penetrating the alveoli and not on the air volume entering the airways. The greater reduction of the breathing level during a post-bed rest exposure to acceleration in comparison with the baselines may have induced deficient alveolar ventilation and more marked changes of external respiration as a whole. Undoubtedly, this was brought about by diminished motor activity during bed rest. In fact, the test subjects who carried out physical exercises in bed rest showed smaller changes in external respiration. The resulting blood redistribution in the pulmonary circulation results in the disturbance of the normal proportions of the blood flow and alveolar air in the lungs which aggravates the lowered oxygen supply to the arterial blood.

Bed rest produces a number of changes in the cardiovascular system. These changes include orthostatic hypotension, change in the vascular tone, decrease in the contractile capacity of the myocardium, disorder in the regulatory mechanisms of haemodynamics, decrease of the circulating blood volume, etc. During exercise and orthostatic tests cardiovascular deficiency is often noted. During an exposure to transverse accelerations the cardiovascular system undergoes far greater stress. Under these conditions oxygen disbalance, blood redistribution and pooling in the lungs and other organs create great difficulties for the cardiovascular system of a healthy young man leading a normal mode of life.

The presence of a longitudinal head-to-feet component of acceleration due to an angle of the couch imposes additional strains on the cardiovascular system because this impairs venous return of the blood to the right heart. Naturally, the development of changes in the contractile capacity of the myocardium, vascular tone, regulation, etc. in weightlessness inevitably affects the cardiovascular function during the subsequent exposure to acceleration as well as the human tolerance as a whole. This was evidently the factor which caused an earlier development of progressive bradycardia, visual disturbances (up to light loss), ECG changes, etc. during an exposure to acceleration which followed bed rest.

Some cardiovascular disorders recorded during the post-bed rest exposure to accelerations shed light on their origin. In our investigations we observed three cases of cardiac and sternal pain which needed medical aid after 62—100 days of bed rest. One of the test subjects experienced pain during an exposure to acceleration which coincided with progressive decrease of heart rate. The two others displayed pain 1.5—2.5 hours after termination of the exposure during the first post-hypokinetic day. It should be noted that during the after period all the test subjects demonstrated instability, lability and cyclicity of arterial pressure and show recovery.

Cardiac pain may be the result of autonomic crises in the men which could have been caused by excitement and fear, and physical stress during an exposure to acceleration as well as disturbed regulation of physiological functions, including the vascular functions. The show recovery of arterial pressure was often followed by a marked decrease of pulse pressure due to an increase in the diastolic and a decrease in the systolic pressure. The cyclic recovery of the arterial pressure when the diastolic pressure tended to remain at an elevated level for a long time was suggestive of vascular spasms. It can be considered that changes in the vascular reactivity, and enhanced susceptibility to spastic reactions may have influenced the coronary vessels and induced cardiac pains. The pains were eliminated by spasmolytic drugs.

Disorders in the regulation of physiological functions, including the vascular bed may have been the cause of the above disturbances. It should be stressed that these observations are of practical importance from the point of view of prediction of the state of health of cosmonauts upon recovery and evaluation of the required medical aid. As is known, vascular activity is associated with the structure of vascular walls. Changes in the elastic properties of vessels of the legs due to prolonged bed rest were found by Pestov [6]. The author reported that the vessels became relatively rigid and partially lost their flexibility and contractility. According to Pestov these changes resulted from the reduction of the absolute value and relative fluctuations of the hydrostatic blood pressure on the vascular walls which are characteristic of the erect posture, running, walking, etc. Therefore, prolonged disuse of compensatory antigravity mechanisms may lead to deconditioning and decline of their functional capabilities. This is the reason why changes in the vascular tone and elastic properties of the vessels of the legs caused diminished blood inflow to the right heart upon transition to the upright position, thus impairing orthostatic tolerance. This is the real meaning of the decrease of the hydrostatic blood pressure for the test subjects changing their recumbency for the erect position.

The reduction of the circulating blood volume which is typical of weightlessness may aggravate circulatory disturbances upon an exposure to acceleration because it diminishes the effective volume of circulating blood. The heart plays an important part in the compensation of circulation during an exposure to acceleration. Blood redistribution and reduction of gas exchange in the lung as well as circulation disorders in other organs are accompanied by general hypoxia and increased load on the heart. Naturally, the poorer functional state of the myocardium and its reduced contractile ability induced by prolonged bed rest influence the effectiveness of the cardiac function during an exposure to acceleration.

It can be assumed that, if the body responds to post-hypokinetic acceleration with a higher heart rate and later with an earlier compensatory disorder (progressive decrease of the heart rate and more marked ECG changes), then the cardiac muscle is exposed to less favourable conditions than during control studies. Nevertheless, the vascular tone and, particularly, its neuro-humoral regulation play the major role in the compensatory reactions. Therefore, it is evident that the study of the contractile capacity of the myocardium, vascular tone and different aspects of their regulation upon a post-hypokinetic exposure to acceleration needs further thorough investigations. The problem of how to maintain high human tolerance to acceleration after an exposure to weightlessness is still very important.

In future the crew members and passengers of the Shuttle vehicles and orbital spacecraft may be subjected to longitudinal $+G_z$ accelerations. In these situations deconditioning and asthenization of the human body may cause significant decrease of the acceleration tolerance. Obviously, an anti-G suit should be recommended to enable endurance of accelerations of this type. It is important to determine permissible levels of $+G_z$ accelerations, taking into consideration the weightlessness effects, and thus to assess the hazards involved.

References

[1] E. L. Beckman, K. R. Coburn and R. M. Chambers, Aerosp. Med. **32**, 1031 (1961).
[2] P. B. Miller and S. D. Leverett, Aerosp. Med. **36**, 13 (1965).
[3] A. R. Kotovskaya, R. A. Vartbaronov and S. F. Simpura, Problems of Space Biology **6**, 106 (1967).
[4] A. R. Kotovskaya, R. A. Vartbaronov and S. F. Simpura, 18th Int. Astronaut. Congr. Belgrade 1967 (p. 123).
[5] L. I. Kakurin, The Influence of Limited Muscular Activity on the Human Organism, Doctoral Thesis, Moscow 1968.
[6] I. D. Pestov, paper presented at 19th Congr. of Int. Astronaut. Fed. USA 1968.

Life Sciences and Space Research XIV — Akademie-Verlag, Berlin 1976

ON THE MECHANISMS OF CHANGES IN SKELETAL MUSCLES IN THE WEIGHTLESS ENVIRONMENT

V. S. Oganov and A. N. Potapov

Institute for Medico-Biological Problems, Academy of Sciences of the USSR, Moscow, USSR

Some characteristics of muscle contraction and mechanical properties of two muscles (*M. soleus* and *M. extensor digitorum longus*) in Wistar rats after 22 days of weightlessness have been investigated. On the second day after return to earth, the following changes were evident: slowing of twitch responses of the muscles studied; shortening half tetanic contraction time (defined by point of interactions of the increasing curve with 50% level of the peak value) in *soleus*; a rise of tension in both muscles as shown by the curve "length-force"; an increase of twitch/tetanus ratio and fatigability in both muscles. During repeated study of muscle properties, on the 26th day after return to earth, there were not any significant changes in values of most the above mentioned indices except the diminished strength of *soleus*.

Shortening of contraction time in slow antigravitary *soleus* muscle is believed to be a sign of adaptive change in its characteristics resulting from unloading under the conditioning of weightlessness. Increased stiffness and diminished strength of muscles are considered as functional signs of atrophic processes, developing in the experiment. Dynamics of functional changes, conditioned, as we believe, mainly by weightlessness and correlated with morphological and biochemical data obtained on analogous material, allows us to postulate a reversible character of the changes described.

Although the countermeasures against the adverse effects of weightlessness have appreciably contributed to the successful accomplishment of Skylab missions, the problem of how to maintain good work capacity of space travellers in future has not yet been solved. Serious attention should be paid to the maintenance of high functional activity of the neuromuscular system [1, 2] which should be adequate for the normal work of the musculo-skeletal system on return to the earth.

The present communication deals with studies of mechanism and methods of adaptation of the intact (unprotected by countermeasures) neuromuscular system to the weightless environment. The paper gives analytic information about weightlessness effects on the functional parameters of skeletal muscles and the state of the neuromuscular system during relative disuse of the skeleto-muscular apparatus of animals.

White rats of the Wistar strain flown aboard the Cosmos 605 biosatellite and tested in hypokinetic experiments were used to examine isometric contraction *in situ* and *in vitro*. Dogs with disused skeletal muscles were also used to study isotonic muscle contraction during locomotion. This was done with a combination of traditional techniques (electromyography, cyclogrammetry) and a newly developed method of dynamomyography to record the *in vivo* strength of individual muscles [3].

The data collected from the rats exposed to weightlessness for 22 days were compared with the information obtained from two control groups. The first group was kept in the vivarium (Control 1) and the second group was nearly synchronously exposed to the flight profile (Control 2). The changes were thought reliable if there were significant differences (according to the Student *t*-test) between the flight group and both controls or between the flight group and Control 2. An examination of the flight rats on the second day of recovery showed a marked weight loss of both muscles under study — *M. soleus* and *M. ext. digitorum longus*; the loss was larger in the first muscle. On the 26th post-flight day weight decrement persisted in the fast muscle only.

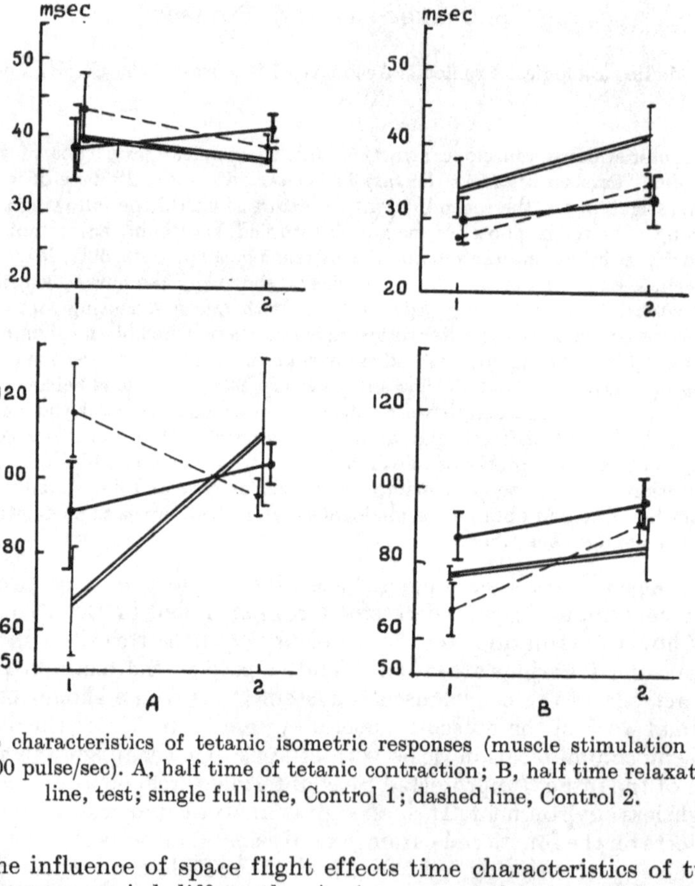

Fig. 1. Time characteristics of tetanic isometric responses (muscle stimulation with a frequency of 100 pulse/sec). A, half time of tetanic contraction; B, half time relaxation; double line, test; single full line, Control 1; dashed line, Control 2.

Under the influence of space flight effects time characteristics of twitch and tetanus responses varied differently. An increase of the time of development of muscle tension and half relaxation time of twitch of *M. ext. digitorium longus* and a similar trend in *M. soleus* were observed only on the second post-flight day. In studies of time characteristics of tetanic responses the only significant change was decrement of time of half tetanic contraction in *M. soleus* of flight rats in comparison with Control on the second post-flight day (Fig. 1).

An investigation of strength characteristics demonstrated that the tension amplitudes in twitch and tetanus stimulations did not differ significantly in the two control groups. Direct measurement of tetanic amplitudes on the 26th post-flight day (this was not done on the 2nd day due to technical difficulties) showed a significant decrease in flight in *M. soleus* (Fig. 2).

In order to assess mechanical properties of muscles, the relationship between the muscle tension and passive stretch, i.e. the length tension curve, was studied. As follows from Fig. 3, the comparison of these curves indicated increased stiffness of flexor muscles on the 2nd post-flight day which was statistically significant for the slow muscle only. By the 26th day the elastic properties of the slow muscle recovered whereas increased stiffness of the flexor *M. ext. dig. long.* became statistically significant (Fig. 3). Fatigue ability of both muscles on the 2nd post-flight day was significantly lowered and on the 26th day did not differ from that in the two control groups.

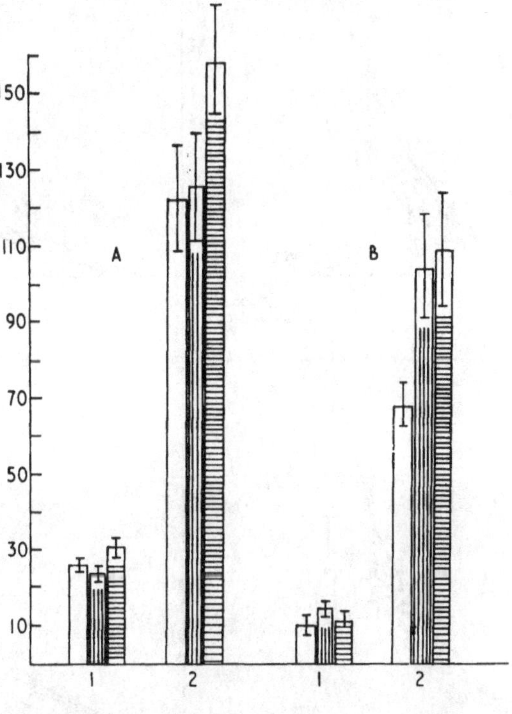

Fig. 2. Amplitude of twitch (1) and tetanus (2) in grammes in *M. ext. digitorum longus* (A) and *M. soleus* (B) on the 26th post-flight day. Unshaded bar, test; vertically shaded bar, Control 2; horizontally shaded bar, Control 1.

Thus, it can be stated that the 2nd day after the termination of 22-day orbital flight revealed statistically significant changes of fundamental characteristics of muscle tension and elasticity: increase of contraction and relaxation twitch responses of the fast muscle; decrease of half time of tetanic contraction of *M. soleus*; increase of muscle stiffness; increased fatigue and weight decrement. Repeated examination on the 26th post-flight day showed significant changes in increased stiffness and weight decrement of *M. ext. digitorum longus* and decreased tetanic amplitude of *M. soleus*.

Previous laboratory investigations demonstrated that, as the result of load decrease on the shoulder muscles due to amputation of the distal part of the forearm, slow and fast muscles also showed dissimilar functional changes. For instance, the slow muscle *cap. med. m. tricipitis brachii* of the operated limb displayed weight losses and strength reduction as well as speeding up of contraction

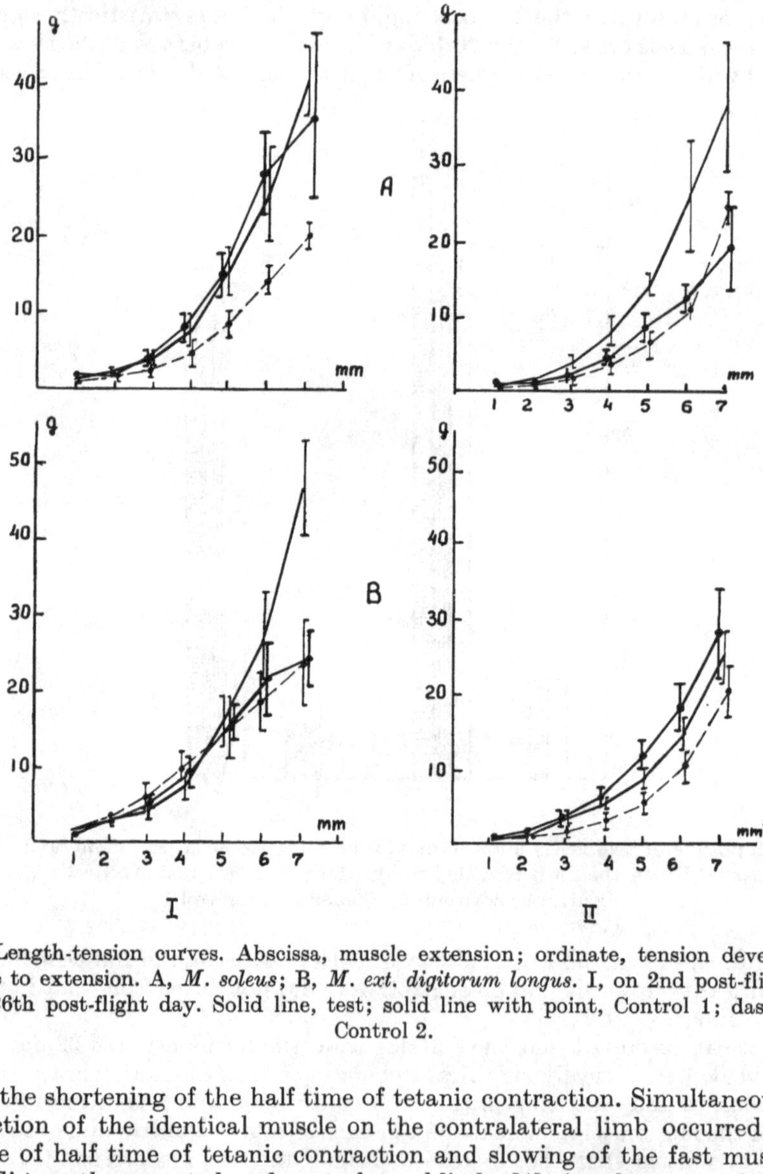

Fig. 3. Length-tension curves. Abscissa, muscle extension; ordinate, tension developed in response to extension. A, *M. soleus*; B, *M. ext. digitorum longus*. I, on 2nd post-flight day; II, on 26th post-flight day. Solid line, test; solid line with point, Control 1; dashed line, Control 2.

due to the shortening of the half time of tetanic contraction. Simultaneously the contraction of the identical muscle on the contralateral limb occurred due to increase of half time of tetanic contraction and slowing of the fast muscle (*M. brachialis*) on the operated and contralateral limbs [4]. Another type of disuse of the musculo-skeletal system, 120-day hypokinesia of animals, showed slow-down of tetanic contraction and relaxation of the slow *M. soleus* in the absence of similar changes in the mixed *M. plantaris* and fast *M. ext. dig. long.* In this case a significant reduction of strength parameters primarily of *M. soleus* takes place [5]. During re-adaptation following hypokinetic exposure the tetanic response of *M. soleus* continued to slow down due to increase of the relaxation time and the contraction of the fast muscles *M. ext. dig. long.* and *M. plantaris* slowed down also; this did not occur during the exposure to hypokinesis.

Changes of time parameters of muscle contraction were also noted in dogs with disused motor system due to hypokinetic procedures. Preliminary results show that during a 16—40-day hypokinetic exposure *M. gastrocnemius* undergoes reversible slow-down: decrease of the time and level of tension during walking and running on a treadmill and slight reduction of its amplitude. In these experiments the total bioelectric activity of posterior spinal roots was recorded [6], and it was found that changes of the muscular contraction of hypokinetic dogs (30 days) were paralleled by changes in the pattern of afferent proprioceptive impulsation (of the synchronization type): reduction of the average frequency of discharges and a less pronounced increase of effective energy of bioelectric activity.

When discussing the above facts the following situation should be kept in mind. Selective and reversible speeding up of tetanic contraction of the slow *M. soleus* simultaneously with strength decrement in the weightless environment is similar (probably, due to similar causes) to the change in the contraction time of the slow muscle on the operated limb in unloading experiments. It can be postulated that these adaptive changes reflect the trend towards maintenance of the dependence of the rate of muscle contraction on the level of strength [7] under the conditions of reduced strength requirements. Our findings agree with the data published in the literature about a similar trend of changes in the neuromuscular functions during disuse associated in any way with strength unloading of muscles.

The slowing of some skeletal muscles in the amputation model as well as in hypokinesis may be caused by changes in the pattern of the muscle function under these conditions resulting in an increase of their tonic activity and decrease of their phasic activity [4]. These data are also in conformity with the information published in the literature about muscle slow-down during hypokinesia.

Among the described effects of actual and simulated weightlessness it is important to note changes which were atrophic in nature and were noted in both slow and fast muscles: weight retardation, change of elastic properties, strength decrease. It should be pointed out that development and involution of these changes during re-adaptation after weightlessness and hypokinesia occur at different times in different muscles. It finally follows from our experiments that stability of reversal of changes during re-adaptation depends in general on the duration of disuse of the motor system. In dog's experiments of relatively short duration as well as in flown rats these changes were readily reversible; in 120-day experiments on rats change of time characteristics of muscles were more stable.

Analysis of the above changes in the parameters of muscular contraction confirms the widely recognized idea that they are brought about by changes in muscle trophics resulting from deficiency of the functional activity of the motor apparatus. This point of view finds support in L. A. Orbeli's theory on the trophic function of the nervous system and in modern concepts of its formation and realization [8, 9].

The above observations on total proprioceptive impulsation show the validity of another assumption based on the statements of M. R. Mogendovich [10] and R. P. Genevskaya [11] that changes in the trophic function may be in their turn associated with an alteration of afferentation which may also take place in the weightless state [12].

It can be demonstrated that this hypothesis and ensuing concepts of the different time constant in the neuroreflex (operational) control and in the regulation of muscle trophics make it possible to describe adequately the general pattern of changes in the human motor function which was noted in short-term flights (2—3 weeks) [1, 13] and also in longer-duration missions [2, 16]. The most typical feature is that disturbances in movement coordination in weightlessness are, as a rule, mild and transient; at the same time decrement of work capacity, physical endurance and other clinical symptoms of asthenization of the neuromuscular system which are to a certain extent related to the trophic effects of the nervous system seem to develop at later flight stages, to be manifest post-recovery and to have greater stability during readaptation. They may also account for disorders of man's movement and posture and changes in muscular functions which are again noted during re-adaptation [13, 15, 16].

Returning to the results of the Cosmos 605 experiments, it can be postulated in agreement with the above concepts that some of the changes in the functional parameters which were similar for slow and fast muscles (decrease in weight and strength, increase in stiffness) may be induced by atrophic processes. The levelling off of these changes tends to be delayed in comparison with time characteristics and does not coincide in stages in slow and fast muscles. This feature together with the selective pattern of changes of time characteristics in the slow muscle can be regarded in the light of the above concepts as a manifestation of the specific realization of the trophic function of the nervous system in metabolically and histochemically different muscle fibres. It is known that muscle histochemistry and related time characteristics of contraction depend strongly on muscle innervation [17, 18]. It is also known that under the influence of specific stimulation time characteristics of muscles change in parallel with alterations of their metabolism and histochemistry (see [19]).

The apparently adaptive pattern of the changes of time characteristics which is also suggested by their reversal suggests other possible methods of their realization. It can be assumed that a part is played by the so-called peripheral factor which implies the mechanism of regulation restricted to the muscle-tissue level [20]. This is also confirmed by the rearrangement of time characteristics and histochemistry of denervated muscle in relation to the imposed level of functioning [20, 21].

Thus, our investigations give evidence that characteristics of individual muscles differing in function and fibre structure may change differentially in the weightless environment. This may influence the pattern of changes and recovery of motor functions. Further study of mechanisms of these changes may have impact on the selection and procedure of applying the appropriate preventive and rehabilitation measures.

References

[1] C. A. Berry, in: Bioastronautics, Data Book, NASA, Washington 1973 (ch. 8, p. 349).
[2] W. E. Thornton and J. A. Rummel, Proc, Skylab Life Sciences Symp. 1974, NASA, 1, 407 (1974).

[3] V. S. Gurfinkel and V. S. Oganov, in: Space Biol. and Aerospace Med. IV Soviet Conf., Kaluga, **1**, 287 (1972).
[4] G. S. Katinas, V. S. Oganov and A. N. Potapov, Fiziol. Zh. SSSR, **60**, 1606 (1974).
[5] V. S. Oganov and A. N. Potapov, Kosm. Biolog. i Med. **7**, 2, 22 (1973).
[6] S. A. Skuratova and V. S. Oganov, Kosm. Biolog. i Med. **3**, 6, 71 (1969).
[7] A. V. Hill, Mechanics of Muscular Contraction, Mir, Moscow 1972.
[8] E. Gutmann, Uspekhi sovremennoi biologii **53**, 323 (1962).
[9] L. Guth, Physiolog. Revs. **48**, 645 (1968).
[10] M. R. Mogendovich, in: Nervous Trophics in the Norm and Pathology, Medizine, Moscow 1970 (p. 48).
[11] R. P. Genevskaya, Neurotrophic Regulation of Plastic Activity of the Muscle Tissue, Nauka, Moscow 1974.
[12] O. G. Gazenko and A. A. Gurjian, in: Problems of Space Biology, **6**, 22 (1967).
[13] N. N. Gurovsky, L. I. Kakurin, A. D. Egorov, V. S. Georgievsky, B. S. Katkovsky, A. P. Polyakova and I. B. Svistunov, Symp. on Space Biology and Medicine 1970, Intercosmos, Budapest 1970 (p. 1).
[14] J. F. Kubis and E. J. McLaughlin, Skylab 1/2 Preliminary Biomedical Report 1973, NASA (p. 283).
[15] C. E. Ross and J. R. Hordinsky, Skylab 1/2 Preliminary Biomedical Report 1973, NASA (p. 31).
[16] J. T. Baker, A. E. Nicogossian, G. W. Hoffler, R. L. Johnson and J. Hordinsky, Proc. Skylab Life Sciences Symp. 1974, **1**, 297 (1974).
[17] A. J. Buller, J. C. Eccles and R. M. Eccles, J. Physiol. **150**, 417 (1960).
[18] F. C. A. Romanul and G. P. Van Der Meulen, Nature **212**, 1369 (1966).
[19] J. P. Van Der Meulen, P. H. Peckman and J. T. Mortimer, Ann. N.Y. Acad. Sci., **228**, 177 (1974).
[20] J. Melichna and E. Gutmann, Pflugers Arch. **352**, 165 (1974).
[21] M. D. Brown, Nature **244**, 178 (1973).

Life Sciences and Space Research XIV — Akademie-Verlag, Berlin 1976

RESULTS OF MEDICAL INVESTIGATIONS CARRIED OUT ON BOARD THE SALYUT ORBITAL STATIONS

O. G. Gazenko, N. N. Gurovsky, A. M. Genin, I. I. Bryanov, A. V. Eryomin and A. D. Egorov

Institute of Medico-Biological Problems, Moscow, USSR

The purpose of the medical investigations carried out in flights of the Soviet orbital station Salyut were: to further studies and clarification of the phenomenology and mechanisms of changes in the main functions of the human body during prolonged weightlessness; to accumulate additional data on human responses at an early stage of adaptation to weightlessness; to assess the effectiveness of countermeasures against adverse effects of weightlessness during and after flight. The orbital station was equipped with a set of countermeasuring devices: a trainer for physical exercises (a treadmill), gravity simulation suit for long wear, bicycle ergometer, anti-G suit to be worn immediately post-flight, drugs, etc. The system of medical control included daily recordings of electrocardiography and respiration and regular physical examinations using provocative tests (physical exercises of a known force, lower-body negative pressure). Specific biomedical investigations were performed before, during and after flight.

As in previous flights, the atmosphere in the habitable modules was similar to that on the earth. Food supply met the requirements for basic nutrients and satisfied individual preferences of the crew members. In addition to the time required to fulfil the flight programme, the work-rest schedule allowed the cosmonauts 8 hours sleep, four meals, 2.5 hours physical training and personal time. The biomedical investigations carried out during and after flight furnished information towards a better understanding of human responses to the effects of space flight factors and the mechanisms underlying their development.

1. Medical Investigations

The USSR has launched the following Salyut stations: Salyut 1 in 1971, Salyut 3 in 1974 and Salyut 4 in 1975. Altogether three stations, carrying seven crew members, were placed in earth orbit. The space mission of the Soyuz 11 spacecraft and Salyut 1 orbital station (Commander, G. T. Dobrovolsky; flight engineer, V. N. Volkov, engineer scientist, V. I. Patsaev) lasted about 24 days. The space mission of the Soyuz 14 spacecraft and Salyut 3 orbital station (Commander, P. R. Popovich; flight engineer, Yu. P. Artyukhin) lasted about 16 days. The space mission of the Soyuz 17 spacecraft and Salyut 4 orbital station (Commander, A. A. Gubarev; flight engineer, G. M. Grechko) lasted almost 30 days.

The basic goals of medical investigations carried out in these flights were as follows: further study of reactions of the human body to weightlessness and other flight factors; assessment of the effectiveness of countermeasures against adverse effects of weightlessness and methods of maintaining high mental and physical performance during and after prolonged space flight; improvement of the on-board systems of medical control.

The manned flights of the Soyuz spacecraft and Salyut orbital stations were performed under favourable radiation conditions. The largest total doses by the end of the flight were recorded by the dosimeters in the working compartment of the Salyut 4 station and amounted to 0.7—1.02 rad. According to personal dosimeters, the total radiation dose received by the crew members was 0.66 rad. Thus, radiation doses were much less than the permissible level as indicated for the 30-day orbital flight (according to USSR standards this is 15 rad) and could not exert an adverse effect on the health state of the cosmonauts.

Life support systems of the Soyuz 11, Salyut 1, Soyuz 14, Salyut 3, Soyuz 17 and Salyut 4 spacecraft maintained the environmental parameters within the assigned levels, which were close to the earth atmosphere (Table 1). It should be noted that partial CO_2 pressure was normally 5—6 mmHg and rose higher for short periods before replacement of regeneration filters.

Table 1

Variations in the Environmental Parameters in Salyut Manned Flights

Parameter	Station		
	Salyut 1	Salyut 3	Salyut 4
Total pressure (mmHg)	880—920	838—872	780—840
Partial pressure of oxygen (mmHg)	235—280	171—200	170—240
Partial pressure of carbon dioxide (mmHg)	0.5—11.0	2.3—9.8	0.24—11.12
Partial pressure of water vapour (mmHg)	6.5—10.0	6—9	6.6—13.8
Temperature (°C)	17—21	20—25	18.4—28.4

The cosmonauts took hot meals three times a day. The caloric value of a daily diet was 2900—3000 cal. The menus were composed so as to take into consideration individual preferences of the cosmonauts and were balanced with respect to nutrient substances. The Salyut crew members also had all the necessary items of personal hygiene. In accordance with the flight schedule, the cosmonauts had 8 hours of sleep, took four meals a day, spent 2.5 hours in physical exercise and were allowed some private time. Their duties took 8—9 hours a day.

In order to prevent adverse effects of space flight on the human body the cosmonauts:

(i) Performed physical exercises on the trainer and bicycle ergometer. The trainer included a treadmill and pulling devices. The trainer was used for walking (no-load work on the treadmill) or running (motor-driven work on the treadmill) and for various pulling exercises. The system of bungee-cords provided a pull by the cosmonaut on the treadmill with a force of 50 kg, applying a load to the long axis of the body and made it possible to exercise practically every muscle group of the body.

(ii) Wore training G-suits (for no less than 8 hours a day). The G-suits reproduced a constant static lead on the musculo-skeletal system and skeletal musculature. This largely counteracted the zero gravity.

(iii) Used lower-body negative pressure for the last five flight days according to a special scheme (about 2 hours in the Salyut 4 flight).

(iv) Used a special suit to prevent circulation disorders upon recovery (the suit was to be donned before re-entry);

(v) Used drugs before re-entry.

In order to provide medical control, an electrocardiogram, seismocardiogram and pneumogram were recorded daily, voice communications and television broadcasts were analysed, and cosmonauts' reports of their state of health were studied. Detailed medical investigations were carried out using special equipment which recorded on a memory device (or transmitted to the earth) the following parameters: ECG (12 leads), kinetocardiogram, sphygmogram (from three points), arterial pressure (tacho-oscillographically and by distal perimetry), phlebogram, pulmonary ventilation, body temperature, electro-oculogram, plethysmogram and dynamogram. Cardiac output and peripheral resistance were derived from calculations (according to Bremser-Ranke).

Physical exercises were used as functional tests: squatting on the treadmill (Salyut 1 and 3) or bicycle ergometry (Salyut 4: pedalling for 5 min at a rate of 60 rev min^{-1} with a power of 440 kg m min^{-1}). Gravity load on the cardiovascular system was simulated by lower-body negative pressure. The test consisted of two stages: -25 mmHg for 2 min and -35 mmHg for 3 min.

Bearing in mind the fact that some of the medical results of Salyut 1 and Salyut 3 flights have been already published, the present communication discusses them briefly in comparison with the data obtained from the 30-day flight of Salyut 4.

2. Health Condition of the Cosmonauts

The Salyut 3 crew members and the Salyut 4 flight engineer had an illusion of back tilt during orbital insertion. The state of health of the cosmonauts did not deteriorate at that time. A few hours later the cosmonauts had a sensation of rush of blood to the head which reached a maximum in the Salyut 3 crew members by the third day of the mission and began to diminish afterwards. The sensation of rush of blood to the head was not burdensome and did not interfere with the approach and docking manoeuvres of the Soyuz-Salyut vehicles. G. M. Grechko (Salyut 4) had the feeling of rush of blood to the head throughout the entire mission. However, as he stated, this did not impair his performance. The Salyut 3 crew members experienced disturbances of spatial orientation upon transition on board the station. With eyes closed they were unable to determine precisely their body position in relation to various objects within the station. Later disorientation appeared periodically when the cosmonauts went to the Soyuz spacecraft. Disorientation disappeared entirely about a week later.

In the early stages of the flight the cosmonauts did not make abrupt head movements: they moved slowly and smoothly. This helped them to avoid distinct vestibulo-vegetative disturbances. The Salyut 3 crew members did not experience continuous nausea, had good appetite, sound sleep and maintained high work capacity. On the 5th—6th day of flight the cosmonauts did not need to restrict their motor activity, they could control and coordinate their body movements well in zero gravity and learnt how to use this state for more rapid and economic movements along the station; abrupt head tilts did not bring about the loading sensation on the vestibular apparatus. It is interesting to mention observations made by P. R. Popovich and Yu. P. Artyukhin on the sensation of rush of blood to the head in the weightless environment. They noted that the sensation — the feeling of heaviness — decreased every time when they pressed themselves against the walls, attached themselves to the pulling system of the trainer or did physical exercises. A. A. Gubarev did not notice a feeling of rush of blood to the head in weightlessness. A day after he boarded the station he experienced heaviness in the head which, however, did not interfere with his work. Nevertheless, he reported that during the first days of orbital flight he had the feeling that the viscera in his abdominal cavity were moving upwards (headwards). Unlike other cosmonauts he had transient disorientation until the 20th day of flight.

During the first eight days G. M. Grechko experienced uncomfortable feelings which were enhanced by abrupt head movements, but had no nausea or vomiting. During the first 5—8 days the Salyut 4 cosmonauts had a reduced appetite and consumed only part of their daily diets. Later they regained their normal appetite and consumed their daily diets in full.

The cosmonauts took no drugs for motion sickness. According to both crew members of the Salyut 4 station, the period of acute adaptation to weightlessness lasted for 8 days. For instance, G. M. Grechko stated on the 8th flight day that he had no more problems, weightlessness became his ally, it was much easier to work, movements required no effort. No vestibular disturbances occurred even during rotation in the chair; in the weightless environment it was easier to rotate than on the earth.

Among other symptoms noted during acute adaptation nasal congestion should be mentioned. The cosmonauts attributed this to a cold in the head. However, this symptom could have been induced by edema of the nasal mucous membrane due to blood redistribution. On the 8th—9th day both Salyut 4 crew members, and on the 11th day the flight-engineer, reported moderate headaches at the end of the working hours which were alleviated by analgesics. On the 18th day both cosmonauts noted moderate pain in the legs which may have been induced by intensive physical exercises made before.

During the first flight days the cosmonauts had some sleep deficiency. They slept about 4—5 hours and awoke early. Afterwards they slept better, for 6—7—8 hours a day. A. A. Gubarev occasionally took sleeping drugs.

3. In-flight Cardiovascular Studies at Rest and during Provocative Tests

As in all previous flights, the cosmonauts showed an increase in heart rate a few hours pre-launch. This was indicative of the stressful effect of the prelaunch period.

In orbital flight the heart rate and other cardiovascular parameters were stabilized at a relatively constant level after several days of variations. The heart rate of the Salyut 1 and Salyut 4 cosmonauts was significantly higher than pre-flight levels during the 24- and 30-day missions.

The indices of the phasic structure of the cardiac cycle at rest were lower than in the baselines, the difference being significant in most cases.

The changes in the phasic structure of the cardiac cycle, in contrast to the changes noted in previous flights, are not indicative of the syndrome of functional hypodynamia (due to insufficient volume loading) and seem to be associated with increased heart rate. Clinical electrocardiography showed no important changes in the shape of the curve. During the Salyut 4 flight electrocardiographic study revealed a slight deviation of the electric axis of the heart to the right from the frontal plane in A. A. Gubarev and reduced amplitudes of the P_1, P_2 and P_3 waves in G. M. Grechko. The latter also displayed reduced amplitude of T waves, though the P/T ratio differed from the baselines. ECG changes in the sternal leads were moderate and may be attributed to the change of the heart position in the chest or to the increased load on the right compartment of the heart.

In-flight arterial pressure at rest in all cosmonauts was close to the pre-flight values. However, during the first examination on the 4th day of flight A. A. Gubarev showed a slight increase in all types of pressure (minimal, mean, lateral and terminal systolic pressure) and G. M. Grechko displayed an indistinct tendency towards a fall in pressure, especially on the 28th flight aday. The data indicate that cardiovascular indices at rest showed no variations which can be interpreted as pathologic.

During LBNP tests in flight the heart rate increased: at −25 mmHg to 18 and 15 beats/min in A. A. Gubarev and G. M. Grechko, at −35 mmHg to 21 and 16 beats/min, respectively. These changes were slightly different from the pre-flight data. However, in G. M. Grechko on the 14th flight day they were more marked than pre-flight.

During LBNP tests arterial pressure either remained essentially unaltered or decreased by 7—10 mmHg (minimal, lateral, systolic) which indicated good tolerance to the exposure. During the LBNP test on the 20th flight day A. A. Gubarev showed a 12—13 mmHg decrease in the lateral and mean pressure.

The bicycle ergometry test was followed by an increase in the heart rate by 33—53 beats/min in A. A. Gubarev and by 29—66 beats/min in G. M. Grechko. The most pronounced reaction was observed in A. A. Gubarev on the 14th day (heart rate increment by 53 beats/min) and in G. M. Grechko on the 12th day (heart rate increment by 66 beats/min). It should be noted that the workload during bicycle ergometry was 420—470 kg m min^{-1} for A. A. Gubarev and 370—440 kg m min^{-1} (on the 4th day — 540 kg m min^{-1}) for G. M. Grechko. During bicycle ergometry maximal arterial pressure increased up to 150 to 170 mmHg in both crew members. On the first minute of the recovery period it decreased to 140—145 mmHg which was in agreement with the pre-flight data.

Summarizing the results of in-flight examinations of crew members with the use of provocative tests, it should be stated that their reactions were very close to those pre-flight. It is also important to note that the reactions did not vary in relation to mission time. This gives evidence that adaptive capabilities of the human body during a 30-day flight remain high enough to maintain adequate regulation of the cardiovascular function at these levels of the provocative tests. G. M. Grechko exhibited a more marked response to the provocative tests in flight. This may be brought about by the less intensive physical exercises he did in flight.

4. Post-flight Examinations

After the 16-day flight the Salyut 3 crew members tolerated well enough both active and passive orthostatic tests. Their orthostatic tolerance was at least not lower than after the 2- and 8-day flights and significantly higher than after the 18-day mission of Soyuz 9. Orthostatic tolerance returned to the baselines in a week.

The LBNP test carried out 3.5—5.5 hours post-flight showed a decrease in the tolerance. For instance, the heart rate of the commander increased pre-flight by 10 beats/min at the most and post-flight by 28 beats/min. During the post-flight LBNP test at −25 and −35 mmHg the flight engineer showed significant variations in the arterial pressure and gradual decrease of the mean pressure. In LBNP tests carried out 3 and 5 days after flight the responses did not differ from the pre-flight values.

During bicycle ergometry tests the Salyut 3 crew members displayed signs of deconditioning after their 16-day mission. In comparison with the pre-flight values they showed a more marked increase in heart rate, decrease of oxygen pulse rate during exercises and increase of oxygen consumption during recovery. These changes were not detected on the 7th day.

The readaptation of the Salyut 4 crew members after their month-long mission was more lengthy and demanding.

Immediately upon landing A. A. Gubarev and G. M. Grechko, like the Salyut 3 crew members, reported subjective heaviness of the body, particularly of the arms, complained about fatigue, showed unsteady gait and experienced vertigo upon tilting the head. In the upright position they had tachycardia and slight dizziness. However, a few hours later, and especially after sleep, the state of health of the cosmonauts improved significantly. On the next day the cosmonauts noticed painful feelings in the gastrocnemius and femoral muscles, as typically occur after intensive physical training. Muscle pains, as well as the other changes, soon disappeared.

Electrocardiographic examination of A. A. Gubarev (12 leads) on the first post-flight day showed a decrease of the T wave amplitude in all the leads and T inversion in the left sternal A and D leads (according to Neb). These changes gradually returned to normal and disappeared on the 12th post-flight day. Electrocardiographic changes were accompanied by unpleasant sensations in the cardiac region during physical exercises.

During the tilt test on the recovery day and on the next day A. A. Gubarev showed poor condition, cold sweat and paleness, and consequently the test was terminated on the 10th min. G. M. Grechko endured the test well. The tilt tests on the 3rd—7th post-flight days were completed by both cosmonauts, though A. A. Gubarev still exhibited decreased orthostatic tolerance.

During the bicycle ergometry test (600 kg m min^{-1}) carried out on the 3rd post-flight day A. A. Gubarev experienced unpleasant sensations in the cardiac region and painful feelings in the gastrocnemius muscles. As a result, the test was discontinued. Heart rate was not more than 120 beats/min during the test. On the 7th day the test was repeated and completed; on the 4th min the heart rate rose to 136 beats/min. It should be noted that during bicycle ergometry on the 3rd—4th day A. A. Gubarev showed reduced amplitude and slight deformation of T waves (in the 4th sternal lead A according to Neb) in the ECG. On the 4th post-flight day G. M. Grechko carried out the ergometry test to the end, although he had unpleasant cardiac sensations at a heart rate of not more than 131 beats/min.

Routine clinical analysis of the blood made in the Salyut crew members post-flight demonstrated an increase in the white blood cell count and a decrease in the red blood cell count and haemoglobin. The content of eosinophils in the peripheral blood was also reduced.

It should be noted that the content of catecholamines in the blood of A. A. Gubarev and G. M. Grechko on the day after return was elevated. On the 7th day after return the epinephrine content returned to normal whereas the norepinephrine content remained increased. On the recovery day there was a decrease in the potassium concentration and an increase in the calcium and magnesium concentration in the plasma. On the 7th day the potassium concentration was still lower than pre-flight. It is interesting to note that immediately post-flight creatine phosphokinase activity was elevated and remained high on the 7th day after return. Total activity of lactate dehydrogenase did not change significantly; however, its fraction-1 was increased.

The results of post-flight examinations of the Salyut 4 crew members indicate certain transient changes in cardiac metabolism, cardiac deconditioning and typical responses to physical exercises which in clinical cardiology are interpreted as myocardiodystrophy or "adrenalgic heart". This is a frequent event in sports medicine: it occurs in people who begin heavy physical exercises without preliminary training.

Post-flight all the cosmonauts showed weight losses. For instance, P. R. Popovich lost 3.8 kg, Yu. P. Artyukhin 3 kg, A. A. Gubarev 2.2 kg, G. M. Grechko 4.8 kg. The weight was recovered on the 2nd or 3rd day after return. The only exception was G. M. Grechko whose weight did not return to normal until the 10th day. The rapid weight recovery was indicative of fluid retention in the body. This was also confirmed by the fact that all the cosmonauts, A. A. Gubarev and G. M. Grechko included, showed a marked reduction of fluid and electrolyte excretion immediately post-flight. Water consumption was higher than diuresis. Only on the 2nd day after return did the fluid-electrolyte excretion approach the pre-flight values.

A comparative analysis of the pre- and post-flight examinations of Soyuz 9, Soyuz 13, Soyuz 16, Salyut 3 and Salyut 4 crew members as well as a detailed study of medical results of Skylab missions suggest that an increase in mission time alters significantly the pattern and duration of readaptation on return to earth. Nevertheless, the changes detected can hardly be attributed to weightlessness alone. They may be brought about by nervous and emotional strain due to the hazardous situation, the unusual and inconvenient environment and by a high sense of responsibility for the successful accomplishment of the mission. It can also be induced by strenuous work of the crew members in experimentation, insufficient rest and sleep, disturbed nutrition or water consumption. It is not easy to ignore these factors and it is more difficult to standardize them, especially taking into account individual variations of cosmonauts. Therefore the results obtained should be interpreted with great caution. However, it is obvious that in space missions of long duration special attention should be paid to the maintenance of the normal function of the cardiovascular system and fluid-electrolyte metabolism.

Short-term(7—8-day)-flights do not cause significant cardiovascular changes post-flight. This can be associated with insignificant de-adaptation of the heart to earth gravity. The initial state of the cosmonaut's health is very important for the prediction of space flight tolerance. Cosmonauts who previously had diseases influencing the cardiovascular system showed the least favorable reactions during readaptation. On return to earth the Skylab astronauts also exhibited reduced tolerance to gravity and physical exercises. However, according to American specialists, Skylab 4 crew members were in a better condition than Skylab 3 astronauts and Skylab 3 crew members felt better than Skylab 2 astronauts. They ascribe this to the fact that Skylab 3 and 4 astronauts exercised more. Therefore, in order to avoid or at least to minimize cardiovascular deconditioning and orthostatic intolerance, it is necessary to use special countermeasures, to develop optimal schedules of work, rest, sleep and nutrition. From this point of view the following are very important: in-flight medical investigations which help to follow up the state of health of cosmonauts, determination of cardiovascular function, responses to tests of a known force and obtaining data that can be used in assessing the effectiveness of countermeasures that are applied. Complete reports of crew members about their state of health are also very significant.

The period of acute adaptation occupies a special place in the effects of space flight factors on the human body. During this period most marked adaptive reactions of the body to zero gravity take place. In this period rearrangement of the organism develops; it begins to function at a new level and to involve many sensory systems. The vestibular analyser, which is subjected to the lack of gravity stimulation and to haemodynamic reorganization, plays an important role in the adaptation. As is known, venous congestion in the cerebral circulation exerts a specific effect on cerebral pressure and enhances susceptibility of the organism to vestibular crises which are typical of circulation disturbances in the vestibulo-basilar artery. The transient pattern of vestibular of vestibular disturbances, and their clear-cut correlation with the general adaptive reaction of the human body to blood redistribution, confirms the involvement of the haemodynamic component in the genesis of vestibular reactions. This is the factor which may explain the discrepancy found by the American researchers between vestibular stability measured by ground-based tests and tolerance to weightlessness in real space flight. Nevertheless, it should be admitted that people naturally resistant to motion sickness prove to be more tolerant to weightlessness than people who have developed high resistance due to special training.

Life Sciences and Space Research XIV — Akademie-Verlag, Berlin 1976

SPACELAB AND ITS UTILIZATION FOR BIOMEDICAL EXPERIMENTS

G. SEIBERT

European Space Research Organisation, Neuilly-sur-Seine, France

The types of investigations possible in the orbital laboratory Spacelab are considered. A summary of services available to the Spacelab user is given, in which the load carrying capacity and the characteristics of the main subsystems (e.g. thermal control, environmental control, electrical power and energy as well as data management) are briefly described.

Life science investigations may be undertaken for two reasons: firstly, to ensure safety and efficiency, and, secondly, for their scientific interest relating to effects of weightlessness or cosmic radiation.

Safety of the crew and their genetic cells in relation to cosmic radiation is considered vital; essential knowledge is also required about the performance of the vestibular balancing mechanism and the related problem of "stomach awareness". The effect of zero-gravity on the cardiovascular system is studied, and the effect of circulatory changes in the brain, possible psychological stress and effects on exercise tolerance measured. Due to their rapid reproduction, important information may be gained from micro-organisms in respect of mutation rates when exposed to radiation. Plants depend, to some extent, on gravity in germination and growth. Of interest here is the relative importance of gravitational and photonic influences.

1. Introduction

The Spacelab is an integral part of the Space Shuttle Transportation System. ESRO, shortly to be renamed "European Space Agency", representing ten participating European countries, is responsible for the design, development and manufacture of Spacelab. NASA is responsible for developing and building the Shuttle itself and for the launch and operation of Spacelab.

2. The Shuttle-Spacelab System

Spacelab is carried to and from orbit by the space shuttle. The shuttle is composed of the fully reusable shuttle-orbiter, a large extensible tank containing the ascent propellant and two partially reusable solid rockets. The orbiter is about of the size of a DC-9 aeroplane. During its orbital stay Spacelab remains with the orbiter, i.e. it remains wholly contained in the large cargo bay of the orbiter. The orbit achievable by thc shuttle-Spacelab system is limited to low earth orbits. Orbit ranges from 100 to 500 naut. miles and inclinations in the range of 28.5 to 104° are foreseen. When the desired orbit is attained, the cargo bay door opens, Spacelab is checked out and prepared for its experimentation programme. The

Spacelab crew can work in the pressurized module of Spacelab which is connected to the orbiter by a pressurized tunnel, but when off-duty, the crew is housed in the orbiter cabin. The tunnel is necessary because the Spacelab, for reasons relating to the centre of gravity, could not be directly attached to the orbiter cabin. The orbiter not only provides living quarters for the crew but also a number of services to Spacelab: alignment and stabilization of the assembly, data transmission to the ground, provision of a certain amount of electric power, and dissipation of some of the thermal energy given off by Spacelab.

After completion of the experiment mission the orbiter re-enters and lands more or less like a conventional aeroplane. Following the landing, Spacelab is removed from the orbiter and refurbished for the next flight cycle starting with the integration of a new set of experiments into Spacelab. A typical shuttle-Spacelab flight profile is shown in Fig. 1. Spacelab is designed for an operational lifetime of 10 years or 50 missions of seven days. This nominal flight duration of seven days is extensible to 30 days.

3. Main Spacelab Characteristics

The Spacelab flight vehicle consists of two basic elements, a pressurized module and an unpressurized pallet, which can be used separately or in combination. Fig. 2 shows a typical example of a module/pallet configuration. The Spacelab concept embodies the principle of modularity so that the size, in terms of pressurized volume or unpressurized length may be varied to meet the needs for a particular flight. Thus, three basic flight configurations are possible: module only, module plus pallet and pallet only. In addition to the variety of configurations, further mission flexibility is obtained through the choice of "common payload support equipment" (CPSE), e.g. airlocks, viewports, feedthroughs, vents to allow access to the unlimited vacuum capacity of space, etc. Thus, the modular elements of Spacelab may be arranged into various flight configurations to suit the needs of a wide variety of users.

3.1. The Module

The module, which allows up to four persons to work simultaneously, offers working conditions close to those found in a laboratory on the ground: (a) an atmosphere composed of a mixture of oxygen and nitrogen at a pressure of one atmosphere, (b) a controlled environment, (c) electrical power, (d) communication and data management, (e) certain standard furnishings such as equipment racks and work benches. A modular approach has also been adopted for the interior design of the module. It will be equipped with quickly removable secondary structure and standard equipment racks. This permits off-line integration and checkout of experiments and easy turn-round of the module.

The basic module contains a core of subsystem equipment but also provides for 8 m^3 of experiment equipment. The aft extension module is dedicated entirely to experiment installation, giving an additional 14.6 m^3, and is configured to permit efficient experiment operation. In this configuration a total cylindrical side wall length of 4.4 m is fully available for experimenters. Also, centre aisle packaging of experiment racks is possible in the extension module that leads to efficient use of the available module.

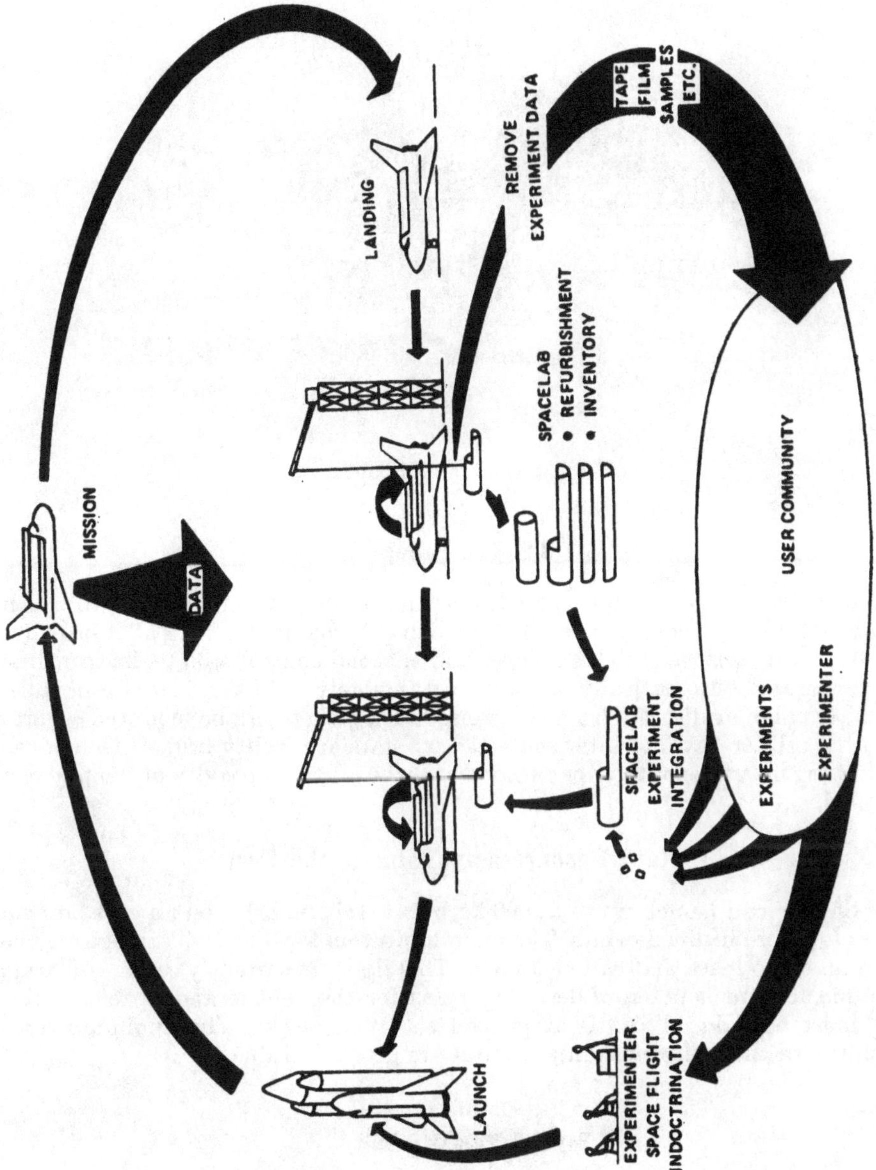

Fig. 1. Typical shuttle — Spacelab flight profile.

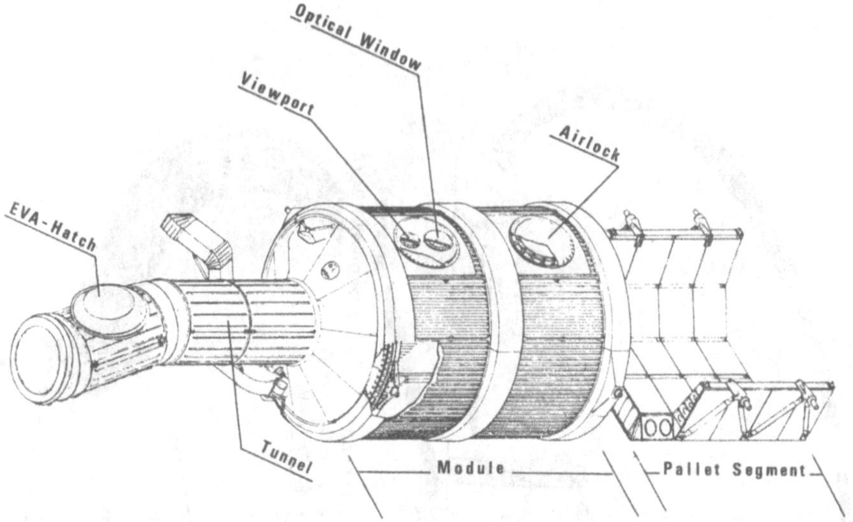

Fig. 2. Spacelab with pallet.

3.2. The Pallet

The pallet is simply a space-exposed structure or observation platform to which instruments such as telescopes and antennae may be mounted. The pallet provides some very basic services (such as power and thermal control) and its instruments may be operated automatically or controlled remotely from the Spacelab module, the orbiter cabin or directly from the ground. The pallets can be mounted separately in the orbiter bay and interconnected by standard utility bridge elements so that viewing interference may be avoided and the centre of gravity of the payload controlled.

4. Main Resources available to the User

The Shuttle can launch up to 29500 kg into a 185 km orbit on an east launch, but less for more inclined orbits. The main limitation for Spacelab arises from the maximum cargo mass that can be landed. This figure is currently set at 14500 kg. When due account is taken of design margins for the shuttle and Spacelab, this figure must be reduced and is at present set at 11340 kg. The minimum mass capabilities of the different configurations are given in Table 1.

Table 1
Payload Mass Capability

Configuration	Module length (m)	Pallet length (m)	Payload Mass (kg)
Long module	6.9	—	5500
Short module + 9 m pallet	4.2	9	5500
9 m pallet	—	9	9100
15 m pallet	—	15	8000

The primary electric power is provided by a dedicated orbiter fuel cell, and Spacelab's electrical power and distribution system provides conditioning and distribution to the subsystems and payload. The unregulated power amounts to 7 kW average and 12 kW peak (15 min every 3 hours) of which about 4 kW average and 9 kW peak are left for the payload. A peaking battery could be added to satisfy peak power demands. Details on voltage lines and electrical energy are shown in Table 2. Electrical energy is dependent on the flight configuration and degree to which mission dependent equipment is used, but the basic supply can be augmented by standard kits of consumables (hydrogen and oxygen), each kit providing up to 840 kWh. The characteristics of the services available to users in the environmental and thermal control and the communication and data management system are also given in Table 2.

Active thermal control is effected by a two-fluid system of water (within the module) and freon (on the pallet). The environmental control system employs LiOH cannisters for CO_2 control. The command and data management system (CDMS) is probably the most sophisticated subsystem. It gives a variety of services including data acquisition and distribution, monitoring, processing, displaying and recording of experiment data. Three identical computers are proposed, one each for subsystems and payload and one for redundancy. The CDMS also provides formating of data for communication to the ground and reviews commands. Limited on-board processing will be possible utilizing software supplied by the experimenters.

5. The Aspect of Man

When considering the Spacelab design concept one must not forget the important aspects of crew accommodation and habitability. It is, for instance, necessary to provide adequate living conditions (sufficient space, acceptable acoustic and height levels, atmosphere regeneration) so that crews who have not had intensive astronaut-type training can work aboard it. In addition, safety constraints must be imposed on the various on-board equipment, and there must be a certain amount of redundancy in the survival equipment.

The Spacelab crew consists of up to four scientific astronauts concerned with the scientific experiments and a normal work schedule foresees two crew members continuously in Spacelab. The crew members are housed in the orbiter cabin when off duty so that a total orbiter crew complement can be up to seven persons. For this purpose the orbiter includes equipment for sleeping, health and recreation, eating, waste management and personal hygiene. To compensate for these constraints, man's presence aboard offers a large number of advantages. Mention has often been made of the technical aspect of human intervention in manned systems: maintenance, repair, replacement of defective parts, etc. The most important aspect is probably the active involvement of the experimenter in the conduct of his experiment: he can, for example calibrate optical systems in flight, manipulate biological specimens, initiate and directly observe particularly tricky experiments, or even modify the programme of an experiment when necessary as a result of the appearance of an unexpected event. In addition, the tunnel permits extra vehicular activity (EVA) in support of pallet mounted experiments via an EVA hatch.

Table 2
Services available to Spacelab Users

Service	Capability of Characteristic
Electrical power	4 kW average, 9 kW peak } unregulated dc 26 V to 32 V dc 115 V/200 V/400 ± 0.1 Hz 3-phase Mission dependent: Regulated 28 V dc (±2%) 115 ± 5% V ac, 60 Hz ± 1% 220 ± 5% V ac, 50 Hz ± 1%
Electrical energy	Configuration dependent but approx. 400, 430 and 590 kWh unregulated dc available for (a) long module, (b) short module + pallet, and (c) pallet only modes respectively. Mission dependent kits (840 kWh) available for additional energy.
Thermal control	Air cooling with separate avionics loop. Liquid loop with provisions for cold plates. Max. capability: Module atmosphere 1 kW at 18–26 °C Avionics bay atmosphere 3.13 kW at 24–50 °C Pallet cold plates 4.85 kW at 24–40 °C (Pallet only mode 6.6 kW at 10–32 °C)
Environmental control	Normal atmosphere N_2/O_2, 1 atm. Air temperature 18–26 °C Relative humidity 30–70% Cleanliness 5 μm filtering
Data management	Separate subsystem and payload computers 16 bit (fixed) and 32 bit (fixed and floating point) operations; 500 K ops/sec, 64 K 16 bit word memory; alpha-numeric display and key board. Mission dependent: Digital data recorder, 30 mbps Video and analog data recorder, 6 MHz.
Communications	Down-link via orbiter and Tracking and Data Relay Satellite (TDRS), provides up to 50 mbps digital or 6 MHz analog and video. Telemetry 20 kbps interleaved with orbiter data. Up-link via orbiter 2 kbps command.

6. Opportunities for Biomedical Research

The considerations on biomedical experiments are based on the European experiment ideas proposed for the first Spacelab flight. Life science investigations may be undertaken for two reasons; firstly, to ensure safety and efficiency, and, secondly, for scientific interest in effects of weightlessness or cosmic radiation. One of the main objectives of space flight research up to date has been to find out about the effects on humans of longer journeys into space. It is necessary to know about these effects before planning a long space flight, for example a "fly past" of Mars, which would take about eighteen months. The two major problems which have appeared so far are the effect of weightlessness on the balancing mechanism, and damage of living cells by space radiation. The conditions of zero gravity, "hard" radiation, and vacuum make possible some basic scientific experiments, which may be grouped as follows:

Human physiology and psychology
Vestibular function (separated from the above due to its importance)
Animal experiments
Effects of hard radiation: on animals; at cellular level
Plant experiments.

A major problem in the Spacelab project is the limited amount of time available for the scientists to do experiments, particularly if one of them has to be a subject. The other difficulty is that no one person can be expert in all the subjects likely to be flown on a mission, so that projected work may have to take account of crew skills. Apart from vestibular problems which will have special consideration, some method of crew selection on health grounds will have to be developed to screen the potential European crew applicants.

6.1. Vestibular Function

Following the American Apollo programme and Skylab flights, and data made available from Russian space flights, it has become obvious that "space sickness" or "stomach awareness" will be a major problem in the immediate future. This effect of vestibular dysfunction due to zero gravity has become more obvious recently because spacecraft are now larger, thus allowing more independent movement of the occupants. Owing to the relatively short flights of Spacelab, the two or three days of partially incapacitating effects on the crew are a major difficulty. The re-adaptation problem during re-entry and landing poses difficulties for the pilot. However, the most disturbing aspect of the situation is the almost total inability of existing test and screening procedures to predict which members of a crew are likely to suffer from space sickness by testing on the ground. If space flight is to become a more routine activity, then it is an urgent matter to conduct basic research into vestibular function. The availability of a reasonable sized laboratory in a zero gravity condition offers a good opportunity to perform dynamic testing, either on animals, or crew members, by use of a small centrifuge or a chair mounted on rails, without the disturbing effect of a constant gravitational field.

6.2. Human Physiology and Psychology Experiments

It has already been demonstrated that there is a shift of blood electrolyte balance, and a loss of calcium from bones, together with a loss of blood plasma volume. There is also a change in red blood cell mass. All these effects clear up within a few weeks of return to earth, but the mechanisms are not well understood. It is quite obvious that the other major effect of zero gravity will be on the circulation system, since the heart is a pump and is occupied in pumping blood against gravity. It could be expected that the compensatory mechanisms adopted by the body would shed some light on the underlying control systems. A "freely floating" human being could be investigated by ballisto-cardiography which is a difficult procedure on earth. Some quite simple measurements need to be made in zero gravity to determine, for example, what happens to venous pressure. The effect of gravity on time estimation by the crew together with any effect on circadian rhythm would need to be ascertained.

6.3. Animal Experiments

It is very convenient to use the scientists themselves for some experiments provided that these are not harmful. However, some experiments must involve some danger, and these could most conveniently be done on animals. By using animals, experiments involving poisonous chemicals, radiating materials and electrode implantation become possible. Animal experiments are the only method by which post-flight dissections can be carried out. Small animals may be a convenient method of obtaining some types of information as has already been done by using minnows for vestibular experiments. Small animals also reproduce rapidly making them suitable for genetic experiments. Facilities will be required on board to provide "life support" functions for animals together with some method for preserving part of whole animals which have been killed for experimental reasons. It is most important that any animal species selected for work in space should have been very thoroughly investigated on earth so that their "normal" reactions are known. Some experiments could be done by pre- and post-flight examinations only, both on humans and animals. Many of these could be non-invasive, for example, time of re-adaptation to gravity.

6.4. Effects of Hard Radiation

The other vital issue involved in prolonged space flight is the effect of radiation on humans, animals, plants and lower organisms. The "hard" radiation (such as HZE particles) found in space cannot be imitated on earth because, although it is not very intense, the various types found naturally and simultaneously in space cannot be produced simultaneously by any earthbound machines. It is important to know about the effects of this radiation on living organisms because the weight limitations on spacecraft preclude complete shielding. Radiation experiments form a good example of the necessity of animal experiments so that a potentially lethal procedure can be carried out. This could take the form of either post mortem examination on return, or procedures during the flight.

One of the most dangerous effects of radiation is in the production of mutations in cells during reproduction phases. When this possibility is considered in relation to man, animals and, particularly in the short term, pathogens, the effects of the introduction of a new disease to earth, either to man or plants could be serious. It is also essential to avoid both introduction of extra-terrestrial organisms to earth and contamination of other planets with earth organisms.

From the point of view of scientific interest, the effects of radiation at cellular level and the mechanisms of lesion production should be investigated. It may prove useful to find the effect of combined "hard" radiation and vacuum on organic chemicals to see if any change is produced which might indicate how life forms evolved from simpler compounds.

The absence of gravity can be used to make easier or more nearly theoretically "perfect" certain physico-chemical processes which are used for biological experimentation, for example electrophoretic separation.

6.5. Plant Experiments

When long space flights are envisaged, it will be essential to know about the growth of plants for food. A major factor in determining the organization and growth responses of plants on earth is the particular orientation of the growing parts in relation to external stimuli such as light and gravity; with respect to the latter, each organ exhibits a precise angle and direction of growth in the gravitational field.

Much of the sublet of plant responses resides in the inherent polarity of the individual constituent cells and in the polarity of movement of growth regulatory chemicals from one cell to another. The redistribution of these chemicals that occurs in an organ when it is displaced from its proper orientation to gravity is recognized as instrumental in reinstating the normal growth pattern. The precise mechanisms involved, however, await elucidation. Whereas experiments can, and are, conducted on earth in uniform gravitational fields by proper rotation of the subject, the condition of zero gravity is available only in space. The combined conditions of cosmic radiation and weightlessness are likewise unique in space.

7. Conclusion

In the 1980's, a new tool, Spacelab, will be at the disposal of the scientific community to perform experiments in space. From the previous short chapters describing the Spacelab system and its capability in terms of resources and services available to the user, one can see that the advantages offered by Spacelab are the large mass and volume permissible, the increased mission opportunities, the manned check-out, supervision and repair of experiment instrumentation. For Europe it will make research possible in fields like life sciences, which were almost ignored by space research due to the limited capabilities of conventional automatic satellites.

The main reasons why life science experiments in space are of interest is that the conditions of zero gravity and the availability of hard radiation over a wide spectrum are unique to the space environment. The areas of experimentation envisaged comprise:

(i) investigations of the known physiological effects of space flights on man: such investigations can include acute experiments on animals;

(ii) exposure of living systems to the space environment with the objective of investigating fundamental biological processes on which gravity or hard radiation is thought to have an influence;

(iii) experiments where the ability to control gravity right down to zero will make it possible to investigate a biological system in a manner impossible on earth.

Radiation Biology

Life Sciences and Space Research XIV — Akademie-Verlag, Berlin 1976

IRRADIATION OF BIO-OBJECTS ABOARD THE COSMOS 690 BIOSATELLITE

Yu. A. Akatov, A. N. Gladilkin, I. V. Ignatov, S. B. Kozlova, A. V. Kolodin, R. A. Kuzin, V. I. Popov, L. N. Seliverstov, V. G. Semyonov, M. A. Sychkov, B. I. Solyanov and V. V. Yurgov

Institute of Medico-Biological Problems, Moscow, USSR

Animals on the Cosmos 690 were exposed to Cs 137 gamma-radiation of 320 curies. The on board emitter was a spherical container made of tungsten alloy with a gamma source placed in the centre. A special dose filter provided a uniform $\pm 10\%$ distribution of the dose field. Animal containers were equipped with thermoluminescent dosimeters. Radiation was monitored by an on-board dosimeter and displayed. The emitter was controlled by commands from the ground. On the tenth flight day the emitter was turned on and bio-objects were exposed for 24 hours. The dose received by bio-objects located in different areas of the biosatellite varied from 200 to 1000 rad. The flight experiment confirmed entirely the reliability of the radiation system.

When arranging irradiation of rats flown aboard the Cosmos 690 biosatellite, the following factors were taken into consideration: geometry of irradiation, weight and size constraints of the radiation device and dosage of irradiation (which was planned to be not less than 700 rad per 24 hours).

On this basis it was concluded that rats should be exposed to irradiation from a point source and not from an extended source. In this case the shielding weight was minimized.

Cs 137 ($E = 0.661$ MeV, $T_{1/2} = 26.6$ year) with an activity of 320 curies was used as the radiation source. The radiation source was placed in the flight emitter which was a tungsten spherical container. The emitter was designed so that the source in the "on" and "off" position remained fixed, whereas the container walls formed a collimator with an angle of 120° in the open position and a spherically shaped storage container in the closed position. The emitter weighed 110 kg. The configuration of the emitter guaranteed that walls of the emitter would not open unexpectedly under the influence of vibration, acceleration or impact.

The geometric scheme of irradiation was as follows. The gamma source Cs 137 placed in the emitter was located in the upper part of the biosatellite along its long axis at a distance of 580 mm from the plane crossing the middle of cylindrical rat cages. Five cages formed a block for investigation and maintenance of animals. Blocks I—V were situated in the direct beam on the lower base plate. The beam was collimated in such a way as to leave blocks VI and VII which were placed on the upper base plate beyond the direct beam.

Measurements showed that the dose gradient in blocks I—V ranged from 100 to 500 relative units. This made it necessary to use a special dose-smoothing filter. The configuration of the filter was computed on the assumption that the dose gradient due to the square of the distance should be compensated for by the radiation absorption of the filter.

The dose-smoothing filter was tested in a biosatellite mock-up. It was shown that the configuration provided an average non-uniformity of the dose field of $\pm 30\%$ with significant deviations of the mean dose in some field points. This was induced by different construction components and equipment located in the

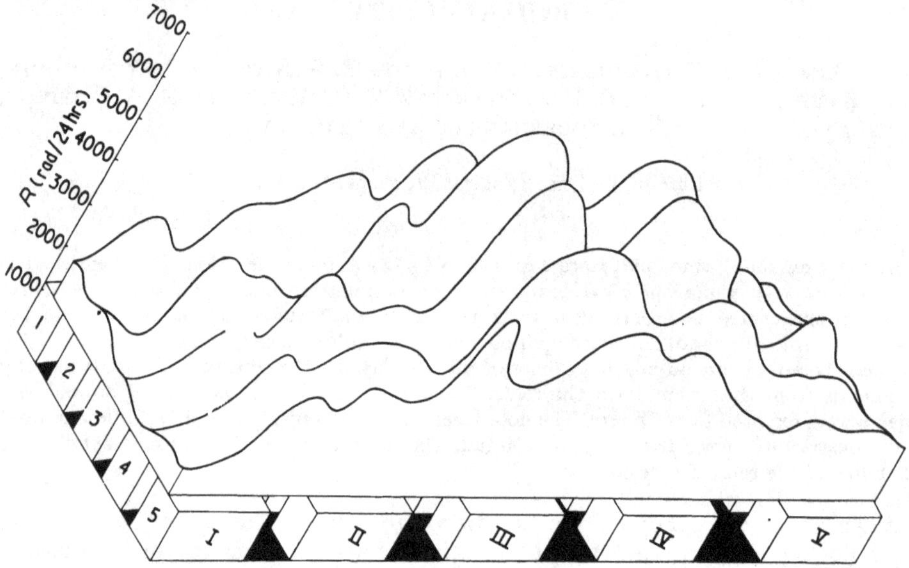

Fig. 1. The dose field in the location of the experimental animals from blocks I—V without a dose-smoothing filter. 1, 2, ... 5, cage numbers; I, II, ... V, block numbers.

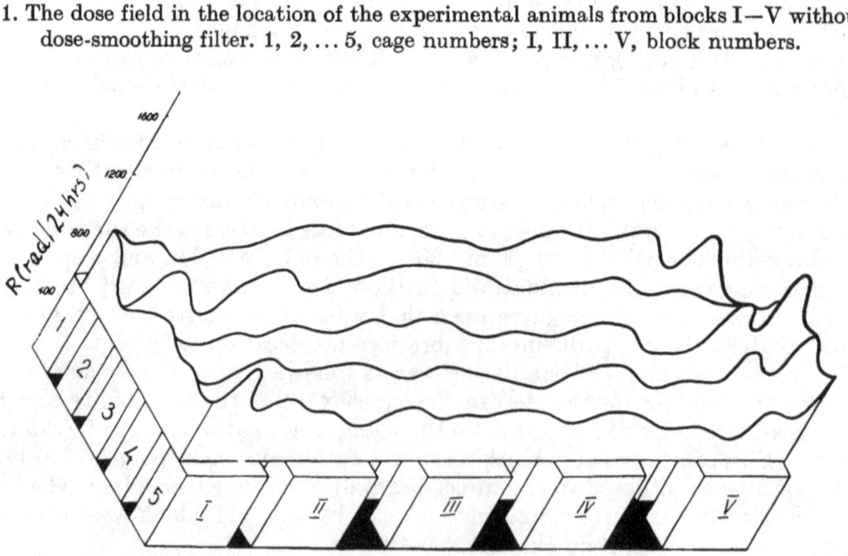

Fig. 2. The dose field in the location of the experimental animals from blocks I—V behind an optimal configuration of the dose-smoothing filter. 1, 2, ... 5, cage numbers; I, II, ... V, block numbers.

direct beam of radiation. The configuration and thickness of the dose-smoothing filter were selected experimentally to provide a mean dose value of 800 rad per 24 hour exposure with uniformity of the dose field.

The dose field was measured by means of ionization sphere chambers and LiF thermoluminescent dosimeters [1]. The ionization chambers and LiF dosimeters were calibrated against reference Cs 137 sources. Measurements of dose fields in the biosatellite without the dose-smoothing filter and behind several filters are given in Figs. 1 and 2.

The spatial pattern of the field where the experimental animals were located (blocks I—V) for the dose-smoothing filter is shown in Fig. 2. In blocks VI—VII the dose was provided by diffuse radiation only. The role of diffusers was played by the upper base plate and other construction components of the biosatellite as well as by dose-smoothing filters and the base plate of the spherical emitter. The mean dose in blocks VI and VII measured by means of ionization chambers was 220 rad/24 hr in the centre of the rat cages.

In order to identify earth effects on the dose field, measurements were performed within the biosatellite when it was on the ground and 2 m above it. The results were very close. This suggests the decisive role of construction components of the biosatellite in the arrangement of the dose field in animal blocks. The spectral structure of radiation responsible for the dose field was also examined during the exposure. Spectral measurements were carried out by a NaI(Tl) scintillation spectrometer of 18 mm diameter and height. The effectiveness of the scintillation spectrometer during non-collimated irradiation was assessed by the method described in [2]. Spectral distributions in blocks III, V and VI measured along the axis of the central rat cages are illustrated in Figs. 3—5.

As follows from Fig. 3, the energy distribution of gamma-quanta in the centre of the dose field (block III, cage 3) has a scattering spectrum with a maximum in the 0.1—0.3 MeV region. The contribution of scattered gamma-quanta into the dose in the centre of the field is about 80%. In the periphery of the dose field (block V, cage 3) the spectrum of gamma irradiation is harder (Fig. 4). The contribution of scattered radiation into the total dose is about 40%. The contribution of gamma-quanta with an energy below 100 keV is not more than 3—5% (the contribution into the integrated flux is about 13%). In blocks VI and VII the maximum of spectral distribution also lies in the 0.1—0.3 MeV region (Fig. 5). The contribution of gamma-quanta with an energy below 100 keV is also not more than 3—5%. Thus measurements of energy spectra demonstrate that the dose field is to a considerable extent formed by scattered gamma radiation. This field is non-uniform with respect to the primary and scattered components: at the periphery the radiation spectrum is harder than in the centre where the filter thickness is maximal and where the primary Cs 137 beam is strongly attenuated.

The dose gradient along the rat body was also measured. For this purpose a rat phantom made of tissue-equivalent plastic was used. The phantom had the shape of a cylinder 85 mm high and 54 mm in diameter. LiF dosimeters were placed within the phantom. Measurements were made in the centre and periphery of the dose field of blocks I—V and of block VI. The dose gradient from the centre to the surface of the rat phantom was $\pm 15\%$. Taking into consideration the fact that at this exposure geometry the dose gradient from the upper to the lower surface of the cage was 35% and that in zero gravity the rat may occupy any position in the cage, no correction was made for the dose gradient along the rat body.

The irradiation environment aboard the biosatellite before, during and after

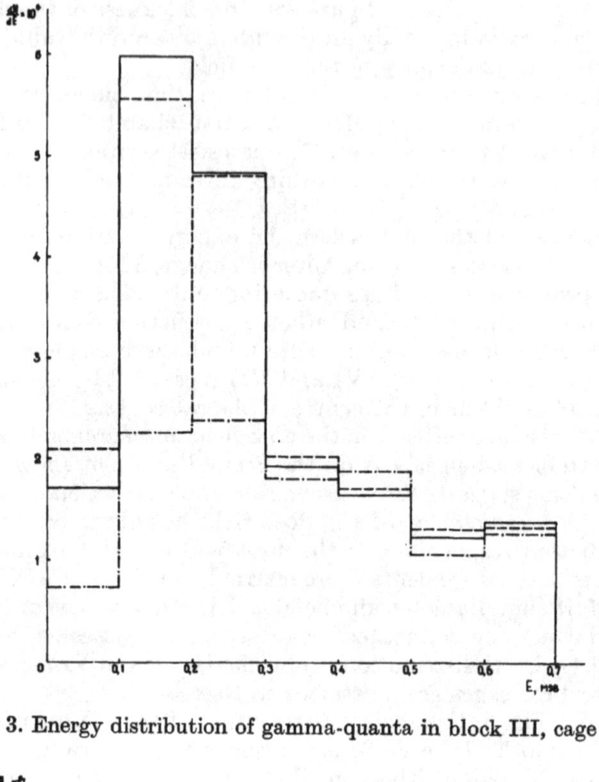

Fig. 3. Energy distribution of gamma-quanta in block III, cage 3.

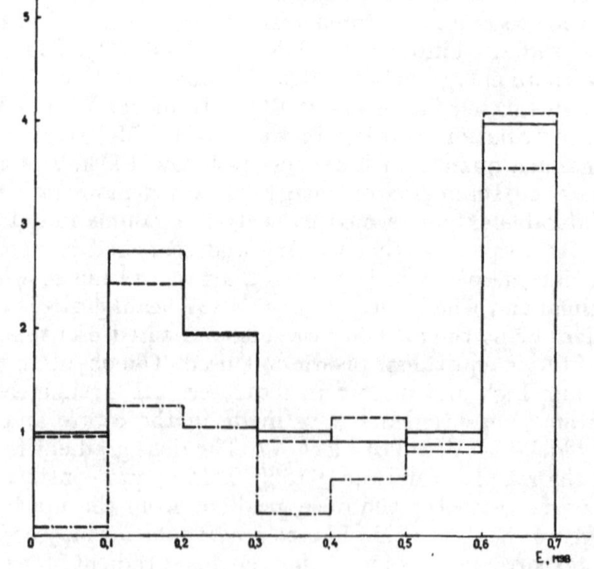

Fig. 4. Energy distribution of gamma-quanta in block V, cage 3.

flight was controlled by thermoluminescent glasses. These dosimeters were fixed to the outer surface ot the rat cages. The irradiation was monitored by means of a dosimeter the readings of which were fed to the control system. According to the flight programme, the radiation equipment was turned on on the tenth mission day. The length of the exposure was 24 hrs. This procedure was followed in the ground-based experiment.

Table 1 gives a summary of dose measurements made by thermoluminescent glasses, LiF thermoluminescent dosimeters and tissue-equivalent ionization

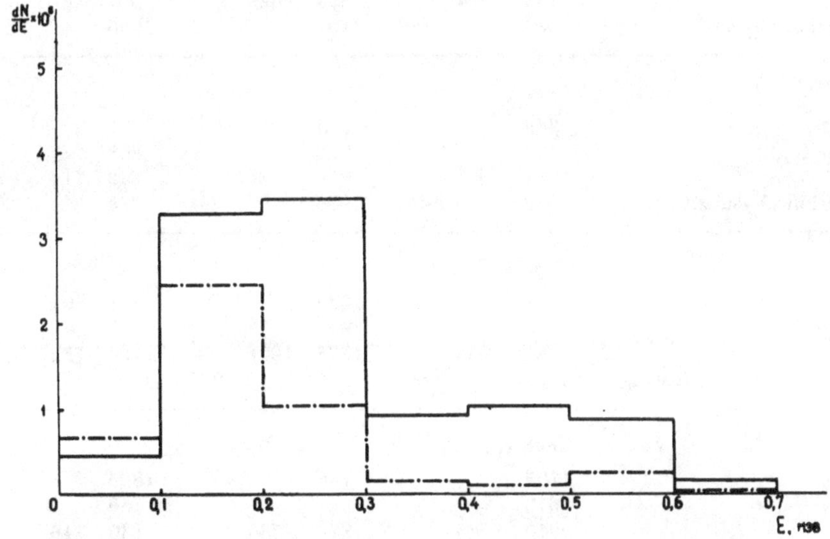

Fig. 5. Energy distribution of gamma-quanta in block VI, cage 3.

chambers. It is seen that doses measured by thermoluminescent glasses (TLG) are slightly higher than those recorded by LiF dosimeters (LiF) and ionization chambers (IC). This can be attributed to the energy dependence of thermoluminescent glasses in the < 200 keV energy range. The mean dose value for the experimental animals kept in blocks I—V was 800 ± 85 rad. The mean dose value for the additional animals kept in blocks VI and VII was 220 ± 25 rad. During the flight experiment the radiation environment within the biosatellite was operationally controlled with the aid of a dosimeter the readings of which were telemetered. The average dose was 20 mrad/24 hr with the emitter closed.

Thus, the radiation and physical investigations allow characterization of the irradiation of animals and other bio-objects flown aboard the Cosmos 690 biosatellite.

The 24-hour exposure of rats and other bio-objects performed on the tenth mission day, as was envisaged by the flight programme, and the absence of radiation hazard throughout the mission give evidence that the designed radiation equipment can function well and reliably.

Table 1

Parameters of the Dose Field (rad/24 hr) measured by

	Block I			II			III		
	TLG	LiF	IC	TLG	LiF	IC	TLG	LiF	IC
	Cage 1								
Before	734	608		955			904	850	
During flight	759			847			924		
After	782	704		934	898		907	841	
Radiation-physical testing	630		882	1107		775	1095		828
	Cage 2								
Before	648			988			737		
During flight	808			902			808		
After	734	601		1059	929		849	714	
Radiation-physical testing	812		699	1065		791	794		806
	Cage 3								
Before	572	744		1075			821	801	
During flight	803			968			869		
After	651	633		1325	1304		796	722	
Radiation-physical testing	832		669	1206		912	901		798
	Cage 4								
Before	1105			785			825		
During flight	913			814			869		
After	940	988		826	735		830	748	
Radiation-physical testing	924		684	720		775	862		843
	Cage 5								
Before	1089	1030		889			1014	953	
During flight	951			973			913		
After	855	794		847	789		995	901	
Radiation-physical testing	930		608	1103		803	1151		859

References

[1] K. K. Schwarz, Z. A. Grant, T. K. Mezh and M. M. Grube, Thermoluminescent Dosimetry, Izd. Zinatne, Riga 1968.

[2] Yu. A. Egorov, Scintillation Method of Spectrometry of Gamma-Radiation and Fast Neutrons, Gosatomizdat, Moscow 1963.

Different Dosimeters in the Centre of the Rat Cages

IV			V			VI			VII		
TLG	LiF	IC	TLG	LiF	IC	TLG	LiF	IC	TLG	LiF	IC
787			532	550		227	197				
803			616			234			220		
813	730		622	633		205	203		239	190	
808		745	666		630	269		181	280		187
985			500			267					
869			550			250			282		
1044	933		712	672		247	223		244	228	
1033		791	708		720	306		236	309		240
1202			972	904		281	266		316		
896			979			262			329		
1084	1013		1181	924		268	305		236	204	
1112		897	1204		800	393		232	316		212
901			1509			305					
858			1122			255			333		
821	679		1255	1354		304	252		213	192	
825		806	1047		989	348		200	277		232
1005			1642	1051		262	215				
968			1221			214			267		
944	772		1530	1572		228	199		157	175	
968		851	1443		1039	297		251	223		203

Life Sciences and Space Research XIV — Akademie-Verlag, Berlin 1976

INVESTIGATION OF RADIATION SENSITIVITY IN MAMMALS UNDER LONG DURATION WEIGHTLESSNESS

Yu. G. Grigoriev, E. A. Ilyin, Yu. P. Druzhinin, L. V. Serova, V. I. Popov, A. D. Noskin, R. A. Kuzin, Yu. I. Kondratyev, M. P. Kalandarova, G. N. Podluzhnaya, B. N. Yurov, V. K. Golov, V. I. Milyavsky and V. V. Verigo

Institute of Medical and Biological Problems, Moscow, USSR

Preliminary results of the radiobiological experiments carried out on the biosatellite Cosmos 690 with a radiation exposure unit on board are presented. The duration of the satellite flight was 20.5 days. On the tenth day of the flight 35 rats were exposed on board the satellite to 220 or 800 rads of gamma radiation. Comparison of data obtained in test and control groups of animals has shown that under the influence of space flight factors a somewhat more severe radiation injury develops than in on-ground conditions.

For predicting the eventual consequences of radiation exposure during space flights it is necessary to take into account the interaction of the organism with a number of other than radiation factors. Of these weightlessness must be first of all considered. In this connection the scientists came to the conclusion that what is needed is an artifical source of radiation on board a biosatellite, that would allow experimentation during conditions of weightlessness with a broad range of ionizing radiation doses. Such spacecraft were the biosatellites Bios 2 and Cosmos 690.

The Cosmos 690 was launched on 22 October 1974, to an orbit with parameters as follows: apogee, 380 km; perigee, 223 km; orbit inclination, 62.8°. According to the flight programme it was successfully landed on 12 November 1974, having been in flight for $20^1/_2$ days. The main task of the studies performed was to obtain data on radiation sensitivity changes in animals and the development of radiation injury under simultaneous exposure to weightlessness and to ionizing radiation under conditions of space flight.

The radiobiological experiment on board Cosmos 690 was carried out mainly on rats. Each of 35 rats was placed in a separate small cylindrical chamber without fixation (Fig. 1). Inside each chamber were installed a manger and a water feeder, a light and a system for ventilation and waste collection. An exposure unit was installed on board the satellite with a gamma radiation source (^{137}Cs). The radiation exposure began on the tenth day of flight and lasted for 24 hours. The absorbed doses in two groups of animals equalled 220 and 800 rads respectively. In addition two rats received 580 rads each, and two others 1170 rads.

Several series of control experiments were carried out on earth. First of all a dose-effect relationship was determined. One group (Control 1) was exposed to radiation and all flight conditions, except weightlessness, were simulated. Control 2 group differed from the former only in that overloads were also excluded. An intact control group of animals was also observed in a simple vivarium. Temperature, humidity and the composition of the gaseous medium were kept within normal limits. After landing, all the animals were delivered to the laboratory. On the 1—2 and 26—27 days following landing, that is, on the 11—12 and the 36—37 days post-radiation, the animals were examined.

During the 20-day flight two rats that received 800 rads died; two rats died in the control 1 group, one that received 1170 rads and the other from intermittent pneumonia. Among the control 2 group no death occurred.

Fig. 1. Chambers for rat containment.

Immediately after the satellite landing the general state of the animals was satisfactory. The rats were not very mobile, muscular tone being depressed. On the first day after landing it was noted that in these rats the weight increase was significantly lower than in similarly exposed control rats (Table 1). During the subsequent observation period the kinetics of weight in test rats exposed to

Table 1

Weight of Flight and Control Animals exposed to 220 or 800 rads of Gamma Radiation

Group and Radiation Dose	Weight (g)		
	Before experiment	After experiment	Difference
800 rads			
Flight	223.5	239.9	+16.4
Control 1	219.2	263.9	+44.7
220 rads			
Flight	223.4	256.2	+32.8
Control 1	223.0	274.9	+51.9
Intact control	213.8	276.5	+62.7

800 rads differed from that seen in other groups. The recovery of weight was less rapid.

Haematological changes in rats exposed on Cosmos 690 did not differ qualitatively from those in the Control 1 and Control 2 experiments. However, quantitatively, maximal injury was observed in test animals exposed to 800 rads. A

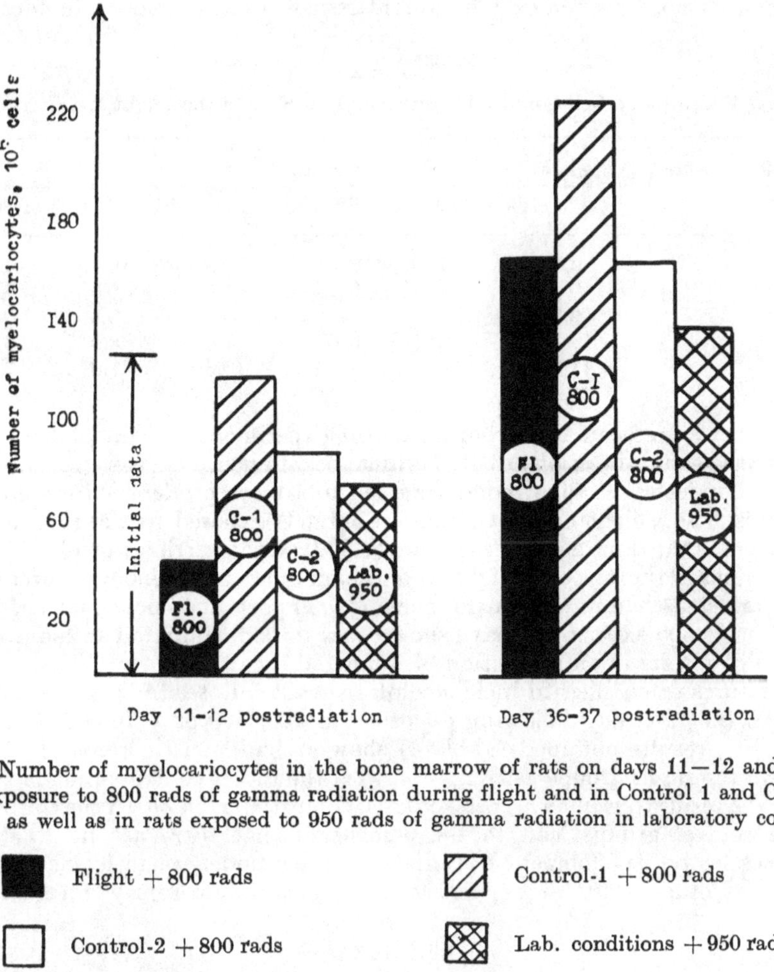

Fig. 2. Number of myelocariocytes in the bone marrow of rats on days 11—12 and 36—37 after exposure to 800 rads of gamma radiation during flight and in Control 1 and Control 2 groups, as well as in rats exposed to 950 rads of gamma radiation in laboratory conditions

high degree of bone marrow depletion and a sharp decrease of lymphocyte count were noted (Fig. 2). The changes observed correspond to the radiation injury induced in control animals by an exposure to 950 rads.

The numbers of circulating white blood cells in rats of the flight and control series is shown in Table 2. A trend toward deeper leucopenia was observed in the flight group exposed to 800 rads; the recovery phases of leucocytes were delayed.

On the day 11—12 following exposure to 800 rads the colony-forming ability of stem-cells in test and control rats was measured. A considerable increase of colony formation was noted in the case of bone marrow transplantation from donors exposed on board the biosatellite. Morphological analysis of the colonies showed no essential impairment in the differentiation process of haematopoietic cells in the test animals.

Analysis of some indicators of the state of the thymus and spleen (organ weight, number of cells) showed that the radiation response, as judged by these indicators, is more pronounced in weightlessness than on earth. In fact, when

Table 2

Circulating White Blood Cell Count (10^3 cells mm^{-3}) in Rats of the Flight and Control Series

Days post-exposure	Flight		Control 1		Control 2	
	220 rads	800 rads	220 rads	800 rads	220 rads	800 rads
11–12	9.1	3.7	7.7	4.8	10.7	4.7
17	9.4	7.1	9.2	6.6	6.6	7.5
21	9.7	6.2	—	—	8.8	6.8
30	13.3	9.3	6.1	7.4	8.0	7.8
36	14.7	12.0	8.3	10.9	7.0	7.2

exposed to 220 rads in the ground control experiment, the involution of the thymus was seen only in one of five animals examined, whereas exposure to an equal dose under space flight conditions led to a more pronounced involution in all animals. The weight of the thymus in animals exposed to 220 rads in flight conditions was the same as after exposure to 800 rads on earth (Control 1, Table 3). In test animals the spleen weight and the number of splenocytes were more severely affected than in the control ones. It is important to note that the results of the Cosmos 605 experiment have shown that in non-irradiated animals weightlessness also results in an involution of lymphoid organs.

The features of changes in incidence of aberrant mitoses in cornea epithelium of rats on the 12th day following exposure to 220, 800 or 1170 rads have been studied. The results obtained (Table 4) show a significant increase of aberrant mitoses in the flight groups as compared with the intact control animals.

Thus, the results obtained on rats exposed to radiation on board the Cosmos 690 and the control groups lead to the conclusion that there are no qualitative differences in the development of radiation injury under weightlessness and the influence of other space flight conditions. The radiation injury developed was

Table 3

Thymus and Spleen Weight in Flight and Control Animals 11–12 days following Irradiation

Organ	Dose (rads)	Weight (mg)		
		Flight	Control 1	Intact control
Thymus	220	131 ± 21	277 ± 43	337 ± 36
	800	94 ± 14	127 ± 13	
Spleen	220	335 ± 56	556 ± 56	685 ± 55
	800	334 ± 28	421 ± 32	

somewhat more severe under weightlessness conditions. According to the available end points the radiobiological effect of exposure to 800 rads in flight corresponded to 950 rads exposure in laboratory conditions on earth.

Table 4

Incidence of Aberrant Mitoses in Eye Cornea Epithelium Cells in Rats of the "Flight" Group 12 days post-exposure and in Control Animals

Flight	220 rads	2.9 ± 0.3
	800 rads	8.8 ± 0.7
	1170 rads	22.1 ± 2.3
Intact control		1.8 ± 0.1
950 rads (laboratory conditions)		6.6 ± 1.4

Life Sciences and Space Research XIV — Akademie-Verlag, Berlin 1976

EFFECT OF IRRADIATION IN THE SPACE ENVIRONMENT ON THE BLOOD-FORMING SYSTEM IN RATS

M. P. Kalandarova, V. V. Verigo, G. N. Podlyzhnaya, G. P. Rodina, L. V. Serova and N. A. Chelnaya

Institute of Medico-Biological Problems, Moscow, USSR

A special experiment was carried out on board Cosmos 605 in order to study the modifying influence of weightlessness on the radiobiological effect. On the tenth flight day 35 rats were exposed to irradiation using an on-board gamma-ray source. They were irradiated for 24 hours at a dose rate of 32 rads/hr with a total dose of 220 ± 25 rads and 800 ± 53 rads. On the 1st and 26th post-flight days haemopoiesis in bone marrow (determined from myelograms and the total count of myelo-karyocytes) was examined and the blood composition was analysed. The total count of thymocytes and spleenocytes was measured in the thymus and spleen. Identical measurements were made in the control animals exposed to irradiation on the earth. The haematological findings indicate an enhancement of the radiobiological effect in the rats irradiated in space flight.

The blood-forming system reparative regeneration was studied in rats exposed to gamma radiation in space, aboard the Cosmos 690 biosatellite. On the tenth day of the flight two groups of rats were irradiated for 24 hours with gamma rays from an on-board radiation source at a dose rate of 32 rad/h; they received total doses of 800 ± 53 rad and 220 ± 25 rad respectively. In order to reveal the modifying influence of weightlessness on the radiobiological effects two control ground-based experiments were also conducted.

One group of rats was irradiated after being kept for ten days in conditions similar to that of space flight. A simulated spacecraft descent was performed on the 11th day after irradiation (Control 1). The second group of rats was irradiated after living for ten days in the same conditions without simulated spacecraft descent (Control 2). On the 1st and 26th days after the 20.5-day space flight (post-irradiation days 12 and 38) the blood-forming function of the bone marrow was evaluated in rats using myelogram data and total bone marrow cell counts. Blood cell counts were also studied.

Since long-term simulation of zero gravity is impossible on earth the experiment was designed to differentiate any modifying influence of weightlessness on the biological effect of radiation. At present the experimental data are undergoing statistical processing, so the purpose of this paper is to report some preliminary results and conclusions.

For the first ten days after irradiation the development of radiation injury took place in the space environment and it was mainly the reparation of haemopoiesis that could be studied after the end of the flight. In all three groups of rats the bone marrow reparation processes were similar. However there were certain differences in the time-course of reparation between the groups. The degree of haemopoietic reparation to a certain extent depended on the conditions under which the radiation injury had been received. Fig. 1 shows changes in the bone marrow myelokaryocyte count, percentage of myeloblasts (mature and immature) and white blood cell count in rats following irradiation. The myelokaryocyte

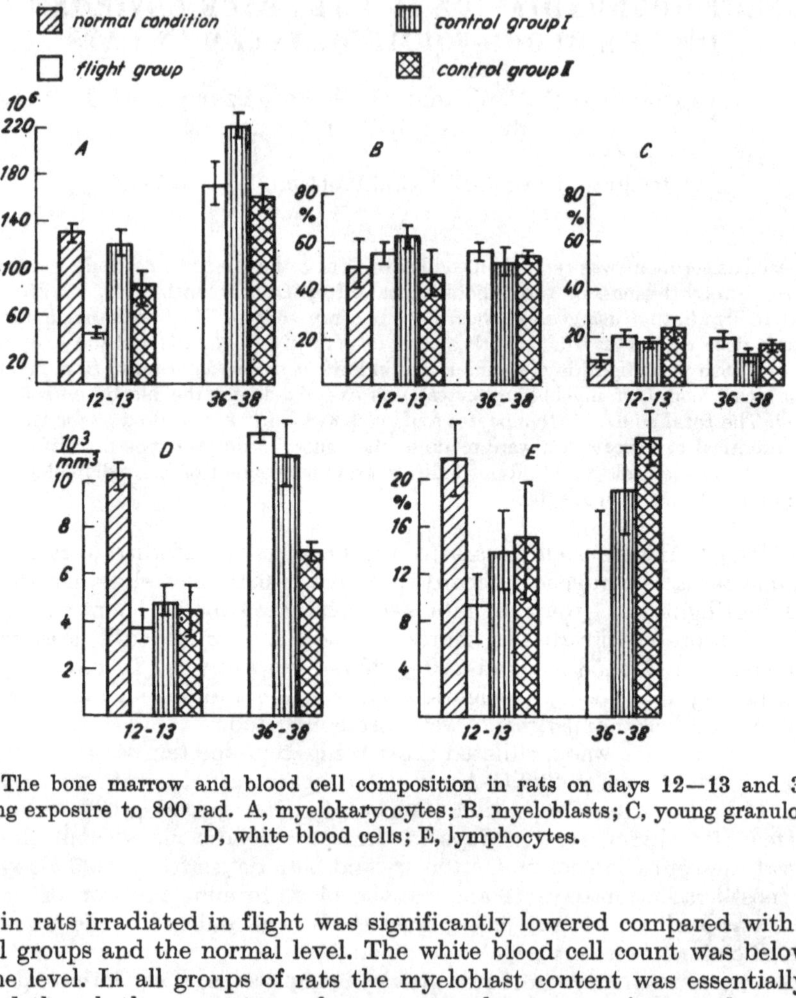

Fig. 1. The bone marrow and blood cell composition in rats on days 12—13 and 36—38 following exposure to 800 rad. A, myelokaryocytes; B, myeloblasts; C, young granulocytes; D, white blood cells; E, lymphocytes.

count in rats irradiated in flight was significantly lowered compared with both control groups and the normal level. The white blood cell count was below the baseline level. In all groups of rats the myeloblast content was essentially unchanged though the percentage of young granulocytes was increased compared with normal. Within 36—38 days following the irradiation the bone marrow myelokaryocyte count and the white blood cell count returned to (or even somewhat exceeded) the baseline level. The bone marrow myeloblast count remained essentially unchanged and the number of young granulocytes continued to be fairly elevated in all three groups of rats.

An analysis of the bone marrow myeloblast content (Fig. 2) revealed some increase in erythroblasts in the experimental (flight) group and in Control 2.

As can be seen in Fig. 2, the red blood cell count was normal in all three groups. The rats irradiated in flight showed a decrease ($P = 0.05$—0.001) in reticulocyte level compared with normal and with both control groups on the 12th day after irradiation. Later there were fluctuations in the reticulocyte count with a trend to recovery in all three groups. Within 36—38 days after irradiation the reticulocyte count returned to baseline level with the exception of the experimental group where it was still increased compared with the control group.

There was a definite relationship between the time-course of lymphocyte changes in the bone marrow and blood, and irradiation conditions. The rats

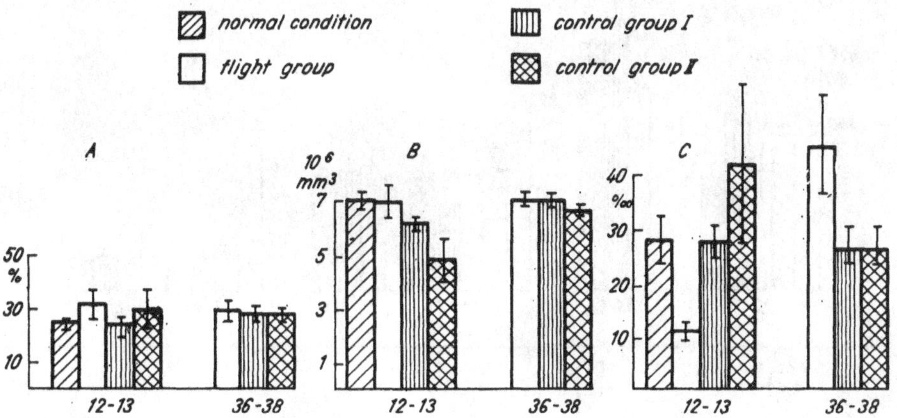

Fig. 2. The bone marrow and blood cell composition in rats on days 12—13 and 36—38 following exposure to 800 rad. A, erythroblasts; B, red blood cells; C, reticulocytes.

irradiated in flight showed a statistically significant decrease in the lymphoid elements during the bone marrow regeneration (post-irradiation day 12—13, $P < 0.01$ compared with normal; post-irradiation day 36—38, $P < 0.005$ compared with normal and $P < 0.01$ compared with control 2).

An examination of the bone marrow reticular stroma demonstrated that, on post-irradiation day 12, rats irradiated in flight showed the maximum increase in the total number of reticulum cells, monocytes and plasmocytes ($P < 0.01$) compared with normal and control 1. On post-irradiation day 36—38 the total number of these cells was not significantly different from normal in all three groups of rats. Changes in the bone marrow and blood following both in-flight and ground-based exposure to 220 rad were qualitatively similar to those exposed to 800 rad in the same conditions. The differences were only quantitative.

A comparative study of the time-course of changes in the total amount of myelokaryocytes after doses of 220 rads revealed alterations similar to those observed after doses of 800 rad received in similar conditions. The most pronounced lowering of the total amount of nuclei-containing elements on post-exposure day 12 was found in rats irradiated in flight. In control 1 this change was minimal (Fig. 3). At the second examination of the bone marrow on post-exposure days 36—38 the number of myelokaryocytes was back to normal in all rats. The changes in myeloblasts, young granulocytes and white blood cells, as well as in the bone marrow erythroblasts following exposure to 220 rad were similar to those found after exposure to 800 rad in the same conditions but in all cases they were less pronounced.

The one exception was the number of lymphocytes which on post-irradiation day 12 was normal in the experimental group and considerably elevated in

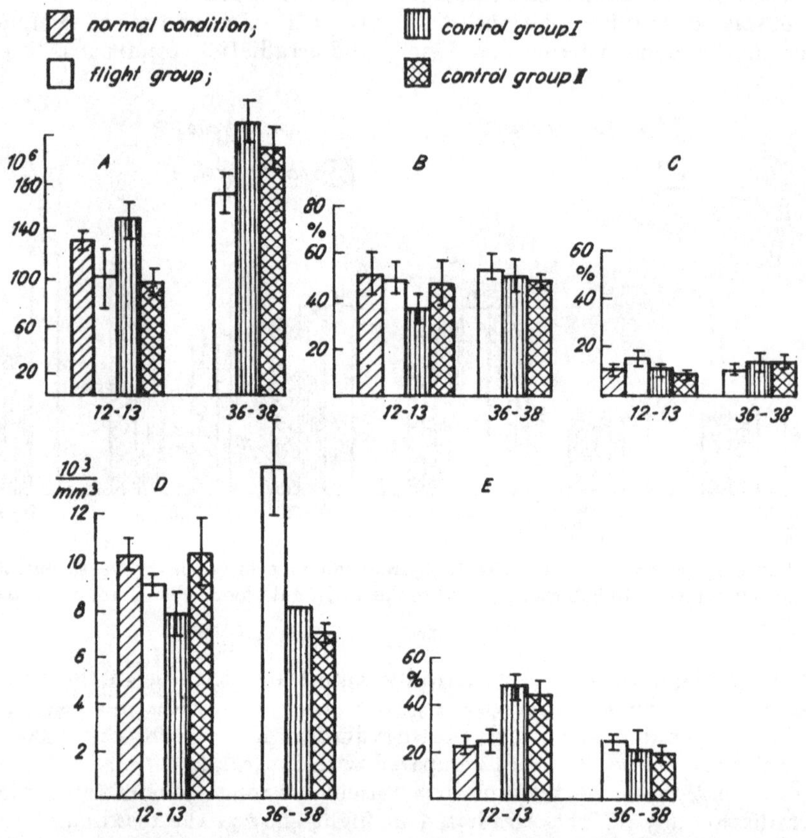

Fig. 3. The bone marrow and blood cell composition in rats on days 12—13 and 36—38 following exposure to 220 rad. A, myelokaryocytes; B, myeloblasts; C, young granulocytes; D, white blood cells; E, lymphocytes.

controls 1 and 2. No changes in the bone marrow stroma throughout the observation period were found in rats of all three groups exposed to 220 rads.

The experimental group irradiated with both doses of 800 and 220 rad showed a greater decrease in most haematological data than controls 1 and 2 on post-irradiation day 12.

In the present experiment it was found that ionizing radiation was the main factor responsible for the injury to the blood-forming system. Weightlessness also contributed to the progress of the acute radiation injury of the blood-forming system. At present it is not yet possible to conclude which aspect of radiation injury is most affected by weightlessness — the extent and duration of haemopoietic inhibition or the rate of reparation processes. As a tentative suggestion it could be said that the reparation processes rate was affected more, since physiological regeneration in these conditions was apparently slowed down, probably due to a decreased demand for blood-forming cells by the body.

Some data obtained in the experiment on rats aboard the Cosmos 605 biosatellite (exposure to space flight factors but no irradiation) indicated that after a 22-day flight there was some inhibition of haemopoietic reparation in rats (a decrease in total myelokaryocyte, erythroblast and lymphocyte counts of the bone marrow on post-flight day 1—2 with recovery within 26 days). The return of most haematological index data to the baseline level by post-irradiation day 36—38 (post-flight day 26) shows the readjustment of haemopoietic reparation rate to normal gravity conditions. A comparative analysis of the results obtained confirms the conclusion that pathologic processes, specifically radiation injury, can be affected and stressed in conditions of prolonged weightlessness.

Life Sciences and Space Research XIV — Akademie-Verlag, Berlin 1976

BIOCHEMICAL CHANGES IN RATS FLOWN ON BOARD THE COSMOS 690 BIOSATELLITE

I. Ahlers[a], E. Mišúrová[a], M. Praslička[a] and R. A. Tigranyan[b]

[a] Institute of General Biology, Šafárik University, Košice, CSSR
[b] Institute of Medico-Biological Problems, Health Ministry, Moscow, USSR

Male rats of the Wistar strain were examined one day and 26 days after the 20-day space flight on board the Cosmos 690 biosatellite. On the 10th flight day the animals were exposed to 24-hour gamma irradiation: some animals were irradiated at a dose of 220 rad and the others at a dose of 800 rad. The data were compared with the results obtained from two control groups one day and 26 days after the experiment: (i) rats kept in the vivarium and fed with a special diet, and (ii) rats from a synchronous experiment (which simulated the flight profile with the exception only of weightlessness).

The studies showed the following changes one day after flight: an increase in the concentration of total cholesterol, glucose, urea and corticosterone in the blood plasma, an increase in the concentration of triglycerides in the plasma, liver and bone marrow, a decrease in the concentration of total cholesterol in the adrenals, a reduction in the DNA and RNA content in the spleen and bone marrow, a decline in the potassium concentration together with an increase in the sodium concentration in soleus muscle. The degree and pattern of the changes depended largely on the dose of irradiation. The parameters returned to normal 26 days post-flight.

The present paper gives the results of biochemical examinations on male rats of the Wistar strain flown 20.5 days on board the Cosmos 690 biosatellite. The purpose was to study the effect of prolonged weightlessness combined with ionizing radiation. On the 10th flight day the animals were exposed to 24-hour irradiation: some of them were irradiated with a dose of 220 rad and the others with a dose of 800 rad.

The examinations were carried out one day and 26 days post-flight. The flight results were compared with the data obtained from two control groups of animals: (1) rats of the same stock kept in the vivarium on a special diet (Control 1) and (ii) rats from a synchronous experiment which simulated all flight conditions (Control 2) one day and 26 days after simulated flight. Blood, bone marrow, skeletal muscles, myocardium, liver, spleen, adrenals, brain, spinal cord, white and brown fatty tissue of the animals were examined. Triglycerides were measured by the method of Eggstein and Kreutz [5], total cholesterol by the method of Zlatkis et al. [6], phospholipids by the method of Bartlett [3], glucose by the method of Bittner and McCleary [4], lactate by the method of Barker and Summerson [2] and corticosterone by the method of Popens et al. [1]. Statistical differences were derived from the comparison of the flight data with the examinations of intact animals and Control 2.

The investigations have shown the following changes.

Blood. The content of total protein in rats one day post-flight did not differ from that of intact (Control 1) animals; the rats irradiated with 800 rad showed a decrease in the albumin content and an increase in the globulin content; they also demonstrated a significant increase in the urea content. It should be noted that no changes of this kind were found in the flight group of rats exposed to 220 rad or in the corresponding group of Control 2. Flight animals displayed no variations in the total content of nucleic acids (DNA + RNA) in the white blood cells; however, Control 2 rats showed a noticeable decrease of nucleic acids one day after simulated flight. This change was correlated with the dose of irradiation: after an exposure to 220 rad the amount of nucleic acids decreased by 35% and after an exposure to 800 rad it fell by 50% compared with that of intact animals. Later the amount of nucleic acids in the white blood cells returned to normal. One day post-flight the rats exposed to a dose of 800 rad showed a significant increase in the glucose concentration which persisted 26 days after flight. One day post-flight the animals irradiated with 220 rad displayed a marked decrease in the lactate concentration and creatine phosphokinase activity as well as an increase in the corticosterone concentration. One day post-flight a noticeable increase in the concentration of triglycerides (similar in the 220 rad and 800 rad groups) and a significant increase in the concentration of total cholesterol (greater in the 800 rad group) were noted. Changes in the concentration of non-esterified fatty acids were different: their quantity increased in the 220 rad animals and fell in the 800 rad rats. In animals of both flight groups the phospholipid concentration one day post-flight did not differ from that of intact animals, and 26 days post-flight increased in 800 rad animals (in comparison with the intact rats and 220 rad animals). These changes returned to normal 26 days post-flight.

Bone marrow. In both groups the concentration of triglycerides increased noticeably one day post-flight and returned to normal 26 days post-flight. At the same time Control 2 animals showed the following picture: one day after simulated flight a difference between animals irradiated with 800 rad and intact animals was found; 26 days post-flight a noticeable increase in both groups was observed. The concentration of phospholipids one day post-flight was increased only in 800 rad animals and 26 days post-flight only in 220 rad rats; no differences were found in Control 2 groups. The DNA concentration was lowered one day post-flight only in the 800 rad animals and returned to normal 26 days post-flight. A similar picture was observed in Control 2 animals. No changes were found in the RNA concentration or the RNA/DNA ratio in flight and Control 2 animals in comparison with intact rats.

Skeletal muscles. The content of phospholipids in the microsomes of femoral muscles was noticeably reduced one day post-flight and returned to normal 26 days after flight; no changes were found in Control 2. One day post-flight in the soleus muscle the content of sarcoplasmatic and myofibrillar proteins was significantly reduced, LDH activity was diminished, the ratio of LDH isoenzymes was altered and activity of SGOT and SGPT transaminase in the sarcoplasmatic proteins was increased. In control experiments such changes (indicating the development of atrophic processes) were not noted. At the same time one day post-flight a decrease of the K/Na ratio in the soleus muscle, an increment of this value in the plantaris muscle and no change of the ratio in the diaphragmatic muscle were found. 26 days post-flight these changes recovered completely.

Myocardium. One day post-flight the structure of protein fractions remained unaltered except a slight decline of the content of T-proteins. The ATPase activity of myosin was significantly reduced, the lowered activity of the enzyme persisting 26 days post-flight.

Liver. One day post-flight the content of triglycerides was significantly increased in both groups of animals and 26 days post-flight it returned to normal; no such changes were found in Control 2. No variations were detected in the concentration of total cholesterol and phospholipids of all flight and Control 2 animals in comparison with intact rats. One day post-flight both groups of animals exhibited a significant decline in the DNA concentration and no variations in the RNA concentration; they also showed a significant increase in the RNA/DNA ratio. 26 days post-flight these parameters returned to normal. Such changes were not found in Control 2.

Factors of the space mission and Control 2 did not influence the activity of alkaline DNAase. In Control 2 one day post-flight a marked (nearly 2-fold) decrease of the total activity of acid DNAase was noted; 26 days post-flight, activities of the enzymes were 20% higher than those in intact animals. After real flight a trend of similar changes was found; however, this effect was not statistically significant. Free activity of acid DNAase remained unchanged after flight experiment and increased by 7—8% in Control 2. Under the influence of irradiation the number of breakages in DNA molecules grew in the flight experiment and in Control 2; this effect was increased in proportion to the dose of irradiation. The number of breakages in DNA molecules was greater in flight than in Control 2 at identical irradiation doses.

Spleen. One day post-flight the DNA concentration decreased by 25% in both groups of animals in real flight and in Control 2; the RNA concentration reduced only in the 220 rad animals of Control 2; the RNA/DNA ratio remained unaltered. 26 days post-flight these values returned to normal both in flight and Control 2 rats. Space flight factors did not influence the activity of alkaline DNAase and Control 2 conditions reduced the enzyme activity by 30%. The total activity of acid DNAase in flight and Control 2 groups did not change, except for a 24% decline 26 days after the termination of Control 2. One day post-flight the free activity of acid DNAase was reduced by 8%; later this effect was absent from both the flight and Control 2 rats. One day post-flight the rate of DNA synthesis decreased and during recovery it increased in the flight animals; in Control 2 these changes were less marked.

Adrenals. The content of total cholesterol in all animal groups varied insignificantly in comparison with the intact rats. The DNA and RNA concentrations as well as the RNA/DNA ratio did not change in all rat groups of the flight or Control 2 experiment.

Brain and spinal cord. The concentration of protein and RNA in neurons and gliacytes of the supraoptic nucleus of the hypothalamus, cerebellum (Purkinje cells and their gliacytes and satellites) and anterior corni of the lumbar cord (motoneurons and their gliacytes and satellites) in experimental rats was similar to that in intact animals. The activity of protamine-peptidehydrolases in the cerebellum, quadrigeminal bodies, motor and optic zones of the cerebral hemispheres in all experimental rats did not differ from that in the vivarium animals.

Fatty tissue. The concentration of non-esterified acids in the white and brown fatty tissue in all experimental animals did not differ from that in the intact rats.

Thus, the detected changes indicate functional rearrangement of metabolism and are reversible.

References

[1] J. POPENS, E. SILINSH and I. VITOLS, Voprosy med. khimii 8, 628 (1962).
[2] S. B. BARKER and A. H. SUMMERSON, J. Biol. Chem. **138**, 535 (1941).
[3] G. R. BARTLETT, Biol. J. Chem. **234**, 466 (1959).
[4] D. BITTNER and M. MCCLEARY, Amer. J. Clin. Path. **11**, 423 (1963).
[5] M. EGGSTEIN and E. R. KREUTZ, Klin. Wschr. **448**, 262 (1966).
[6] A. ZLATKIS, B. ZAK and A. J. BOYLE, J. Lab. Clin. Med. **41**, 486 (1953).

Life Sciences and Space Research XIV — Akademie-Verlag, Berlin 1976

STUDY OF THE BIOCHEMICAL INDICATORS OF CHRONIC IRRADIATION IN RATS

L. D. SZABÓ, A. B. BENKÓ, L. GYENGE and T. PREDMERSZKY

National "Frédéric Joliot-Curie" Research Institute for Radiobiology and Radiohygiene, Budapest, Hungary

Daily urinary excretion of pseudouridine, creatinine and creatine of chronically irradiated Wistar rats was estimated. The irradiation conditions were: ^{60}Co gamma source, dose-rate 10 rad/day, total dose 200, 400 and 600 rad. Control groups were kept under similar conditions. Urine samples were taken three times after the end of the irradiation period. It was found that: (1) pseudouridine excretion seems more suitable for indicating radiation injury than the creatine/creatinine ratio in chronic irradiation in rats; (ii) there are significant changes in dose dependence of pseudouridine excretion in the post-irradiation period; (iii) a new method for pseudouridine estimation gives closely similar data to those of earlier investigations.

The demand for a biochemical indicator of radiation injury which is reliable over a wide dose range arose almost simultaneously with our growing knowledge about the injury itself. Many papers have dealt with this problem and in recent years this interest was evoked by the developing atomic industry and space research [1]. In these fields the doses of radiation to which people are accidentally exposed are usually absorbed chronically or, more exactly, the doses are distributed homogeneously or inhomogeneously over a long period. The known biochemical indicators are valid almost exclusively for the case of acute irradiation with relatively high dose rates. Unfortunately the data obtained under these conditions are sometimes controversial, especially in the lower dose range.

Experiment show also that the presence or amount of substances used as indicators often depends on the quality of the radiation together with the dose rate. The cause of this phenomenon is rather complex: it involves the fact that the organism reacts with different intensity to different types and intensities of the damaging radiation as well as that repair and recovery mechanisms can occur over a prolonged period.

All the biochemical methods in the literature which were more or less suitable in the case of acute irradiation proved to be completely useless in studying chronic radiation damages. This is understandable if one considers that substances that are excreted by an irradiated organism after an acute dose, as products of cell disintegration, can be effectively re-utilized by an organism in a state of active recovery. On a theoretical basis one can predict that the best substance for indicating both acute and chronic radiation injury must be one that is completely unable to participate in metabolic processes. Drahovsky and others [2] wrote first on pseudouridine as a possible bioindicator because they got a straight line for dose dependence of urinary pseudouridine excretion in rats. There have been several works on pseudouridine in recent years [3, 4].

No metabolic process is known at present leading to possible re-utilization in mammals of pseudouridine released from decaying t-RNA molecules. This property suggests its use as a biochemical indicator even for non-acute irradiation.

Our first aim was to develop a method for the determination of this substance that can meet the requirements of screening tests.

Methods given in the literature [2, 4] involve two- or three-fold chromatography, the last one being usually very time-consuming and requiring a lot of

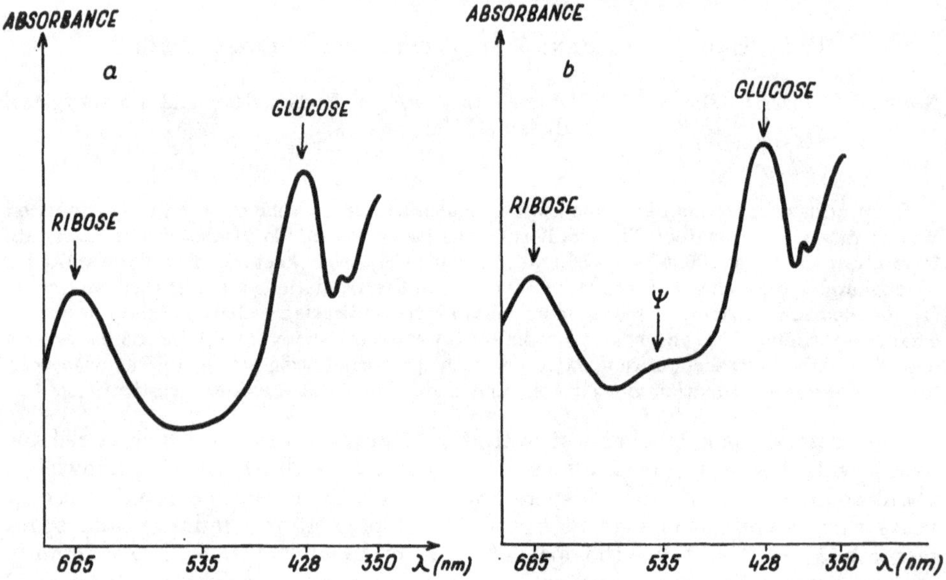

Fig. 1. a, Visible spectrum of a mixture of glucose and ribose after developing by the orcinol reaction. b, Spectrum of the orcinol product of a urine sample. The pseudouridine (Ψ) peak provides a base for estimation (when corrected by suitable factors depending on ribose and glucose peak heights).

materials. Our simple method is based on the orcinol colour reaction which gives a product with pseudouridine with maximal light absorption at 535 nm (unlike other ribose-containing substances and glucose, which develop colours with absorption maxima at 665 and 428 nm, respectively [5] (Fig. 1). Pseudouridine content can be calculated from the light absorbance of a urine sample in suitable dilution boiled with orcinol solution, measured at the above wavelengths. The calculation factors needed can be obtained from pure uridine, pseudouridine and glucose samples. The only purification step is adding 2 volumes of 96% ethanol to urine samples and removing the precipitate before final dilution. Nucleosides remain quantitatively in the supernatant. This method was presented at the 9th FEBS Meeting in 1974 [6], together with our first results on gas-chromatography of urinary pseudouridine. The latter seemed to be quite promising in spite of several difficulties. Meanwhile a quantitative gas-chromatographic method was published by Chang et al. [7]. We shall continue to develop our method, which, in respect of the pre-treatment of the samples is slightly different from the published one.

Using our spectrophotometric method for determination we obtained the curve shown in Fig. 2a for urinary pseudouridine in rats after acute irradiation. Pseudouridine content on this and next curves is presented as a percentage of the control value for standardized data. Dots represent mean values of several experimental results. Single measurements show sometimes more than 15% deviation

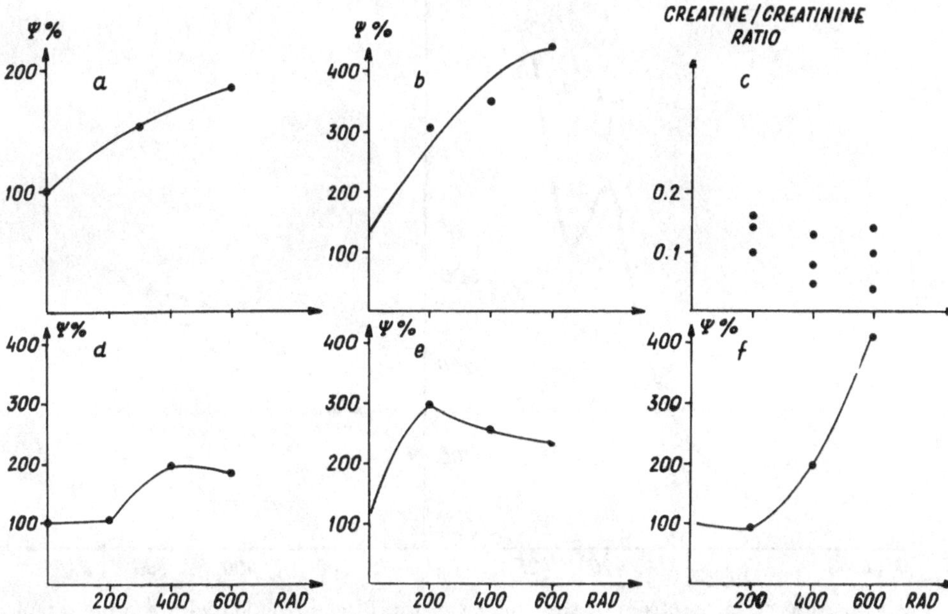

Fig. 2. Values as percentage of control (except c). a, pseudouridine content of daily urine of rats plotted versus dose of acute irradiation; b, pseudouridine content of the urine collected from rats the day after the end of chronic irradiation at several total doses; dose intensity was 10 rad/day; c, creatine/creatinine ratio estimated from the same material as before; dots represent single measurements; d, pseudouridine content of the urine collected from the same rats as in b and c on the 3rd day after ending exposure; e, pseudouridine content of the urine collected on the 4th day after ending exposure; f, pseudouridine content of the urine collected on the 10th day afteı ending exposure.

from this value. It cannot be excluded that these differences are partly due to the limited accuracy of our rapid estimation method. Considering the magnitude of the effect examined we believe, however, that it does not alter the significance of the changes observed.

Chronic irradiation of Wistar rats was performed in a gamma-ray field with a dose intensity of 10 rad/day. Total absorbed dosses were 200, 400 and 600 rad, respectively. Control groups were kept under similar conditions and, in the factors examined, they did not differ significanly from the animals kept as controls. Groups consisted of 8 or 7 animals.

Urine was collected in metabolic cages during the first 24 hours after the end of the irradiation. The curve in Fig. 2b represents amounts of pseudouridine excreted daily as a function of dose of chronic irradiation. For the sake of comparison Fig. 2c shows the creatine-creatinine ratio [8] in the same samples. Dots represent values obtained from single measurements. The construction of a continuous curve from them was not possible. Figs. 2d, e, f represent pseudouridine excretion 3, 4 and 10 days after the end of the irradiation, respectively. These unusual shifts in pseudouridine excretion may be due to acceleration of the

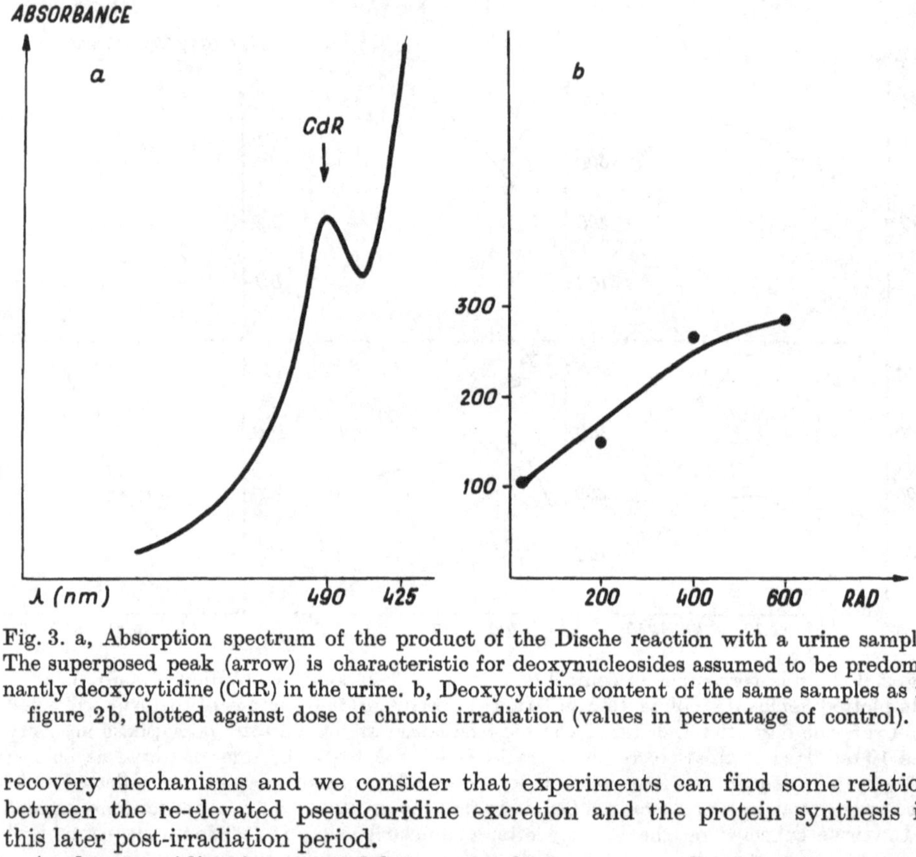

Fig. 3. a, Absorption spectrum of the product of the Dische reaction with a urine sample. The superposed peak (arrow) is characteristic for deoxynucleosides assumed to be predominantly deoxycytidine (CdR) in the urine. b, Deoxycytidine content of the same samples as in figure 2b, plotted against dose of chronic irradiation (values in percentage of control).

recovery mechanisms and we consider that experiments can find some relation between the re-elevated pseudouridine excretion and the protein synthesis in this later post-irradiation period.

As deoxycytidine is excreted by man only in very small quantities, making measurements difficult and unreliable, it may have not such great importance for space research. There are, however, reports concerning human material especially in cases of acute therapeutic irradiation [9]. Rats excrete several milligrams per day of deoxycytidine (CdR)* so one can easily measure it in the urine of these animals. We have a method for estimation of this substance similar to the pseudouridine determination described. It includes the routine ethanolic purification, suitable dilution and colour formation with cysteine HCl as described by Dische [10]. For evaluation spectrophotometric measurement of the product is needed at two wavelengths, usually 490 and 425 nm (Fig. 3a). Calculation factors are obtained by similar developing of a pure CdR standard sample and a

* CdR = Cytosine-deoxy-Riboside

urine sample previously treated with active carbon in order to remove, among others, nucleosides including CdR. So, by determining the absorption ratio E_{425}/E_{490} of pure CdR and that of the interfering peak, approximate calculation of the CdR content becomes possible. Curve b of Fig. 3 shows CdR excretion determined from the same samples as was pseudouridine. A significant dose dependence can be observed.

In the case of rodents this seems to be a promising way for indicating chronic radiation injury, and with developing new estimation techniques it may be applicable to humans as well.

References

[1] L. D. SZABÓ and F. ANTONI, Manual on Radiation Hematology, Tech. Rep. Series, No. 123 IAEA Wien 1971 (p. 287).

[2] D. DRAHOVSKY, A. WINKLER and J. SKODA, Nature London **201**, 411 (1965).

[3] P. V. BYRNE and I. V. CHAPMAN, Int. J. Rad. Biol. **24**, 641 (1973).

[4] A. Z. EISEN, S. WEISSMAN and M. KARON, J. Lab. Clin. Med. **59**, 620 (1962).

[5] S. P. COLLOWICK and N. O. KAPLAN, Methods in Enzymology XII, Academic Press, New York, London 1967 (p. 107).

[6] A. B. BENKÓ, L. GYENGE and L. D. SZABÓ, paper presented at 9th Fed. Europ. Biochem. Socs. Meeting 1974, Budapest, Hungary.

[7] Y. CHANG-SAI, D. B. LAKINGS, R. W. ZUMWALT and C. W. GENRKE, J. Lab. Clin. Med. **83**, 816 (1974).

[8] G. B. GERBER, P. GERTLER, K. I. ALTMAN and L. H. HEMPELMANN, Rad. Res. **15**, 307 (1961).

[9] Z. DIENSTBIER, M. ARIENT, B. ZICHA and J. POSPISIL, Biochem. Ind. of Radn, Proc. Injury in Man Panel Series, IAEA Wien, 1971 (p. 33).

[10] Z. DISCHE, Proc. Soc. Exptl. Biol. Med. **55**, 217 (1944).

Life Sciences and Space Research XIV — Akademie-Verlag, Berlin 1976

PIONEER 10 AND 11 JOVIAN ENCOUNTERS: RADIATION DOSE AND BIOLOGICAL LETHALITY

M. W. Miller, G. E. Kaufman and H. D. Maillie

Dept. Radiation Biology and Biophysics, University of Rochester, Rochester, NY, USA

In their recent Jupiter flybys Pioneer 10 and Pioneer 11 passed through a belt of intense particulate radiation. For Pioneer 10 the radiation dose on the craft's outer surface was at least 5×10^5 rads from electrons plus 1.0×10^6 rads from protons; the radiation dose inside the craft (0.3 cm aluminum) was approximately 4.5×10^5 rads. For Pioneer 11 the surface dose was at least 1.3×10^5 rads from electrons plus 3×10^5 rads from protons; the interior radiation dose was approximately 1.2×10^5 rads. Significant survival of microbial spores would be possible at these calculated doses; however, even the interior dose of Pioneer 11 would be lethal to man and most multicellular biological organisms.

The planet Jupiter, like the planet Earth, is circled by a roughly toroid-shaped belt of trapped ionizing particles. This "trapped radiation belt" (TRB) or "Van Allen radiation belt" consists mainly of electrons and protons, with smaller fluxes of α and other particles. Prior to the Pioneer 10 Jupiter flyby, Jupiter's TRB had been estimated to extend from 5 to 2 Jupiter radii (R_J), and it was thought that the maximum flux of 1- to 30-MeV electrons was, at most, 2.6×10^7 cm^{-2} s^{-1} [1]. The Pioneer 10 data, which will be discussed below, show that the TRB begins at $20R_J$ and extends in at least to $2.8R_J$ [2], and show a maximum flux of 3- to 20-MeV electrons of 5×10^8 cm^{-2} s^{-1} [3]. Clearly, the radiation levels surrounding Jupiter had been underestimated. Preliminary reports in *Science* concerned the charged particles encountered by Pioneer 10 (25 January 1974) and Pioneer 11 (2 May 1975) during their recent Jupiter flybys. This report summarizes the particle fluences encountered by Pioneer 10 and Pioneer 11 during their flight inbound from $108R_J$ to periapsis and outbound from periapsis to $105R_J$, estimates the radiation dose received as a result of these encounters, and considers the implications of such exposures on the survival of biological organisms.

There were 15 detectors for energetic electrons and protons on board Pioneer 10, as listed in Table 1. A similar array of detectors was on board Pioneer 11. Seven detectors (detectors A to G) counted electrons in the range from 0.41 to 50 MeV, and eight detectors (detectors H to O) counted protons in the range from 0.44 to 150 MeV. The particle fluences encountered during various portions of the Pioneer 10 flight are given in Table 1. These fluences were calculated by numerical integration of the flux data given in the *Science* reports [3—6]. Some of the detectors (A, C, D, I and J) apparently became saturated when the spacecraft was within the TRB ($20R_J \rightarrow$ periapsis $\rightarrow 20R_J$). The fluences listed for these de-

Table 1

Energy Ranges and Estimated Fluences (per cm²) for Pioneer 10 Energetic Particle Detectors for the Jovian Inbound and Outbound Passage ($108\ R_J \rightarrow$ periapsis $\rightarrow 105\ R_J$).

Detector symbol	Energy range (MeV)	Inbound		Outbound					Reference
		$108R_J$ to $20R_J$	$20R_J$ to $2.8R_J$	$2.8R_J$ to $20R_J$	$20R_J$ to $35R_J$	$20R_J$ to $80R_J$	$35R_J$ to $80R_J$	$80R_J$ to $105R_J$	
Electron fluences									
A	0.41–1.0	7.8×10^{10}	6.1×10^{11}*	3.9×10^{11}*	2.4×10^{9}		1.9×10^{9}		[5]
B	>3		7.9×10^{12}	6.1×10^{12}					[3]
C	>6	3.0×10^{9}	6.1×10^{9}*	7.7×10^{10}*	9.5×10^{7}		5.3×10^{7}		[5]
D	6–30	3.5×10^{9}	2.0×10^{11}*	1.7×10^{10}*	2.5×10^{7}		3.3×10^{7}	1.8×10^{7}	[3]
E	>20		5.3×10^{12}	3.8×10^{12}					[6]
F	≧21	1.1×10^{8}	1.9×10^{11}	1.1×10^{11}	4.1×10^{6}				[4]
G	>50		1.4×10^{11}	5.7×10^{10}					[6]
Proton fluences									
H	0.5–1.8	1.6×10^{9}	1.0×10^{10}	1.0×10^{10}		4.0×10^{8}		1.0×10^{8}	[3]
I	0.44–2.0	1.1×10^{10}	6.1×10^{10}*	1.0×10^{11}*		7.6×10^{8}			[5]
J	> 1.23	1.5×10^{9}	3.3×10^{11}*	2.1×10^{11}*		6.1×10^{7}			[5]
K	3.3–21	1.3×10^{8}	1.0×10^{11}	6.8×10^{10}		5.8×10^{6}			[5]
L	5.6–21	1.3×10^{7}	3.1×10^{10}	3.4×10^{10}		1.0×10^{6}			[5]
M	16.2–21	3.9×10^{5}	8.9×10^{9}	6.8×10^{8}		3.5×10^{4}			[5]
N	> 30		1.2×10^{10}	8.3×10^{9}					[3]
O	70–150		3.0×10^{8}	1.3×10^{8}					[6]

* Detector apparently saturated.

tectors are minimum estimates based on the assumption that during the saturation period the flux was at least equal to the last measurable inbound flux.

To calculate absorbed radiation dose, electron and proton fluences were first determined for the various energy ranges which could be distinguished by the detectors on Pioneer 10. Since there were several overlaps in the energy ranges of the various detectors, it was frequently necessary to subtract appropriate detector fluences to obtain an approximate energy spectrum (for example, the proton fluence in the 3.3- to 5.6-MeV range was obtained by subtracting the fluence for detector L, counting 5.6- to 21-MeV protons, from the fluence for detector K, counting 3.3- to 21-MeV protons). In cases where the degree of overlap could not be determined (for example, between detector D, counting 6- to 30-MeV electrons, and detector F, counting ≧ 21-MeV electrons), minimum estimates of fluence were used. The absorbed radiation dose in rads for biological materials was then calculated for each energy range as the product of particle fluence and the linear energy transfer rate (LET) [7] in water. Where LET varied significantly over a given energy range, an intermediate value was used.

Table 2 shows the calculation of energy spectra and radiation dose for electrons and protons within Jupiter's TRB ($20R_J \rightarrow$ periapsis $\rightarrow 20R_J$). The estimated surface electron dose was 4.9×10^5 rads. We estimate that the electrons detected during the flight inbound from $108R_J$ to $20R_J$ and outbound from $20R_J$ to $105R_J$ added 3.4×10^3 rads, for a total surface electron dose of 4.9×10^5 rads. The estimated surface proton dose within the radiation belt was 1.0×10^6 rads.

Table 2
Estimated Radiation Dose from Jupiter's Radiation Belts ($20R_J \rightarrow$ periapsis $\rightarrow 20R_J$)

Energy class (MeV)		LET (MeV cm² g⁻¹)	Fluence (cm⁻²)	Surface dose (rad)	Range in aluminum (cm)
Electron dose					
0.41–1.0	(A)	2.5	1.0×10^{12}	3.9×10^4	0.04–0.16
3–20	(B–E)	2	5.0×10^{12}	1.6×10^5	0.54–3.4
20–21	(E–F)*	2	8.8×10^{12}	2.8×10^5	3.4
21–50	(F–G)	2	1.0×10^{11}	3.3×10^3	3.4–7.3
> 50	(G)	2	2.0×10^{11}	6.3×10^3	> 7.3
Total				4.9×10^5	
Proton dose					
0.44–2.0	(I)	3×10^2	1.6×10^{11}	7.7×10^5	0.0003–0.002
3.3–5.6	(K–L)	1.3×10^2	1.1×10^{11}	2.2×10^5	0.006–0.016
5.6–16.2	(L–M)	6×10^1	5.5×10^{10}	5.7×10^4	0.016–0.14
16.2–21	(M)	2×10^1	9.6×10^9	3.0×10^3	0.14–0.2
30–70	(N–O)	1.4×10^1	2.0×10^{10}	4.5×10^3	0.3–1.6
70–150	(O)	8×10^0	4.3×10^8	5.5×10^1	1.6–5.7
Total				1.0×10^6	

* Because of uncertainty in the absolute calibration of the various detectors, the designated energy ranges are only approximate; the "20- to 21-MeV" fluence probably represents a considerably broader energy range.

We estimate that protons in the $108R_J \to 20R_J$ and $20R_J \to 105R_J$ portions of the flight added 5.7×10^4 rads for a total surface proton dose of 1.1×10^6 rads.

Within the spacecraft the radiation dose was considerably reduced since the spacecraft skin absorbed some of the electrons and most of the protons. A "typical" spacecraft skin, a layer of aluminum approximately 0.25 cm thick, would absorb the 0.41- to 1.0-MeV electrons and all but the > 30-MeV protons. Thus, immediately behind the skin the electron dose would be 4.5×10^5 rads and the proton dose would be 4.5×10^3 rads. Some areas within the spacecraft would be further shielded by interior structures. We estimate that the maximum shielding would be approximately 1 cm of aluminum, and that the "> 20-MeV" electrons (detector E) would penetrate at least that thickness of shielding. Thus, we estimate that all interior areas of Pioneer 10 received a radiation dose of at least 2.8×10^5 rads.

The above calculations represent minimum dose estimates, since we limited ourselves to particles detected by the instruments on board Pioneer 10. In this we excluded particles for which there were no detectors (for example, electrons with energies below 0.41 MeV or between 1 and 3 MeV) or for which the detectors saturated (for example, protons with energies between 1.8 and 3.3 MeV). These excluded particles would contribute heavily to the surface dose. Moreover, the production of secondary particles and X-rays, which have been neglected in our calculations, would add to the dose calculated for primary particles.

Investigators at the Jet Propulsion Laboratory (JPL), Pasadena, California, have been conducting research to determine the effect of planetary TRB's on the survival of micro-organisms associated with non-sterile spacecraft [8, 9]. Bacterial sub-populations from the Mariner Mars 1971 spacecraft (nine spore-forming isolates and three non-spore-forming isolates (vegetative bacterial cells)) were exposed to 2- to 25-MeV electrons [8] and to 2-MeV protons [9]. A 300-krad dose of electrons resulted in a mean survival fraction of 0.05 for the spore-formers and 0.007 for the non-spore-formers; a dose of 450 krad of electrons resulted in a survival fraction of 0.01 for spore-formers and 0.003 for non-spore-formers [8]. A dose of 1.1 Mrad of protons yielded a survival fraction of 0.2 for spore-formers and approximately 0.01 for non-spore-formers [9].

On the basis of our calculated surface dose (490-krad electrons plus 1.1-Mrad protons) and the JPL findings, we would predict a survival fraction of less than 0.002 for spore-formers on the surface and a survival fraction of less than approximately 3×10^{-5} for non-spore-formers on the surface. Since the actual surface dose was probably considerably higher than our calculated dose, the survival of spore-formers was probably well below 0.002 and the non-spore-formers were probably virtually eliminated. Thus, the outer surface of Pioneer 10 was significantly decontaminated by the radiation exposure.

We estimate that within the spacecraft the total dose was 280—500 krad, due mainly to electrons. This dose would have resulted in a spore survival of approximately 0.05—0.01. Thus a significant fraction of whatever spore-formers were present would have survived the Jovian radiation dose within Pioneer 10. Non-sporeformers would have had a survival fraction of approximately 0.003—0.007.

Our preliminary analyses of the recent reports [10—16] concerning charged particle detection aboard Pioneer 11 indicate rad exposures sufficient to cause significant exterior microbial decontamination. Pioneer 11's hyperbolic trajectory through the radiation belts and close pass to Jupiter (periapsis at 1.6 R_J) resulted in an overall exterior dose of at least 1.3×10^5 rads from electrons and 3×10^5 rads from protons, for a total of 4.3×10^5 rads. This dose would result in a surviving fraction of 0.04 for spore-formers and 0.01 for non-spore-formers. The craft's interior radiation dose is estimated to be approximately 1.2×10^5 rads, being comprised of 9×10^4 rads from electrons and 3×10^4 rads from protons. The survivability of spacecraft organisms at these internal exposures is estimated to be approximately 0.3 for spore-formers and 0.1 for non-spore-formers.

For almost all "higher" forms of life, such as seeds, plants, algae, worms, insects, and others, the radiation dose inside Pioneer 10 and Pioneer 11 would have been supralethal. For man and other mammals the interior dose far exceeded the lethal level. Thus, Jupiter's radiation belts pose an extreme hazard to any manned mission passing through them. However, were a space vehicle to approach Jupiter along the polar axis, the radiation belts would not be encountered and the radiation hazard might be considerably less.

Acknowledgments

This paper is based on work performed under contract with the US Energy Research and Development Administration at the University of Rochester Biomedical and Environmental Research Project and has been assigned Report No. UR-3490-760.

References

[1] A. J. Beck, Jet Propulsion Lab. Q. Tech. Rev. **1**, 78 (1972); NASA SP-8069 (1971); Proc. Jupiter Radiation Belt Workshop, Tech. Mem. 33-543, Jet Propulsion Lab., Pasadena, Calif. (1972).

[2] A. G. Opp, Science **183**, 302 (1974).

[3] J. A. Simpson, D. Hamilton, G. Lentz, R. B. McKibben, A. Mogro-Campero, M. Perkins, K. R. Pyle, A. J. Tozzolino and J. J. O'Gallagher, Science **183**, 306 (1974).

[4] J. A. Van Allen, D. N. Baker, B. A. Randall, M. F. Thomsen, D. D. Sentman and H. R. Flindt, Science **183**, 309 (1974).

[5] J. H. Trainor, B. J. Teegarden, D. E. Stilwell, F. B. McDonald, E. C. Roelof and W. R. Webber, Science **183**, 311 (1974).

[6] R. W. Fillius and C. E. McIlwain, Science **183**, 314 (1974).

[7] Linear Energy Transfer Rep. 16, Int. Comm. on Radiation Units and Measurements, Washington, D.C. 1970.

[8] D. M. Taylor, C. A. Hagen, G. M. Renninger, G. J. Simko, C. D. Smith and J. A. Yelinek, Life Sciences and Space Research XI, 33 (1973).

[9] D. M. Taylor, C. A. Hagen, J. Barengoltz, C. Smith and R. Renninger, in: Planetary Quarantine Semi-Annual Rev., Space Res. Tech., Publication 900—636, Jet Propulsion Laboratory, Pasadena, Calif. 1973 (p. 2—1).

[10] C. F. Hall, Science **188**, 445 (1975).

[11] J. D. Mihalov, et al., Science **188**, 447 (1975).

[12] E. J. Smith, et. al., Science **188**, 451 (1975).

[13] J. A. Simpson, et al., Science **188**, 455 (1975).

[14] J. Van Allen, et al., Science **188**, 459 (1975).

[15] J. H. Trainor, et al., Science **188**, 462 (1975).

[16] R. W. Fillius, et al. Science **188**, 465 (1975).

Life Sciences and Space Research XIV — Akademie-Verlag, Berlin 1976

CYTOGENETIC ANALYSIS OF SEEDS OF *Crepis capillaris* (L) Wallr. EXPOSED ON BOARD THE EARTH ARTIFICIAL SATELLITE COSMOS 613

E. N. VAULINA, L. N. KOSTINA and A. L. MASHINSKY

Institute of General Genetics, USSR Academy of Sciences, Moscow, USSR

We studied the effect of space flight factors on air-dry seeds of *Crepis capillaris* (L) Wallr. and on radiation injury of seeds exposed to γ-radiation (3 kr, 525 r/min) before and after the flight on the satellite Cosmos 613. Space flight factors induced little increase (which was statistically insignificant) in the rate of chromosome aberrations in cells of the root meristem of *Crepis capillaris* sprouts, but they enhanced the effect of preliminary irradiation of seeds and decreased their radiosensitivity. The modification of radiation injury was statistically significant.

The earlier investigations performed on the automatic station Zond 8, the Soyuz 9 spacecraft and the orbital station Salyut showed that space flight factors increased the rate of chromosome aberrations in non-irradiated seeds [1]. The long-term exposure of space flight factors increased the effect of pre-flight irradiation of seeds and did not affect the result of their post-flight irradiation. The present work continues these investigations. The aim of the work was to study the effect of space flight factors on the radiosensitivity of seeds. For this purpose seeds were exposed to acute gamma-irradiation ^{137}Cs (3 kr, 525 r/min) before and after the flight. The time of exposure of seeds on board Cosmos 613 lasted for 61 days. On their return to the laboratory *C. capillaris* seeds were germinated at 24 °C in Petri dishes on filter paper soaked in a 0.01% solution of colchicine. The sprouts were fixed in a mixture of alcohol and acetic acid (3:1) and stained with acetocarmine. Chromosome aberrations were scored on squashed preparations examined at metaphases. The results are presented in Table 1. The space flight factors slightly increased the rate of chromosome aberrations. It should be noted that the spectrum of chromosome aberrations also changed (Fig. 1). Thus although the control material was characterized by an equal number of isolocus breaks and chromsome aberrations and by the absence of chromatid aberrations, the flight material had a large number of chromatid aberrations (43% of the total number of abnormalities). The number of chromosome aberrations and isolocus breaks, the latter in particular, was smaller in the flight material than in the control (14% against 50% of the total abnormalities). The change in the type of aberrations under the action of space flight factors was observed earlier both in air-dry seeds of *C. capillaris* [2, 3] and on other objects [4, 5]. No prevalence of chromosome aberrations was noted in barley [6]. The previous analysis [7] of the effect of dynamic factors on air-dry seeds of *C. capillaris* showed that these factors significantly increased the rate of chromosome aberrations

Table 1

The Effect of Space Flight Factors in combination with Gamma Irradiation (3 kr) on Air-dry Seeds *Crepis capillaris* (L) Wallr.

Variants	No. of meta-phases	No. of cells with aberrations		Significance of difference (p)	Aberrations					
					Isolocus breaks		Chromatid aberrations		Chromosome aberrations	
		No.	% ± sd		No.	% ± sd	No.	% ± sd	No.	% ± sd
Control	1493	4	0.27 ± 0.14	< 0.05	2	0.13 ± 0.12	—	—	2	0.13 ± 0.12
Flight	1857	7	0.38 ± 0.14		1	0.05 ± 0.057	3	0.16 ± 0.12	3	0.16 ± 0.12
γ_2	1084	273	25.18 ± 1.31	> 0.001	84	7.57 ± 0.81	6	0.55 ± 0.24	216	19.23 ± 1.21
Flight + γ_2	2755	535	19.42 ± 0.76		162	5.85 ± 0.45	13	0.47 ± 0.11	422	15.32 ± 0.68
γ_1 + Flight	957	320	33.44 ± 1.52	> 0.001	123	12.85 ± 1.08	6	0.63 ± 0.27	314	32.81 ± 1.52
γ_1	1377	349	25.34 ± 1.17		78	5.66 ± 0.62	3	0.22 ± 0.14	311	22.59 ± 1.12

sd = standard deviation

γ_1, γ_2, gamma irradiation of the control material before (γ_1) and after (γ_2) the action of space flight factors.

(1.84 ± 0.17 compared with 0.30 ± 0.10 in the control). In addition to this the types of chromosome abnormalities found in flight experiments differed from that in the control with simulated dynamic factors: the number of chromosome aberrations increased and the rate of chromatid and isolocus breaks decreased (Fig. 2). In seeds exposed to space flight the distribution of chromosome abnormalities by the type of aberrations was different. Radiation shifts the spectrum of chromosome abnormalities in the direction of increasing the number of

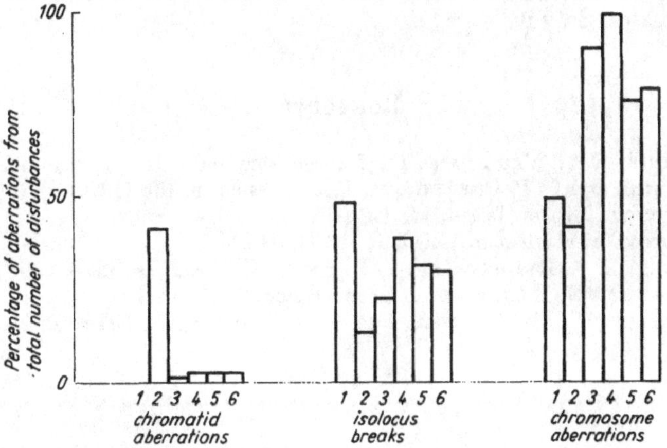

Fig. 1. Distribution of chromosome abnormalities according to their types in different variants of experiment: 1, control; 2, space flight factors (SFF); 3, γ-irradiation of control material before SFF exposure; 4, γ-irradiation of seeds before SFF exposure; 5, γ-irradiation of control material after SFF exposure; 6, γ-irradiation of seeds after SFF exposure.

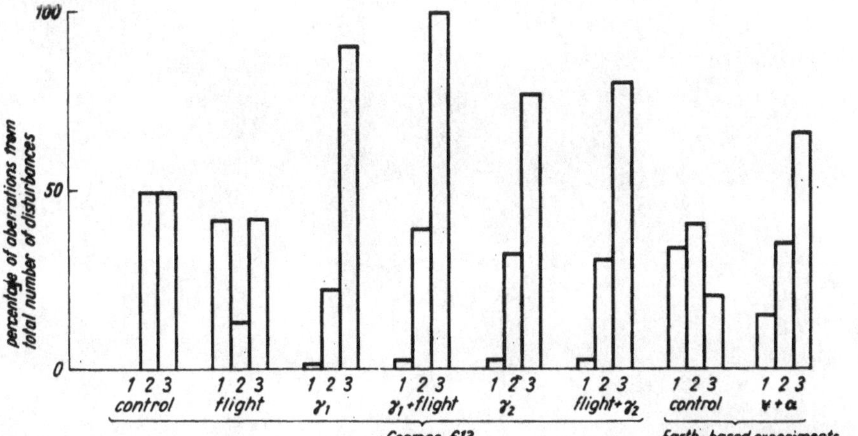

Fig. 2. Distribution of chromosome abnormalities according to their types: 1, chromatid aberrations; 2, isolocus breaks; 3, chromosome aberrations. (From [7].)

chromosome aberrations. Therefore changes in the distribution of chromosome abnormalities cannot be accounted for either by the effect of dynamic factors or by radiation. This effect seems to be produced by weightlessness.

Space flight factors modified the effect of gamma irradiation (Table 1). They decreased the sensitivity of seeds to post-flight gamma irradiation (from 25.18% to 19.42%) and increased the effect of pre-flight irradiation (from 25.34% to 33.44%). The flight factors did not affect the spectrum of chromosome aberrations; in all irradiation experiments the spectrum was similar and typical of gamma irradiation. Similar results were gained by us earlier in experiments on the orbital station Salyut.

References

[1] N. P. Dubinin, E. N. Vaulina et al., Life Science and Space Research XI, 105 (1973)
[2] L. G. Dubinina and O. P. Chernikova, Kosm. Issled. 8, 156 (1970).
[3] R. N. Platonova, Kosm. Issled. 11, 478 (1973).
[4] V. V. Hvostova et al., Kosm. Issled. 1, 186 (1963).
[5] N. L. Delone, N. A. Rudneva and V. V. Antipov, Kosm. Issled. 3, 480 (1965).
[6] K. P. Garina and N. I. Romanova, Kosm. Issled. 8, 158 (1970).
[7] E. N. Vaulina and L. N. Kostina, Life Sciences and Space Research XIII, 168 (1975).

Life Sciences and Space Research XIV — Akademie-Verlag, Berlin 1976

ROLE OF CERENKOV RADIATION IN THE EYE-FLASHES OBSERVED BY APOLLO ASTRONAUTS

P. J. McNulty[a], V. P. Pease[a] and V. P. Bond[b]

[a] Clarkson College of Technology, Potsdam, N.Y., USA
[b] Brookhaven National Laboratory, Upton, N.Y., USA

Visual phenomena in the form of colorless flashes of light were observed by astronauts in deep space when their eyes were closed and adapted to darkness. We describe in this paper laboratory experiments and calculations which indicate that many of these flashes are the result of visible light generated within the astronauts' eyeball in the form of Cerenkov radiation when a relativistic HZE particle traverses it. The sensitivity to Cerenkov radiation measured for three subjects exposed to pulses of pions and muons and the visual phenomena observed were found to be consistent with the reports of flashes observed at rates as high as 2 per minute on Apollo missions 11 through 17.

1. Introduction

These investigations were initiated as a result of the observations by astronauts on Apollo 14 of flashes of light when the spacecraft was dark and their eyes were shut [1]. The possibility that some cosmic ray particles might be visible to astronauts was first suggested by Tobias [2] who argued as early as 1952 that since the dark-adapted retina can detect X-ray ionizations [3] it is conceivable that very densely ionizing HZE particles in space would produce small flash-like light sensations.

The most dramatic aspect of the Apollo light flashes is the variety of visual events. All were colorless except for one that appeared "blue with a white cast, like a blue diamond" [4]. Pinsky et al. have recently summarized the Apollo observations [4]. For convenience they categorize them as stars, streaks, and clouds which occur 66%, 25%, and 8% of the time, respectively. The actual variety of phenomena is, however, much more impressive. The star-like flashes include point flashes, and larger phenomena sometimes referred to as supernova that are similar in appearance to a photographic flashbulb that has flashed across a dark arena several hundred feet from the observer. Occasionally, there were double stars. The streaks were sometimes reported as sharp lines and at other times as diffuse more like bands of light. Some of the streaks had gaps. As with accelerator produced streaks there was often a sense of apparent motion. The cloud-like flashes were large flashes that did not seem to have any distinct shape. They ranged from crescent shaped flashes in the far periphery to events that filled a major portion of the field of view. Some clouds filled the entire peripheral field but left the central portion of the field of view dark.

The average dark-adapting time before the first event was detected in the dedicated Apollo light-flashes sessions was 19.3 minutes. The average post dark-adaptation event rate was one event every 2.9 minutes per observer.

A number of workers [5—8] reported seeing star-like flashes and small streaks when exposed to neutrons with kinetic energies between five and a few hundred MeV. Helium and nitrogen nuclei [9] near the end of their range have been shown to result in flashes and streaks when incident on the posterior portion of the retina in bursts at close to grazing angles. All of these studies were carried out under conditions such that Cerenkov radiation could be ruled out as a possible mechanism.

The investigations described in this paper were initiated to determine the role that Cerenkov radiation plays in producing visual sensations. All experiments were designed to insure that Cerenkov radiation was the physical mechanism behind the observed phenomena. There were distinct advantages in concentrating on Cerenkov light. Independent calculations by Fazio et al. [10] and Madey and McNulty [11] indicated that there were more than enough HZE particles in deep space capable of generating above threshold quantities of Cerenkov radiation to explain the Apollo count rates. The Cerenkov model was simple and free of adjustable parameters. Once the Cerenkov contribution was understood the role played by other mechanisms could be clarified. Details of some of the experiments described in this paper have appeared elsewhere [12—15].

2. Cerenkov Radiation

When a relativistic charged particle traverses the vitreous humor it generates an electromagnetic shock wave that propagates in the forward direction at an angle of about 41° with the particle's trajectory. The shape of the Cerenkov wave front is a cone with the particle at the apex. The Cerenkov photons at high velocities are evenly distributed over the visible frequencies. However, the relative scotopic detection efficiency of the retina is strongly dependent on frequency [16] and it is the absorption of photons that trigger visual sensations. It is necessary to correct for the retinal spectral sensitivity.

Once the number of visible photons generated per unit length by a HZE particle is known the probability that the particle will trigger a visual sensation can be calculated. The dark-adapted peripheral retina integrates signals over areas, known as summation units, as large as 300 μm in diameter. Vision studies suggest that if some threshold number of photons are absorbed within a summation unit a flash of light will be detected [17].

When a Cerenkov particle traverses the eye the summation unit most likely to be triggered is the unit through which the particle passes. In the calculations by Madey and McNulty [11] only the summation unit nearest the trajectory is considered when calculating count rates. Once the average number of photons that will enter the retina within this unit are known, the probability that the threshold number of photons will be absorbed can be calculated. The assumption is made that if the number of photons (2—20) known to be required to trigger a unit with a flash of light [18] is absorbed the unit will be triggered and the particle detected. The computer scans the spectrum of HZE particles in space and determines the probability that each particle at a given velocity will be detected. In Fig. 1 the total count rate for deep space is expressed as a function of the number of photon absorptions assumed to be required for detection.

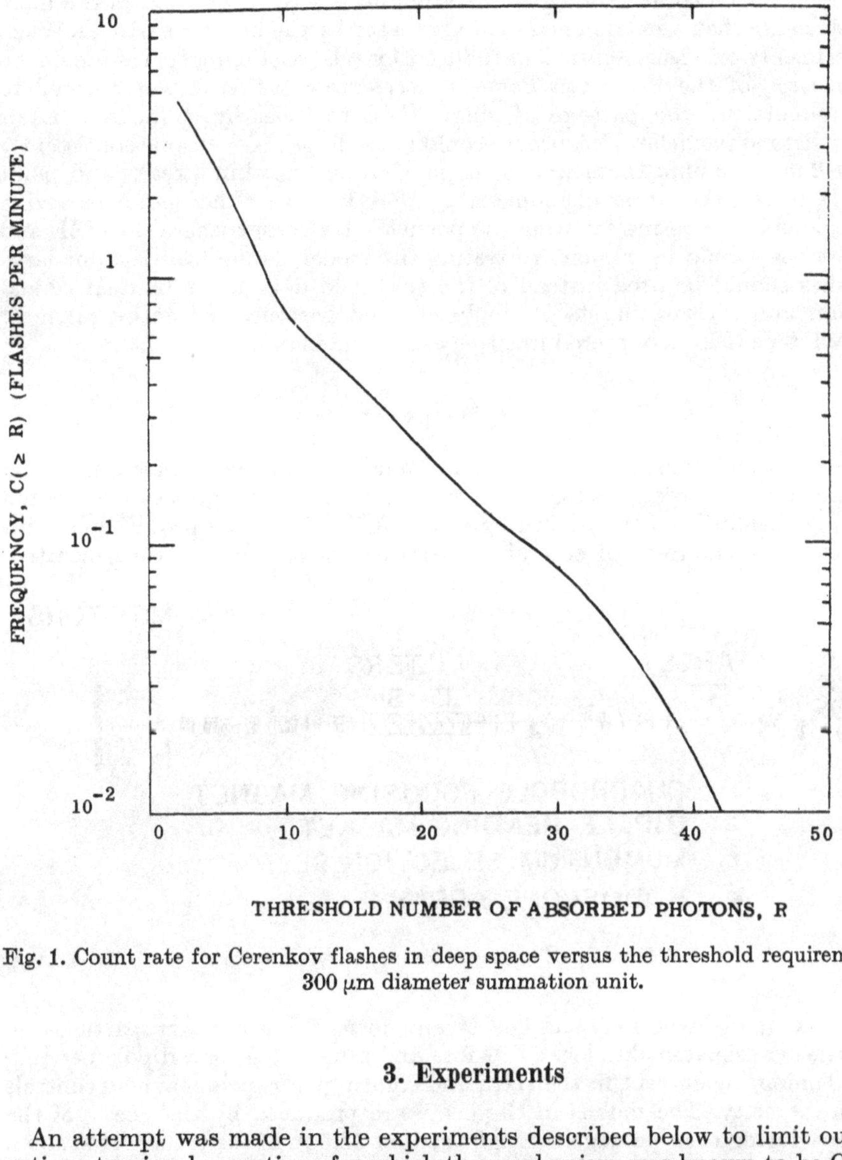

Fig. 1. Count rate for Cerenkov flashes in deep space versus the threshold requirement for a 300 μm diameter summation unit.

3. Experiments

An attempt was made in the experiments described below to limit our observations to visual sensations for which the mechanism was known to be Cerenkov radiation. The physical process was understood and calculations could be made with a model that had no adjustable parameters. When the count rates to be expected in deep space were calculated as in § 2.1 the values obtained were comparable with the Apollo observations for retinal sensitivities that agreed with optical data. Moreover, Cerenkov flashes could be obtained with the least dosage to the retina.

The principal disadvantage was that accelerator facilities could provide only a small range of the relativistic HZE particles found in space. Threshold phenomena could be studied at existing accelerators but we could not obtain the more dramatic flashes that were assumed to be generated by the heavier particles. When our experiments on visual sensations induced by relativistic nitrogen were ended by the closing of the Princeton Particle Accelerator (PPA) it was decided to attempt simulating the passage of single HZE particles by collimated bursts of singly charged particles. Because we could not collimate our beams consistently to below 2 mm × 3 mm, there was no hope of observing thin streaks and point flashes. However, the larger phenomena are determined by the photons arriving at the regions of the retina far from the particle's trajectory where the HZE and burst patterns should be similar. In testing the model the optical data for large area flashes should be used instead of the threshold data for individual retinal summation units. Using bursts of singly charged particles the entire range of Cerenkov intensities encountered in space can be simulated.

3.1. Muons

The muon beam transport system used in our experiments is shown in Fig. 2. A flipping copper target intersected a proton beam that had been extracted electromagnetically from the main ring of the AGS. When the proton pulses hit the target, large numbers of secondary particles, mostly pions, are generated.

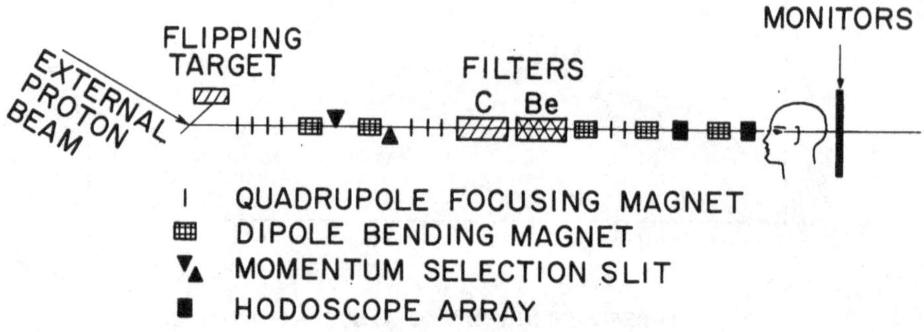

Fig. 2. Schematic of the muon beam transport system.

The first six quadrupole magnets collect and focus the secondary particles on the momentum selection slits located before and after the second dipole bending magnet. Tuning the current in the sixth and eighth quadrupole magnets controls the beam intensity. The muons in the beam are produced by the decay of the beam pions during their passage through the first 13.5 m of the transport system. Pions undergo strong interactions while muons do not. Because of the strong interactions, a beam pion is roughly 10^4 times more likely to interact in the filter material than a beam muon.

It was just this absence of strong interactions that made muons preferable for our initial studies of Cerenkov radiation. The beam that emerges downstream of the 6.3 m of beryllium filter and 2.4 m of carbon filter has a π/μ ratio of 10^{-6}. Just before the beam reaches the observer, two dipole magnets turned on their sides combine to displace the beam upward in the vertical plane by 25 cm. This separates the beam muons from any secondary particles generated upstream. Two hodoscopes define the trajectory of each muon at locations before and after the last bending magnet. Each muons's momentum, 6—9 GeV/c, is defined by computer-monitored changes in its trajectory upon being deflected by the final dipole magnet. In passing through 6.3 m of beryllium and 2.4 m of carbon filter the beam muons undergo considerable multiple scattering. This scattering enlarges the cross-sectional area of the beam. Unfortunately the beam cannot be collimated and focused within reasonable distances. With optimum tuning the full-width-half-maximum dimensions of the muon beam at the exposure site is ± 13.2 cm horizontal and ± 6.3 cm vertical. The full-width-half-maximum angular spread is ± 7 mrad with 90% of the beam within ± 15 mrad.

All sessions involving human exposures were continuously monitored with an ionization chamber by a member of the Health Physics staff. Exposures were also recorded by counters, thermoluminescent dosimeters, chips, film badges, and large area medical X-ray films. The protocol limited the total direct beam exposure for any observer to less than 100 mrad/year. The largest dose accumulated by an observer during the AGS muon exposures was 8 millirad. Visual sensations can be obtained with pulses of less than 1 microrad.

Exposures were carried out in a fail-safe mode; that is, even if, by some combination of accidents and errors, the maximum proton flux hits the target while the beam line magnets were tuned for maximum intensity, the local dose to the subject would be considerably less than 50 mrad.

Subjects were dark adapted for 30 minutes before each session. Each subject maintained the alignment of his eye to the beam by biting into his own dental impression plate. The orientation of the beam particles to the eye could be altered by changing the orientation of either the head or the eye to the beam.

After the subject was aligned, he signalled when ready for a beam pulse. The beam pulse arrived after a fore period that was randomly varied between 1 and 3 sec. Beam tuning and recording of pulse amplitudes were carried out in a separate room beyond the observer's hearing range. Both the observer and the experimenters wore headsets that included both earphones and voice activated microphones. All communications were carried out through this channel. The observer responded yes, no, or doubtful to each pulse. If the response was yes, the visual sensation was described in terms of size, shape, color, brightness and location in the field of view. All visual phenomena observed so far in this experiment were reported to be colorless. The threshold number of particles in a pulse required for a visual sensation is defined to be the number at which the subject observed 50% of the pulses. In determining threshold, the number of particles in successive pulses were randomized about the expected threshold intensity which was estimated from pre-trial data. To insure that each subject would adopt reliable subjective criteria for reporting visual sensations, approximately 40% of catch tests were mixed into each run. A catch test consisted of a simulated pulse during which no particles were delivered. "Yes" or "doubtful" responses were not elicited by any of the catch tests.

The simplest test to determine whether the observed visual sensations are due to Cerenkov radiation involves the location of the visual phenomena in the observer's field of view. The levels of ionization and excitation generated by a muon burst are essentially the same where the beam enters the eye as where it exits. The intensity of Cerenkov radiation arriving at the two locations is dramatically different, however. This difference is explained as follows: The visible Cerenkov photons generated during each muon's passage through the vitreous are emitted in the forward direction at an angle of about 41° with the muon's trajectory. The photosensitive elements, which for scotopic vision are the outer segments of the rod cells, lie flush against the pigment epithelium on the inside wall of the sclera. They form the first transparent layer the particles encounter when entering the eye and the last when exiting. The only photons that will pass through the photoreceptors located at beam entrance must either be generated during the muon's passage through the layer of rod outer segments or arrive as the result of scatterings and reflections elsewhere in the eye. Clearly the region of the retina surrounding the beam exit from the eye receives much more Cerenkov light than the corresponding region about the beam entrance. The observer's eye is oriented to the beam such that the regions of the retina located at beam entrance and exit are equally sensitive to light. When the observer is exposed at such low intensities that the bursts are barely detectable, any visual phenomena resulting from Cerenkov radiation would appear in the field of view that corresponds to the region of the retina surrounding beam exit. This test has been performed on three subjects for a variety of locations on the retina. In all cases the visual phenomena were first detected in those regions of the field of view that correspond to beam exit. In no case did the visual phenomenon first appear in an area that corresponded to beam entrance.

A more dramatic demonstration of Cerenkov radiation is obtained by exposing a subject such that the muons enter the rear of the eye and exit the front. At intensities considerably above threshold, the visual sensation appears as a bright ring of light with a dark center. The visible Cerenkov radiation generated during their traverse of the transparent media in the eye is emitted in the forward direction at an angle of about 41° with the particle's trajectory. Only the regions of the extreme peripheral retina are directly exposed to the Cerenkov light. The posterior portions of the retina at beam entrance receive only a small amount of light scattered off the lens and cornea. The result is a bright ring of light with a dark center. The bright portions of the ring extend all the way to the extreme periphery. There appears to be a gradual transition from the inner portion of the lighted ring to the dark center.

Following our report of the observation of bright flashes with dark centers at the AGS, Pinsky and his co-workers at the NASA Johnson Spacecraft Center briefed the Apollo 17 crew to look for similar phenomena while in space. The Apollo 17 crew confirmed observing flashes with dark centers. Similar visual phenomena have been reported by crew of Skylab 2.

3.2. Nitrogen Ions

The facility for these investigations at the Princeton Particle Accelerator (PPA) is described in [14]. The important feature was that the synchrotron could be operated near its maximum intensity while the subjects were exposed to beam rates as low as 1 particle per minute. Failures of the equipment or changes in the timing of extraction would only result in a reduction of the beam intensity. The beam particles were separated from any secondary particles by a dipole magnet before they arrived at the subject. Each nitrogen nucleus entered the subject's retina with energies of about 503 MeV/amu and, therefore, with velocities just above the Cerenkov threshold.

The alignment of the subject was similar to the procedures used for muons and pions. For final alignment of the retina the subject stared at a diode emitting a very dim red light, which served as a fixation point. The diode was masked to

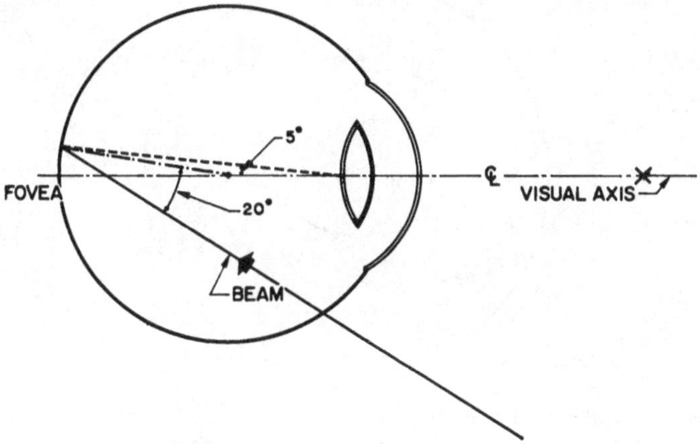

Fig. 3. Orientation of the subject's eye to beam particles for nitrogen exposures at PPA.

produce an image in the shape of a cross that would subtend a visual angle of 2°30′ which corresponds to a length of 0.72 mm at the subject's retina, and thereby provides a convenient scale by which to measure the size at the retina of any induced visual phenomena. Beam particles or heavy fragments with an atomic number of 3 or greater that emerged from the rear of the subject's head were detected by a counter-telescope arrangement, and visual sensations were marked when the subject pressed a button. To allow the subjects time to comment on their observations, the average time interval between beam particles was kept at roughly 1 minute.

For this investigation four subjects were exposed to nitrogen nuclei with energies of 531 MeV per atomic mass unit. The alignment of the left eye and the location of the fixation cross were arranged to produce the relative orientation of the retina to the beam shown in Fig. 3. On entering the eye, the nitrogen nuclei missed the lens and arrived at the temporal retina at a point that corresponded to a visual angle of about 7° from the fovea. This location has a sensitivity to visible light that is typical of much of the peripheral retina, and any visual sensations induced would appear conveniently close to the fixation cross. The beam particles entered the retinal layers at an angle of about 20° with the normal.

In the first round of exposures, four subjects received a total of 277 nitrogen nuclei, 27 of which were accompanied by a visual sensation of a dim streak or a flash about 5° to 10° to the right of the fixation cross. Subject L.P. reported that he required a rather long interval between first observing a faint streak and finally signaling the flash. Subjects V.B. and V.P. had a similar lag; for these two subjects 10 of the 13 coincidences between the counter signal and the observed signal required between 4.5 and 7 seconds. No streaks were observed in the period of 1 to 3 hours each of the four subjects spent dark-adapted but not in the beam.

A typical streak is shown schematically in Fig. 4; it appeared to be colorless, somewhat curved, very thin, and no longer than twice the length of a stroke of the fixation cross or 1.44 mm at the retina. Sometimes the streaks appeared brighter at one end.

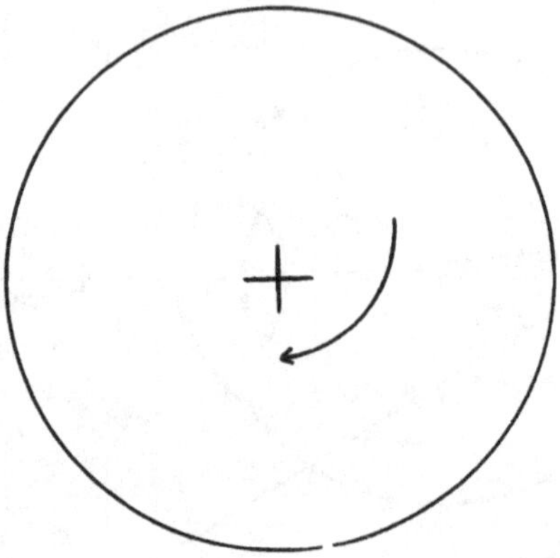

Fig. 4. Typical streak observed relative to fixation image. Arrowhead indicates the direction of apparent motion.

3.3. Pions

A schematic representation of the facility setup for the pion exposures at the AGS is shown in Fig. 5. The beam transport system consisted of quadrupole and dipole magnets set to collect, focus, collimate, and transport 725 $\pm$ 80 MeV/c pions to the eye. The source of the pions is a small Cu target, F-20, in the main ring of the AGS. To trigger the delivery of a beam pulse, a portion of the circulating proton beam is electromagnetically deflected on to the F-20 target. This produces a large number of secondary pions transported as a pulse along the beam line to the subject's eye. The number of particles in each pulse is controlled

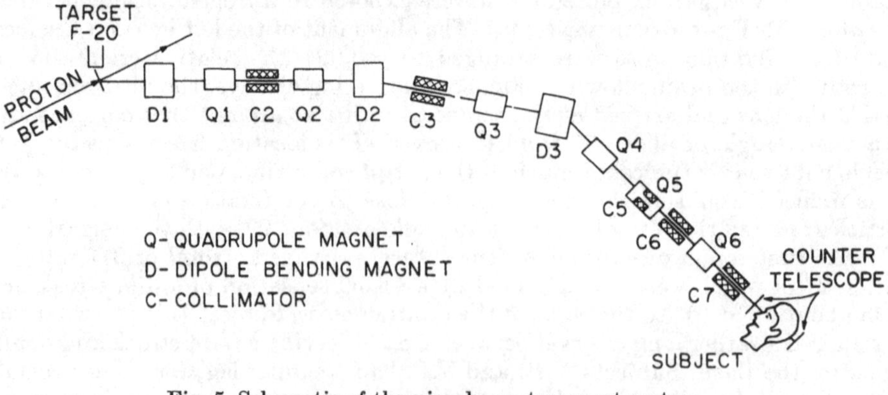

Fig. 5. Schematic of the pion beam transport system.

by the current settings of the upstream magnets. The beam particles entering the eye are monitored by a counter telescope made up of a small scintillation paddle just before the eye and a larger one behind the head. Signals from these counters are then gated to record only particles that pass through the eye during the 0.1 ms beam pulse. Measurements integrated over a large number of beam pulses show that both the muon and pion beam pulses are uniform in intensity to within $\pm 5\%$ across dimensions corresponding to the cross-sectional area of the eye.

The alignment procedures, the methods for taking data and the precautions taken to ensure fail-safe operation were similar to those described above for the muon exposures. The principal advantage of pions over muons is the smaller pion beam diameters that can be obtained by collimation. By adjusting the collimator shown in Fig. 5 just upstream of the eye one could obtain beam diameters as small as 1 mm in diameter, although most data were taken with beams having 2 mm $\times$ 3 mm cross-sectional dimensions or greater.

As with muons the visual sensations were found in the appropriate regions of the field of view for Cerenkov radiation.

Also the number of particles required for visual sensation was found not to vary significantly with the cross-sectional area of the beam spot on the retina, despite as much as a 50-fold increase in dose within the spot.

4. Results and Discussion

In order to determine the role of Cerenkov radiation in the Apollo eye-flashes, the experiments described above were designed to test the model and to ascertain the extent to which the Apollo phenomena could be induced using Cerenkov radiation. These questions are treated separately in what follows.

4.1. Retinal Sensitivity to Cerenkov Radiation

The frequency of seeing curves for three subjects exposed to bursts of pions distributed over a 1.3 cm $\times$ 5.1 cm cross-sectional area at the retina are shown in Fig. 6. In calculating the frequency of seeing curves, the "doubtful" responses were combined with the "no" responses because roughly 40% of the doubtfuls were in response to catch tests. Catch tests are represented in Fig. 6 as exposure to zero pions. The particles that passed through the eye were counted by a 1.3 cm $\times$ 2.5 cm section of a scintillation counter placed in the small space between the collimator and the eyeball. The shapes of the three curves are typical of our threshold measurements and very similar to the analogous curves obtained with optical light sources [18]. There are negative responses to small numbers of pions, then a sharp rise to 100% detection.

No differences were noted in the visual phenomena or the frequency of seeing curves obtained with muons and pions. A comparison is made in Fig. 7 of the frequency of seeing curves obtained for subject V.P. exposed to muons and pions under the same conditions six months apart. The muon and pion curves agree both in shape and absolute value within the day to day variations obtained when the same subject is exposed to light.

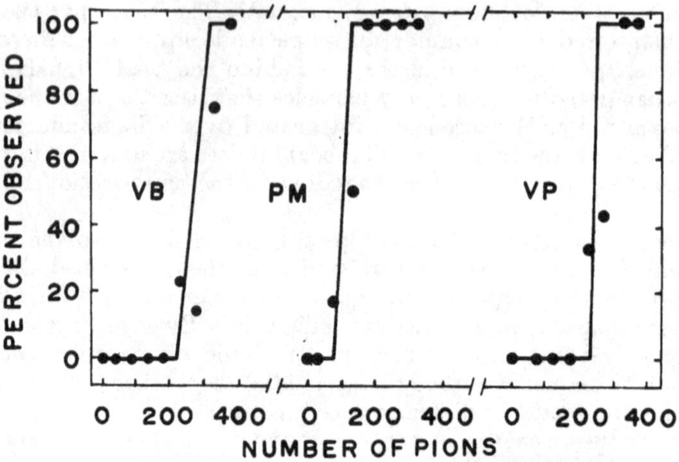

Fig. 6. Frequency of seeing curves for three subjects exposed to 1.3 cm × 5.1 cm bursts of pions.

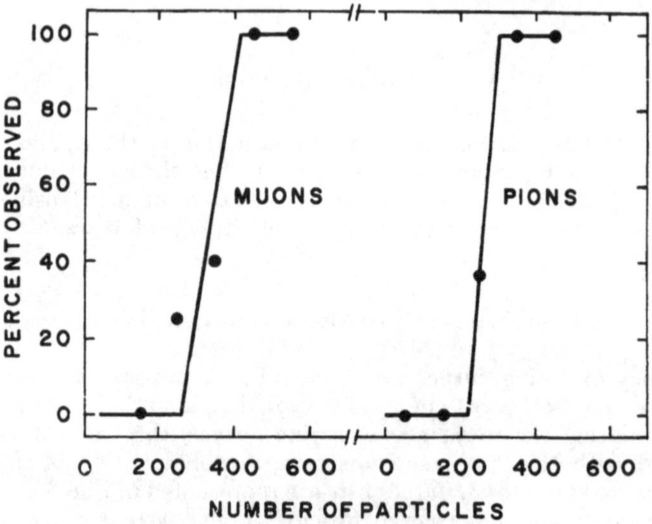

Fig. 7. Comparison of frequency of seeing curves for subject V.P. exposed to muons and pions six months apart under similar conditions.

4.2. Comparison with Optical Thresholds

Testing the Cerenkov model by a direct comparison of the absolute threshold measured for subjects exposed to particles with those measured using optical sources is made difficult by the essential differences between the sources of light in the two cases. Cerenkov radiation is continuously emitted along the particle's trajectory at an angle of about 41°. No correction for absorption in the lens and cornea is needed because the relevant Cerenkov photons are generated within the vitreous and retina. The Cerenkov photons are evenly distributed through the visible frequency range. On the other hand, optical measurements normally involve an external source of either nearly monochromatic light or light that has the spectral characteristics of black-body radiation. Moreover, we have not been able to find published optical threshold measurements with sufficiently large fields of view and pulse durations shorter than 0.1 sec, the time interval over which the retina is believed to integrate signals.

We are planning to measure optical thresholds for large visual fields. In the meantime, a comparison of sorts can be made with the data of Denton and Pirenne [18] for binocular exposure of 23 observers to a steady field of white light covering a visual field of 47°. According to those authors, the average threshold measured in their experiments was equivalent to 27300 photons $cm^{-2} s^{-1}$ of 510 nm light incident on the same field of view. Using the same conversion to equivalent 510 nm photons at the retina the range of measured thresholds corresponds to 12850 to 160590 photons cm^{-2}. If we take the short interval during which the retina integrates signal into account by considering the continuous exposure as a series of 0.1 sec "looks", these optical thresholds range from 1285 to 16059 photons cm^{-2} per "look".

The PIVS data can also be represented in terms of an equivalent number of 510 nm photons at the retina. Assuming a median path length for a beam particle traversing the vitreous of 1.4 cm for an eye that has an interior diameter of 2 cm, the number of equivalent Cerenkov photons of 510 nm incident per cm^2 on the retinas of the three subjects plotted in Fig. 6 for large area exposures becomes 3604, 9192, and 9912, respectively. The details of the calculation are given in [15]. These values for the threshold number of 510 nm photons are well within the range of optical thresholds (1285 to 16059 photons cm^{-2}).

4.3. Comparison with Apollo Count Rates

Once the detection efficiency of a subject's eye to a HZE particle of a given charge and speed is measured one can estimate what that subject's count rate would be if exposed to all the Cerenkov cosmic-ray nuclei in deep space. The measured detection efficiency can be used to establish the required number of photon absorptions within a retinal summation unit. Given this threshold requirement the total count rate in deep space can be calculated (see Fig. 1). The count rates obtained do not appear to be dependent on the size of the summation unit used in the calculations. A detection efficiency of 10% for 503 MeV/amu nitrogen nuclei as measured at the PPA results in count rates of 1.7· per minute in deep space which is comparable with the highest Apollo rates.

Even the pion and muon data can be used to establish a lower limit on the Cerenkov count rate on Apollo. Although the distribution of Cerenkov light that falls on the retina within a 2 mm × 3 mm beam spot of singly charged particles will be quite different from the intensity distribution near a single HZE particle's trajectory (i.e. the area most likely to be triggered), the intensity distribution at locations on the retina far from the particles' trajectories will be similar. The pion and muon bursts can, therefore, be used to estimate the count rates for the larger flashes. The thresholds measured at the AGS varied from 100 to 300 particles per burst which corresponds to Z values of from 10 to 18. Detection efficiencies of 50% for Z values of 10, 16 and 18 respectively correspond to approximate count rates in deep space of 0.6, 0.2 and less than 0.1 flash per minute.

4.4. Simulation of Apollo Phenomena

When $N = Z^2$ singly charged particles are confined to a 2 mm $\times$ 3 mm beam pulse and the pulse duration is less than 0.1 sec, the superposition of N individual patterns should appear similar over most of the retina to the Cerenkov pattern generated by an individual particle of charge Z. As mentioned above, the important difference in the two patterns occurs in the region of the retina covered by the beam spot where the illumination is more smeared out than for a single HZE particle.

As expected with 2 mm $\times$ 3 mm or larger beams spots at the AGS it was not possible to simulate the point flashes and thin streaks observed on Apollo. To the extent that we have been able to compare the descriptions given by Apollo astronauts with our own observations, both groups have observed similar large visual phenomena including cloud-like flashes, crescent shaped flashes, rings of light with dark centers, and broad streaks. Thin streaks were reported when the same subjects were exposed to individual relativistic nitrogen nuclei at the PPA. The length of the streaks compared with the particle's projected path length through the retina would appear to require that these flashes are Cerenkov. Unfortunately, these nitrogen observations were the last experiment run at the PPA before it closed for lack of funding and the time was not available to carry out the tests to rule out completely the possibility that the streaks could have been produced by some mechanism other than Cerenkov radiation.

5. Conclusions

Our experiments and calculations on visual phenomena induced by Cerenkov radiation are all consistent with the assumption that Cerenkov radiation is an important and perhaps the dominant mechanism behind the Apollo eye-flash phenomena. Cerenkov radiation has been used to induce visual sensations similar in description to all the larger Apollo flashes one of which was observed first in our experiments. The count rates calculated for Cerenkov flashes in deep space based on either optical thresholds given in the literature or our measurements agree with those obtained in the Apollo observations. In fact, the wide range of thresholds for subjects exposed to light may explain the variation in count rates among astronauts and for the same astronauts on different days.

Until the same observers have experienced Cerenkov flashes, track-effect flashes of the type studied in other laboratories, and the eye-flashes in deep space, one cannot be totally sure which phenomena are the same. Nor can we rule out a sizable contribution to the Apollo rates due to track effects, particularly of the point flash and thin streak varieties.

Acknowledgments

This research was carried out under the auspices of the US Atomic Energy Commission.

References

[1] P. K. Chapman, L. S. Pinsky, R. E. Benson and T. F. Budinger, Proc. Nat. Symp. Natural and Manmade Radiation in Space 1971, NASA-TMX-2440, 1972 (p. 1002).
[2] C. A. Tobias, J. Aviation Med. **23**, 345 (1952).
[3] D. J. Kimeldorf and E. L. Hunt, Ionizing Radiation: Neural Function and Behavior, Academic Press, New York and London 1965.
[4] L. S. Pinsky, W. F. Osborne, J. V. Bailey, R. E. Benson and L. F. Thompson, Science **183**, 957 (1974).
[5] J. H. Fremlin, New Scientist **47**, 42 (1970).
[6] W. N. Charman, J. A. Dennis, G. G. Fazio and J. V. Jelley, Nature **230**, 522 (1971).
[7] T. F. Budinger, H. Bischel and C. A. Tobias, Science **172**, 868 (1971).
[8] C. A. Tobias, T. F. Budinger and J. T. Lyman, Nature **230**, 596 (1971).
[9] C. A. Tobias, T. F. Budinger and J. T. Lyman, Proc. Nat. Symp. on Natural and Manmade Radiation in Space 1971, NASA-TMX-2440, 1972 (p. 1002).
[10] G. G. Fazio, J. V. Jelley and W. M. Charman, Nature **228**, 260 (1971).
[11] R. Madey and P. J. McNulty, Proc. Nat. Symp. Natural and Manmade Radiation in Space 1971, NASA-TMX-2440, 1972 (p. 757).
[12] P. J. McNulty, Proc. 8th Int. Conf. Nucl. Phot. and Solid State Track Detectors, Vol. II, Inst. of Atomic Phys., Bucharest 1972 (p. 598).
[13] P. J. McNulty, Nature **234**, 110 (1971), AFCRL Rep. No. 71-0377 (1971).
[14] P. J. McNulty, V. P. Pease, L. S. Pinsky, V. P. Bond, W. Schimmerling and K. G. Vosburgh, Science **178**, 166 (1972).
[15] P. J. McNulty, V. P. Pease and V. P. Bond, Science **189**, 453 (1975).
[16] G. Wald and P. K. Brown, Science **127**, 222 (1958).
[17] S. Hecht, S. Shlaer and M. H. Pirenne, J. Gen. Physiol. **25**, 819 (1942).
[18] E. J. Denton and M. H. Pirenne, J. Physiol. **123**, 417 (1954).
[19] M. H. Pirenne and F. H. Marriott, in: Psychology a Study of Science Vol. I, McGraw Hill, New York 1959.

Life Sciences and Space Research XIV — Akademie-Verlag, Berlin 1976

STUDY WITH A MULTI-THRESHOLD HZE-PARTICLE DOSIMETER USING PLASTIC DETECTORS

R. BEAUJEAN, W. ENGE, W. HERRMANN and K.-P. BARTHOLOMÄ

Institut für reine und angewandte Kernphysik der Christian Albrechts Universität, Kiel, FRG

During the Apollo 16 and 17 missions two units of the Biostack experiment were exposed to cosmic radiation. In this experiment plastic detector sheets were used for recording and tracing the heavy ions. In some of these sheets the integral energy loss spectrum was measured. The measurements were performed in two different cellulose nitrate materials and in Lexan polycarbonate under 4 g cm^{-2} and 20 g cm^{-2} absorber thickness. The individual materials have different energy loss thresholds for the registration of heavy ions. The measured number of particles per cm^2 with an restricted energy loss REL greater than REL$_0$ follows a power law $a \cdot \text{REL}^b$ with $b = -2.18 \pm 0.1$ while the value of a depends on the exposure time and the absorber thickness. Calculations show that more than 70% of the fluence in the measured REL region is coming from particles with $Z \geqq 20$.

1. Introduction

The contribution of the highly ionizing high-energy component (HZE) of the cosmic ray particle spectrum to the absorbed dose rate during long term space missions is not alarming. It has been estimated at roughly 10 rad/year with 35% coming from particles with $Z > 2$ and 10% from particles with $Z > 20$ [1]. But the conventional dose rate is not an adequate parameter to describe the hazard from HZE particles. The energy loss of these particles is concentrated in the vicinity of the particle's track. In this core, with a diameter less than 1 μm, the energy deposit may be high enough to cause damage to cellular components.

The effects of such heavy particle high-LET radiation on different biological objects were studied in the experiments Biostack I aboard Apollo 16 and Biostack II aboard Apollo 17. The HZE particles were monitored using nuclear emulsions and plastic track detectors. In two plastic detector sheets the integral energy loss spectrum was measured to give an impression of the relevant flux of heavy particles [2]. In addition to earlier measurements [2, 3] we have now studied the tracks in six plastic detector sheets and checked the methods for obtaining the integral energy loss spectrum. The results are compared with the calculations.

2. Measurements

2.1. Method

Heavy ions passing a plastic detector sheet produce latent tracks. When the ionization level exceeds a certain value which depends on the plastic material the latent track can be revealed by means of etching in NaOH. Starting from the surface, cones were etched along the particle's track. The cone length is a measure for the energy loss. A suitable parameter, which eliminates the charge dependence on the etching process fairly well, is the restricted energy loss REL, which includes in the ionization process only electrons with energies less than a threshold value ω. The REL values for this work are calculated according to Benton [4] using $\omega = 1000$ eV.

Different kinds of plastic detectors give different responses for a constant REL value. The exact relationship between the track etching rate v_t and REL is established in a calibration procedure on tracks of known ions. Knowing this relationship the REL values can be deduced from cone length measurements. The measurements in this work were performed in three different kinds of detectors: a, 100 μm thick Kodak-Pathe cellulose nitrate, b, 250 μm thick Daicel cellulose nitrate and c, 250 μm thick Lexan polycarbonate; these were etched in 6 n NaOH at 50° for 2 h, at 40° for 4 h and at 70° for 8 h respectively.

The scanning and the cone length measurements were performed under a Leitz optical microscope at 125 × magnification. To get approximately the same number of particles in the different detectors the scanned areas for Kodak CN, Daicel CN and Lexan were 16 cm², 25.5 cm² and 45.36 cm² respectively. All tracks (except of stopping ends) were included in the analysis. For etched-through holes the REL values deduced from a cone length half-way through was taken as a lower limit.

2.2. Results from Apollo 16

The response of the Daicel CN sheet was obtained from an in-flight calibration based on light and medium cosmic ray ions [5], while for the Lexan sheet two slightly different calibrations from balloon-borne and accelerator experiments were used. Fig. 1 shows the integral REL spectrum measured in Biostack I under 4 g cm^{-2} and the influence of the difference in the two Lexan calibrations. The shape of the measured spectrum in the individual detectors is caused, a, by scanning and registration efficiency loss at the low REL side due to the threshold of the detector and, b, by the increasing number of etched-through holes at high REL values. The straight line is fitted to the measurements and agrees with a power law $26.5\ REL_0^{-2.18}$ for the number of particles/cm² with $REL \geqq REL_0$.

2.3. Results from Apollo 17

In Biostack II four sheets were studied. The response of the detectors was obtained as follows: (i) in-flight calibration with light and medium cosmic ray ions for Kodak and Daicel cellulose nitrate (the Daicel material in Biostack II is less sensitive than that in Biostack I): (ii) for Lexan the same track etching rate $v_t = 1.26 \exp(0.31\ REL)$ was used as in Biostack I. Fig. 2 shows the result for

Biostack II. The longer mission time of 301.5 h compared with 265.9 h of Apollo 16 gives rise to a higher fluence under the same amount of absorber (4 g cm^{-2}), while an increasing absorber thickness reduces the REL spectrum (20 g cm^{-2}). The power law 33 $\mathrm{REL}_0^{-2.17}$ was fitted to values calculated by the procedure described in the next section. The calculation itself does not yield an exact power law.

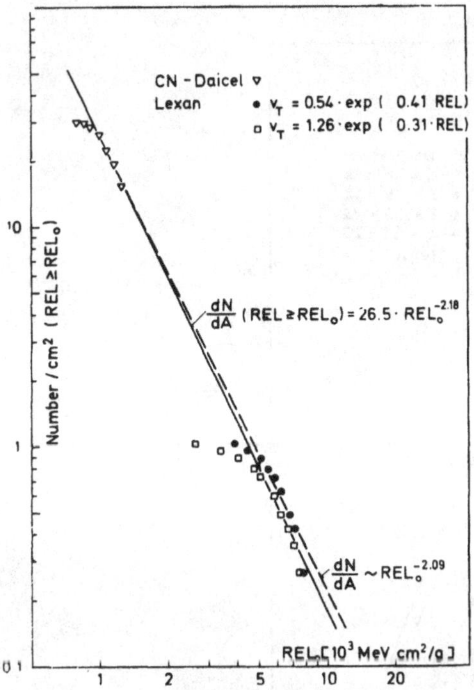

Fig. 1. The integral REL spectrum measured aboard Apollo 16 in plastic track detectors under 4 g cm^{-2} absorbers.

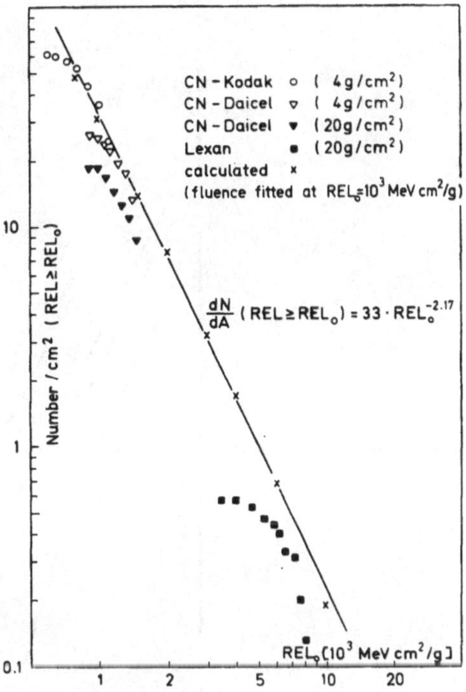

Fig. 2. The integral REL spectrum measured aboard Apollo 17 under 4 g cm^{-2} and 20 g cm^{-2} absorbers.

3. Calculations

In spite of many uncertainties concerning the energy spectrum due to solar modulation, shielding and fragmentation, we tried to calculate the integral REL spectrum and compare the calculations with the measurements. The following assumptions were made: (i) The relative abundances for the individual charges were taken from [6]. (ii) The shape of the external cosmic ray energy spectrum is the same for all elements. The flux maximum at 600 MeV/nuc is 3×10^{-3} particles m^{-2} s^{-1} sr^{-1} (MeV/nuc)$^{-1}$ for carbon [7]. (iii) The shielding was 2 g cm^{-2} for the spacecraft plus variable vertical depth in the Biostack cylinder which has a diameter of about 14 g cm^{-2}. The total absorber is assumed to consist of cellulose nitrate and all calculations are for 2π steradians. (iv) The fluence was corrected for loss due to nuclear fragmentations. The production of low charge elements from higher charge ions is not included. The mean free path length was taken as $\lambda = 56.5 Z^{-0.522}$ g cm^{-2} [1].

With these conditions the number of particles/cm² with an energy loss $\mathrm{REL}_0 \leqq \mathrm{REL} \leqq \mathrm{REL}_{max}$ was calculated by numerical integration by the following procedure: (a) the upper limit REL_{max} depends on the charge Z and was related to a residual range $R_1 = 50$ μm; (b) taking the range-REL relationship REL

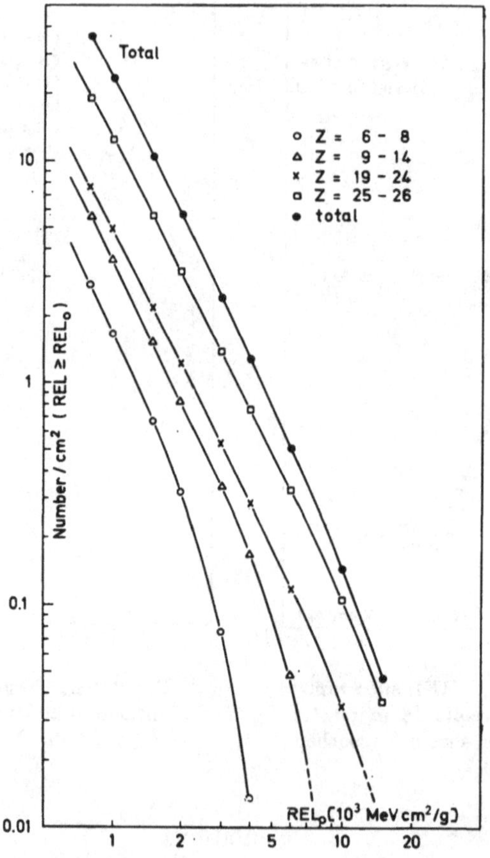

Fig. 3. The calculated contribution of different charge groups to the total integral REL spectrum under 4 g cm^{-2} absorbers.

$= f(Z, R)$ for cellulose nitrate, the lower limit REL_0 was converted to a range R_0; (c) the absorber thickness D as a function of the incident angle was added and yielded R_1^* and R_0^*; (d) taking the range-energy relationship $R = f(Z, E)$ for cellulose nitrate, R_1^* and R_0^* were converted to energies E_1 and E_0; (e) the energy spectrum was integrated from E_0 to E_1; (f) the number of particles obtained was corrected for fragmentation loss; (g) this corrected number was integrated over incident angles $8° < \alpha < 90°$; (h) summing up the individual charge contributions yields the total number of particles:

$$\frac{dN}{dA}\,(\mathrm{REL} \geqq \mathrm{REL}_0) = \sum_{Z=6}^{26} \int_{8^\circ}^{90^\circ} \sin\alpha \cos\alpha \exp\{-D(\alpha)/\lambda(Z)\} \int_{E_0}^{E_1} H(E, Z)\, dE\, d\alpha$$

where E_0 is a function of $[Z, R_0(\mathrm{REL}_0) + D(\alpha)]$ and E_1 is a function of $[Z, R_1(\mathrm{REL}_{max}) + D(\alpha)]$. The calculations yielded a strong dependence of the REL spectrum on the range-REL relationship used. This is connected with the fact that the main contribution in the spectrum comes from iron group particles. Using the correct R-REL function has the consequence that a tail of high energy iron ions is predicted to produce etchable tracks. This increases the REL spectrum around REL $= 10^3$ MEV cm^2 g^{-1} (the lower limit for iron is REL $= 800$ MeV cm^2 g^{-1}) and the fitted power law has an exponent $b = -2.45$.

The measured exponent $b = -2.18 \pm 0.1$ however indicates that these high-energy iron group particles were not registered either in Kodak or in Daicel cellulose nitrate. We therefore used a power law function REL $= 1.2Z^{1.545}R^{-0.49}$ as a compromise between the oxygen and the iron curve, which cuts off the high-energy tail of ions with $Z \geqq 20$. The results obtained with this function for the Apollo 17 flight time under 4 g cm^{-2} are shown in Fig. 3. The fluence at REL $= 10^4$ MeV cm^2 g^{-1} is nearly the same as the correct R-REL function while the fluence around 10^3 MeV cm^2 g^{-1} is reduced approximately by a factor 2. The true fluence is slightly higher than the calculated one because the calculation does not include particles from the rear due to the unknown shielding. This contribution can be estimated from the calculated fluence ratios under 4, 8 and 18 g cm^{-2} absorbers, which are 1:0.68:0.48. Under 4 g cm^{-2} shielding in addition 30—50% of particles may come from the rear, under 20 g cm^{-2} shielding this additional contribution is 40—60%. Taking this into account the calculation is in good agreement with the measurements and predicts the right attenuation under increasing absorber thickness. The calculation of Fig. 3 is plotted in Fig. 2 with a fluence of 31 particles cm^{-2} fitted at REL $= 10^3$ MeV cm^2 g^{-1}.

4. Discussion

The calculations indicated that the integral REL spectrum is dominated by the iron group nuclei in the REL region studied. The contribution of nuclei heavier than iron can be neglected because of the low abundance, while the contribution of medium charge particles becomes important at much lower REL than can be detected in plastic detectors. In the REL region studied the shape of the spectrum can be described fairly well by a power law with an exponent roughly independent of the absorber thickness. Increasing the absorber decreases the fluence, and this tendency is enhanced by nuclear fragmentation.

For future dosimetry it seems to be important that the very sensitive cellulose nitrate detectors are calibrated with iron group ions to give the correct slope of the REL spectrum. Calibrations obtained at low-energy ions with medium charges may predict a slightly different response for high-energy iron nuclei. Most important for integral measurements is the REL threshold of the detector above which nearly 100% of the particle are registered. Knowing the different threshold values for the detectors used the integral REL spectrum can be easily obtained without any tedious cone length measurements by counting the total number of tracks in the individual detectors, plotting these numbers at the proper REL threshold value and then fitting a power law to the measurements. This counting can be automated.

For personal HZE particle dosimetry small packages of plastic detector sheets with different thresholds can measure the relative fluence under complex shielding, while the absolute fluence is determined with a larger calibrated arrangement. This calibration should be performed on high-energy argon or krypton ions from accelerators.

Acknowledgments

Part of this work was supported by the Deutsche Forschungsgemeinschaft.

References

[1] HZE particle effects in manned spaceflight, National Academy of Science, Washington 1973.
[2] G. HORNECK, R. FACIUS, W. ENGE, R. BEAUJEAN and K.-P. BARTHOLOMÄ, Life Sciences and Space Research XII, 77 (1974).
[3] E. V. BENTON and R. P. HENKE, Techn. Rep. No. 26, University of San Fransisco, Dept. of Physics, 1973.
[4] E. V. BENTON, Rep. USNRDL-TR-68-14 (1968).
[5] W. ENGE, R. BEAUJEAN, K.-P. BARTHOLOMÄ and K. FUKUI, Life Sciences and Space Research XII, 51 (1974).
[6] M. GARCIA-MUNOZ, rapporteur paper presented at 13th Int. Cosmic Ray Conference, Denver 1973.
[7] P. MEYER, R. RAMATY and W. R. WEBBER, Physics Today 27/10, 23 (1974).

Life Sciences and Space Research XIV — Akademie-Verlag, Berlin 1976

LESIONAL EFFECTS OF PRIMARY COSMIC HEAVY IONS ON RAT BRAIN

A. PFISTER[a], C. NOGUES[b] and R. KAISER[c]

[a] Laboratoire d'Histologie, Faculté de Médecine Necker—Enfants Malades, Paris, France
[b] Laboratoire d'Histologie, Centre de Recherche de Médecine Aéronautique, Paris, France
[c] Centre de Recherches Nucléaires, Strasbourg-Cronenbourg, France

Heavy ions were detected with nuclear emulsions plates fixed on the skulls of 20 rats which were exposed to cosmic rays at an altitude of 32000 metres. Eight cases are described of correlations between ions tracks and brain lesions.

The passage of heavy ions seems to cause functional rather than destructive alterations in the cells. The metabolic disturbances give a dark aspect to the neurons. The lesions generally appear in wide areas around the track and this fact suggests a physiopathological phenomenon of amplification.

An evaluation of this biological hazard during flights of long duıation at high altitude will be possible when the mecanism of action of heavy ions on nervous tissue is better known, and particularly if experiments carried out in accelerators confirm the small number of results obtained in flight.

1. Introduction

A histopathological study has been carried out to investigate the possibility of a lesional effect of cosmic rays on nervous tissue. Because of the inability of neurones to regenerate it is important to evaluate the risk which threatens crews in long duration missions at high altitude or in space. There have been very few observations on this. W. Haymaker described straight lines or areas of cell depopulation in the cerebellum of monkeys carried to high altitudes in balloons [1]. However, he could not be certain that heavy ions were responsible for these lesions. Recently, he found no anomalies on the brains of pocket mice flown in the last two Apollo missions [2].

Our research began in 1971 and concerns studies on rats which were exposed during high altitude balloon flights. These animals had a nuclear emulsion plate fixed on the skull in order to record spatial correlation between cosmic particles and brain tissue. After recovery, the rats were sacrificed and their brains serially sectioned for histological examination.

2. Material and Methods

In our experiments there were three important basic requirements: (i) animals must tolerate the flight in good condition; (ii) heavy particles must be detected on the photographic emulsion and correlated with the brain with sufficient accuracy; (iii) histological techniques of fixation and sectioning must be perfect.

2.1. Preparation of Containers for Animal Transport

Two types of container were used. The earlier consisted of small, hermetically closed, stainless steel boxes (30 cm × 30 cm × 65 cm) with a double wall. They were thermically insulated by a polystyrene moulded block. Inside, an oxygen source regulated the pressure and composition of the gaseous mixture ($P_B = 550$ mmHg, $P_{O_2} = 100$ mmHg), silica gel and soda lime assured water vapour and carbon dioxide resorption. The temperature inside a box containing two rats was measured in a decompression chamber cooled to −56 °C and was found to remain between 20 and 26 °C for 12 hours. Animals were immobilized by a tightly adjusted cloth sling which was fixed by coil springs to an iron frame screwed to the container.

The second type, which is employed now, has a cylindrical section and is made of aluminium. It can transport six rats. Oxygen is delivered continuously and the internal pressure is maintained at a constant level by a safety valve.

2.2. Ionographic Detection of Heavy Ions

Detection was carried out with Ilford K 5 nuclear emulsion plates (10 mm × 21 mm and 1 mm thick) hermetically sealed in a black plastic envelope. The detector was attached to the rat skull one week before the flight and the skin repositioned over it. When the animals were sacrificed photographs were taken to record the respective positions of plate and skull.

Processing and interpretation of the exposed nuclear emulsions were done by the Corpuscular Physics Group of the Nuclear Research Center at Strasbourg. Their results gave us the spatial coordinates of each heavy ion particle which came through the plate. Knowing these data, the geometrical relationships of plate to skull and the retraction factor of the brain tissues, it is not difficult to extrapolate the trajectories (presumed linear) of heavy ions through the sagittal histological sections of the encephalon.

2.3. Histological Techniques

Since the expected lesions were very small, we had to use histological techniques which avoid artifacts and consequent misinterpretation. 10 to 15 days after the flight, rats were anaesthetized and killed by an intracardiac perfusion of fixative liquid. We proceeded as follows: (i) Nembutal 30 mg kg^{-1} intraperitoneally; (ii) catheterization of the aorta through the left ventricle, opening of the right auricle; (iii) perfusion through the catheter with 0.5 ml of heparin, 0.5 ml of papaverin and then about 30 ml of saline solution until a complete washing out of the circulatory system was obtained; (iv) perfusion of about 80 ml of the fixative liquid (formol, 1: acetic acid, 1: methanol, 8); (v) animals were decapitated, photographic references taken and detector plates removed; (vi) brains were dissected out and post-fixed for two days in the same fixative, after which they were cut in half sagittally, both parts embedded in paraffin and serially sectioned at 5 μm. For staining: hemalum-eosin, Luxol fast blue-cresyl violet or Bodian methods were used.

3. Results

20 rats were flown in six flights (Table 1). For each experiment control rats were sacrificed which had been exposed to the same constraints as the others without leaving the ground.

Table 1
Details of the Exposures

Dates	Number of rats	Flying height (mb)	Duration (hr)
8 Sept. 1971 } 9 Sept. 1971 }	2	7	12
8 Sept. 1972	3	7	2
18 Jun. 1973	4	6	6
19 Sept. 1973	5	5	8
25 Jun. 1974	6	7	6

The ionographic results are given in Table 2; about 50% of the detected tracks had impact angles too small to reach the brain. In other cases, the distance between plate and nervous tissue (4 mm) did not allow the detection of some particles which arrived laterally and could have come through the brain. Therefore, the histological studies were limited to the well defined tracks.

Table 2
Details of the Heavy Ions Detected

Date	Total number of tracks	Surface of detectors (cm^2)	Duration of the flight (hr)	Tracks (cm^{-2} hr^{-1})
Sept. 1972	7	3.36	2	1.04
June 1973	16	4.20	6	0.63
Sept. 1973	26	8.50	8	0.38
June 1974	21	5.76	6	0.60

Each heavy ion track was systematically investigated along a linear extrapolation of the coordinates coming from the nuclear emulsion plate. Two thirds of these tracks did not induce any microscopically evident damage in the brain. In the case of eight other tracks, two types of lesions were observed. Most often, the lesion consists of the existence of blackened neurones with contracted nucleus and hyperchromatic, hyalin cytoplasm. These cells are found randomly along the region of the track, isolated or in small clusters in zones which could perhaps be related to endings or disintegration stars. Dark cells are surrounded by glial cells (satellitosis) and sometimes we noticed neuronophagocytotic aspects. Silver impregnations show a decrease in the number of nervous fibres and synaptic buttons in these regions (Fig. 1).

Another type of damage was detected, in one case, along the trajectory of a class M ion, which passed through the cerebellum from right to left. In the beginning of the track, we observed some isolated dark cells in the granular layer. Further along there was a conical region with a degenerate aspect (about 0.5 mm^3 in volume) and finally a large zone of dispersal where we found numerous dark cells. In the most altered focus, we noticed oedema, rarefaction of nervous fibres and the

Fig. 1. Elongated dark neurones cluster situated along the presumed track of an heavy ion.

disappearance of about 30% of granular and Purkinje cells. Remaining cells often show various cytological abnormalities such as chromatolysis and contracted nucleus (Fig. 2).

In addition, we observed other anomalies which could not be linked with a known heavy ion track. These lesions consisted of isolated dark cells, foci of gliosis and limited areas of depopulation. We detected only one cerebellar linear lesion similar to the observations of Haymaker.

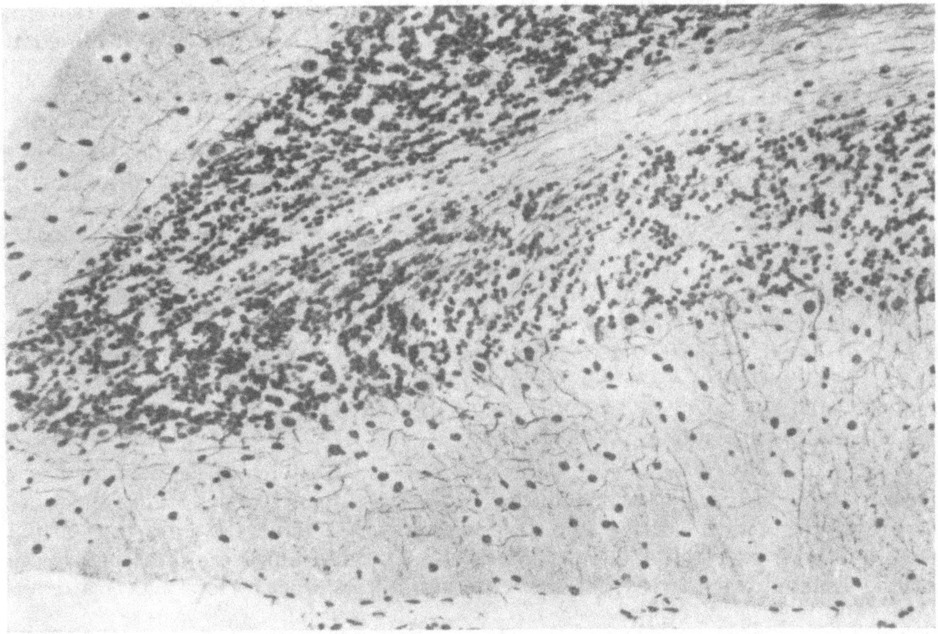

Fig. 2. Limited cellular depopulation region in the cerebellum of a rat. This lesion is situated on the trajectory of a heavy ion.

4. Discussion

1. The passage of heavy ions in a tissue seems to cause functional rather than destructive alterations in the cells. The consequences are not important for tissues which have a strong regenerative capacity, but in the case of nervous tissue (and melanocytes in the skin) the damage is not repaired by newly formed cells. The altered neurons may have a survival of more or less long duration but keep a disfunctioning dark aspect. Glial satellitosis and neuronophagocytosis are often associated and it is likely that the damaged cells die and disappear in a relatively short time. Morphological changes are probably not immediately evident because metabolic disturbances are progressive; in the same way, after a variable period, abnormal cells would disappear and histological damages would be difficult to ascertain. There is, therefore, certainly an optimal interval of time during which lesions are most apparent. We have found this period to be situated around 15 days after irradiation.

We do not think that dark cells are specific for heavy ions effects: this aspect is seen in many other circumstances such as anoxia, cyanide intoxication and often when nervous tissue has been mishandled before fixation. This latter cause of artifact may be dismissed in our experiments because fixation was done by vascular perfusion. Moreover, controls never had such systematized lesions.

2. Amplification of the lesional effect. The distribution of dark cells along a heavy particle trajectory is not tightly concentrated in the core of the track, as we expected at the beginning of our research. Alterations can be found about 100 μm

distant from this core and even further away in some regions. If our observations are correct, the radius of lesional effect of a heavy ion must not be considered as the small ionization cylinder (only a few micrometres) described by Schaefer [3] but as the wide range (more than one centimetre) of delta-ray dispersal suggested by the Soviet physicists [4].

Other arguments may also be proposed to explain this amplification. Damaging effects may be enhanced at the endings of the heavy ion trajectory or when nuclear interaction stars appear. It is also possible that cytotoxic substances are created in the irradiated tissue and are active at a distance. The amplification effect of stars and heavy particles on the skin of black mice exposed at high altitude has already been described [5, 6]. There remains a discrepancy between our results and those obtained at ground level with deuteron microbeams: Curtis [7] and Haymaker [8] found that 400000 rads were necessary to damage brain tissue! Planel et al. [9] saw the same phenomenon with eggs of *Artemia salina*: one heavy ion hit was sufficient in space to inhibit one egg, but 300000 rads was necessary to achieve a similar result in the laboratory.

3. The heavy ions of primary cosmic radiation seem to act differently on living matter than do other known ionizing radiations. Each particle must be taken separately into consideration in order to evaluate the biological hazard of cosmic rays. It is not possible to draw definite conclusions from these facts before our results are confirmed by experiments carried out with heavy ions accelerators.

References

[1] W. HAYMAKER, O. T. BAILEY, E. W. BENTON, F. S. VOGEL, W. ZEMAN et al., Aerospace Med. **41**, 9, 989 (1970).
[2] W. HAYMAKER, W. ZEMAN et al., Aviation, Space and Environ. Med. **46**, Sect. II, 613 (1975).
[3] H. J. SCHAEFER, J. Aviation Med. **25**, 338 (1954).
[4] E. J. KUDRYACHOV, A. M. MARENNYI, V. J. POPOV and O. M. MECHERYAKOVA, Kosm. Biol. Med. **5** (1970).
[5] H. B. CHASE and J. S. POST, J. Aviation Med. **27**, 533 (1956).
[6] A. PFISTER, G. DELTOUR, R. KAISER and L. MIRO, AGARD Conf. Proc. No. 25, **16**, 1 (1967).
[7] H. J. CURTIS, Radiation Res. Suppl. 7, 250 (1967).
[8] W. ZEMAN, H. J. CURTIS, K. L. GEBHARD and W. HAYMAKER, Science **130, 1760** (1959).
[9] H. PLANEL, J. P. SOLEILHAVOUR, R. KAISER and Y. BLANQUET, Life Sciences and Space Research XII, 85 (1974).

Life Sciences and Space Research XIV — Akademie-Verlag, Berlin 1976

EFFECTS OF ^{40}AR ON THE NERVE CELLS OF *Drosophila*

J. MIQUEL[a], E. V. BENTON[b] and G. WELCH[c]

[a] NASA Ames Research Center, Moffett Field, Calif., USA
[b] University of San Francisco, San Francisco, Calif., USA
[c] Lawrence Berkeley Laboratory, University of California, Berkeley, Calif., USA

Despite the potential hazard of galactic radiation to the crews of space vehicles, the effects of heavily ionizing particles on nerve cells remain a controversial subject. To obtain background information relevant to this problem we have investigated the fine structural changes occurring in the brain of adult *Drosophila melanogaster* at 35 days after irradiation with ^{40}Ar from the Berkeley Super-HILAC. The energy of the particles was 7.5 MeV/amv (stopping beam) and the peak energy deposited at the end of the particle range was approximately 3000 keV μm^{-1}. Just before irradiation the flies were etherized and attached to Lexan dosimeters using a harmless adhesive. The particle flux into a particular *Drosophila* brain was ascertained by track counting near the particular fly head and beam penetration was determined by locating the position of the beam Bragg peak in tissue-equivalent plastics prior to the irradiation of the flies. Our preliminary results show that even at the lowest fluence used, namely 6×10^4 tracks cm^{-2} (equivalent to approximately one hit in one target area encompassing about 67 nerve cell bodies), ^{40}Ar irradiation resulted in brain injury consisting of nerve cell swelling with a reduction in the number of ribosomes and glial hypertrophy. These observations lend support to the view that exposure to galactic radiation in prolonged space flight many result in some injury to brain cells.

A full paper appears in *International Journal of Radiation Biology*. **29**, 101 (1976).

Life Sciences and Space Research XIV — Akademie-Verlag, Berlin 1976

THE BIOSTACK AS AN APPROACH TO HIGH LET RESEARCH

H. Bücker, R. Facius and M. Schäfer

Arbeitsgruppe für biophysikalische Raumforschung, Universität Frankfurt, Frankfurt/Main, FRG

By simple geometric and dosimetric arguments the advantage of an experimental approach to high LET radiation research is demonstrated. The Biostack is capable of recording individual hits of heavy ions on single biological targets. This improved method is compared with the common experimental methods for studying biological effects with high LET radiation and is suggested as an methodological improvement in fundamental research.

During the Apollo 16 and 17 missions the various biological specimens in the Biostack received a total dose between 600 mrad and 900 mrad. This dose has been measured by LiF thermoluminescence dosimeters located within the Biostack box on both ends and in the middle of the stack. The absorbed dose due to the heavy ions was determined from measurements in nuclear emulsions [1]. The results showed that an average dose of 3 mrad was produced by HZE particles. However, this value of 3 mrad obviously is an inadequate way to describe the biological effects of HZE particles, more generally called high LET particle radiation. To illustrate this we can compare dose-effect curves for various biological objects, ranging from the procaryotic single cellular system of the spores of *Bacillus subtilis* to the highly differentiated diploid human kidney cells (Fig. 1). In the linear dose scale of this figure the value of 3 mrad had to be placed at a distance of 30 Å from the origin if we equate 10 krad to 1 cm. However, biological effects of the HZE particles on *Artemia salina* eggs, *Bacillus subtilis* spores and on other objects could be detected [2], corresponding to dose values of some tens of kilorads in the dose-effect curves shown, even though they also have been obtained from irradiations with densely ionizing particles [3—6]. In discussing this we want to demonstrate the range of applications in which we consider the Biostack localization methods to be most suitable for radiation biological experiments.

To indicate that we use the term dose simply as the quotient of energy/mass we denote this "dose" as D'. In Fig. 2 a situation is shown which might be considered as representative for an ideal arrangement for many radiation biological experiments. A monocellular layer of optimal packed cells is exposed to some kind of ionizing radiation of a certain mean LET until a predetermined dose has been deposited. This dose is measured by some kind of integrating dosimeter such as film dosimeter, Fricke dosimeter, ionization chamber, thermoluminescence dosimeter or others. After various kinds of post-irradiation treatment the number of cells is counted which reach or do not reach a specific biological endpoint (for example, colony-forming ability, emergence, hatching or other endpoints). The ratio N/N_0 of the number N of cells reaching this endpoint to the number N_0 of the cells doing so without irradiation is taken as a measure of the radiation effect. Now we estimate the number N_p of particles with mean LET L (keV/μm) necessary to deposit the dose D' (rad) in an object of spherical shape with radius R(μm).

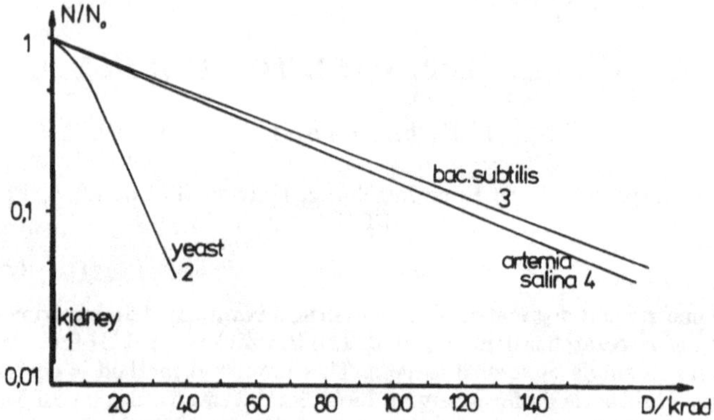

Fig. 1. Dose effect curves for high LET irradiation
1) clone-formation; $\alpha \sim 100$ keV μm^{-1}, [3]
2) colony formation; $\alpha \sim 160$ keV μm^{-1}, [4]
3) colony formation; $\alpha \sim 51$ keV μm^{-1}, [5]
4) emergence; ^{12}C 170 keV μm^{-1}, [6].

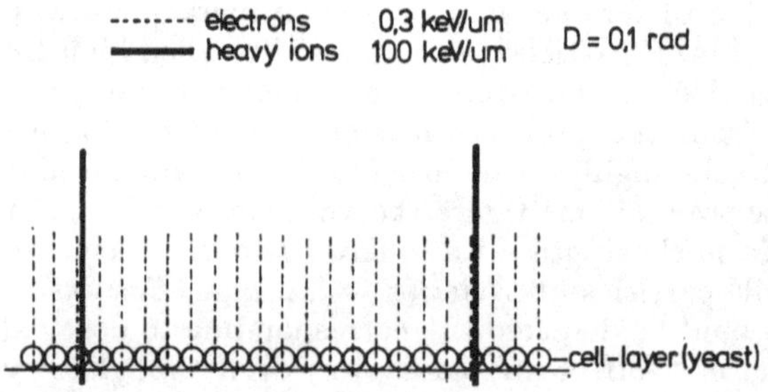

Fig. 2. Irradiation of yeast cells (5 μm diameter) with different LET values.

With a mean path length $l = 4/3R$ we find:

$$N_p = 6.24 \times 10^{-2} \times \pi R^2 \times D'/L$$

$N_f = N_p/\pi R^2$ then is the average number of particles per μm² which is necessary to yield the dose D' to the cell. $d = 1/\sqrt{N_f}$ may be considered as the mean distance in μm between two of those particles assuming a homogeneous beam. In Fig. 2 this mean distance d is indicated by the spacing of the lines representing an irradiation with different LET values. When depositing 0.1 rad by electrons of 0.3 keV/μm into a cell of 5 μm diameter (e.g. yeast cells) the average number of particles per object would be one. Using heavy ions with a LET of 100 keV/μm instead, only every 333rd cell will be encountered by a particle.

In Fig. 3 the mean distance d of particles within a homogeneous beam is plotted against the dose to be deposited for various LET values. According to our assumption of spherical objects this is independent of the object size. The horizontal dashed lines represent some typical sizes of biological organisms with the following significance: below the dose at which the curve for the mean distance crosses the size of the irradiated object the mean particle distance will exceed the size of the object e.g. 2.5 krad for spores of 1 μm diameter when irradiated with carbon ions of 160 keV/μm. Thus only a decreasing fraction of cells will actually be affected by an impact of a heavy ion. The range of LET values which had been measured

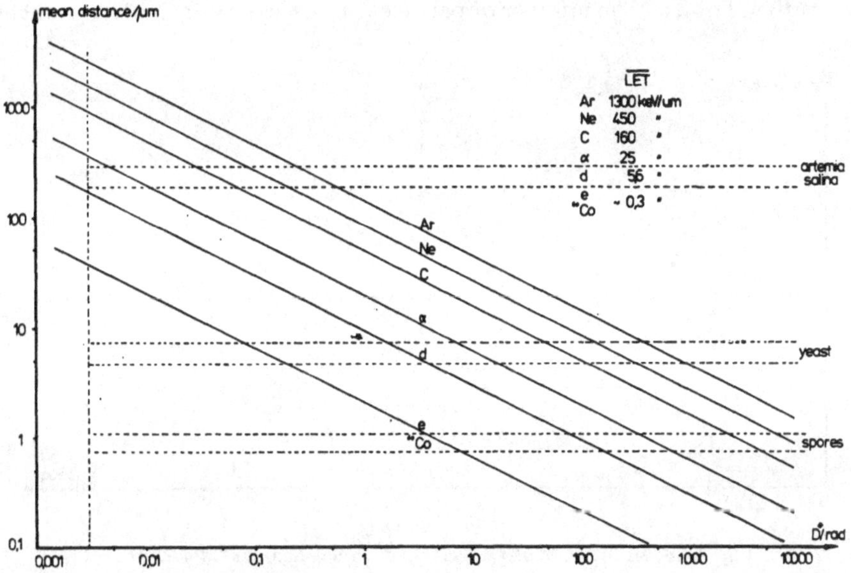

Fig. 3. Mean particle distance in a homogeneous beam against "dose" D'.

in the Biostack is roughly represented by the upper three curves in Fig. 3. The vertical dashed line corresponds to the dose of 3 mrad due to the HZE particles. At this dose the mean distance of the particles would vary between 1 mm and 3 mm which is in fair agreement with the track densities actually measured in the plastic detectors. In Fig. 4 the resulting mean number of particles affecting an object of a given size is plotted against the dose D' which is to be deposited by heavy ions with various LET values, covering the range of cosmic particles. The dashed horizontal line is drawn to find the dose below which this mean number of particles per object falls below one. From then on (for purely statistical reasons) it would be improper to investigate the radiation effects by the above indicated procedure. The population being analysed would contain more and more cells which never experienced the impact of even one particle.

Only experimental set-ups which allow one to keep track of the individual particles, together with the individual cells being hit, are suitable if radiation effects are to be investigated under these circumstances. The common procedure of studying dose effect curves is not applicable for investigating radiation effects in the range of extremely small dose values (3 mrad). At the beginning of the 1950's this principle of irradiating single cells with individual ions was proposed by Baarli and was later tried several times [7, 8]. Our arguments are based on rather crude simplifications which are now slightly refined in order to be applicable to experiments with high LET radiation beyond the low dose restriction of the cosmic radiation. First of all we have to consider the constraints which are already present in the classical dose concept. In general terms these may be described as the necessity of a homogeneous deposition of energy within the irradiated biological object. If we take r in Å as the radius of a cylinder where we assume the dose is deposited by a heavy ion and R in μm as the diameter of an irradiated object, the number of particles M_p necessary to achieve an overall

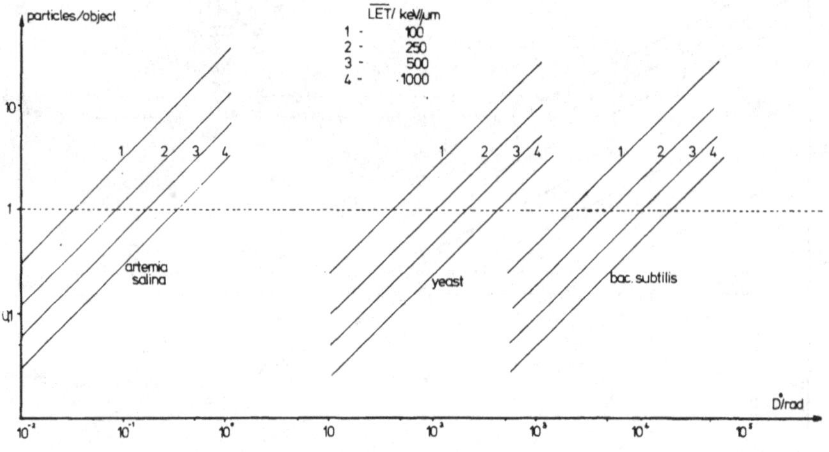

Fig. 4. Mean number of particles per object against "dose" D'.

homogeneous energy deposition within the object would be $M_p = 10^8 \times (R/r)^2$. Even if we allow r to be as large as 100 Å and consider objects of 1 μm in radius, a magnitude of 10^4 is obtained for M_p. Thus the dose values above which the investigations of radiation effects in terms of the common dose concept would become meaningful, had to be derived from Fig. 4 by multiplying the dose at which one ion is striking the target cell with this factor M_p. This results in considerable threshold dose values if small objects are to be irradiated with high LET particles. In Fig. 5 results for the irradiation of yeast cells with heavy ions are shown, to demonstrate that at least at the lower dose values the requirements of even the classical dose concept are hardly met. Whether measurements of this kind can be used to derive, for instance, values for initial and final slopes in order to use them in some recent models of radiation effects at least remains doubtful.

More sophisticated calculations result in threshold doses of the same order of magnitude below which investigations of radiation effects in terms of mean dose are no longer considered to be valid. These limits can be estimated from the stochastic distribution of specific energy and depend on the LET distribution of the particle, the target size and the variance of the stochastic distribution of the specific energy which is considered to be tolerable [9—11]. Thus we are faced with the fact that in order to work properly with high LET radiation by the common procedures, we need large doses. The current knowledge about the purely physical aspect of such an irradiation in terms of microdosimetry might be considered sufficient to describe the physical situation. But the question arises what insight into the chemical and biological consequences of high LET radiation can be gained from living organisms that have come into a machine gun volley. In order to add to the understanding of the biochemical and biological processes involved an appro-

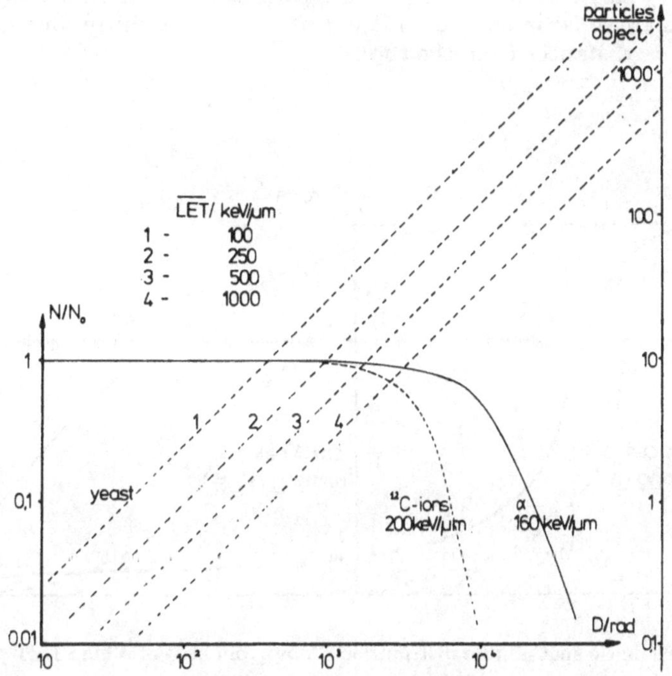

Fig. 5. Yeast-colony formation; α-particles [4], ^{12}C [10].

ach should be used which allows for the investigation of individual biological objects being affected by individual heavy ions. If possible the choice of objects should be extended to the very roots of biological systems, e.g. enzymes, viruses and single-celled organisms. Apart from these spatial limits, however, we have to realize that at present Biostack flight experiments are only possible with biological objects in the resting state.

Various theoretical models for describing the interaction of high LET radiation with living systems also need to be considered. The "Punktwärmehypothese" was proposed by Dessauer in 1923, but only in the beginning of the 1960's has this idea been revived in the form on the thermal spike model. However, when estimating the temperature increase which is produced by a heavy ion within a single cell this results in an increase of only about 5—10 °C. Calculations from Tobias' group in

Berkeley (as reported in the summer LET workshop 1974) resulted in a temperature increase of 7 °C. They also detected shock waves in piezo-crystals irradiated with single heavy ions. These shock waves could lead to torn membranes in biological cells.

The analysis of results obtained by a Biostack experiment which allows the individual recording of hits could be performed in the following way: In the case of a space experiment with a highly heterogeneous radiation environment the physical parameters (charge and energy) of the heavy particle may be determined from the track in the recording detector. In accelerator experiments at least the charge would be known. From these values the radial dose distribution around the path of a heavy particle may be calculated according to one of the various different models which are so far in use [12—15] and which produce fairly consistent results in intermediate distances from the core.

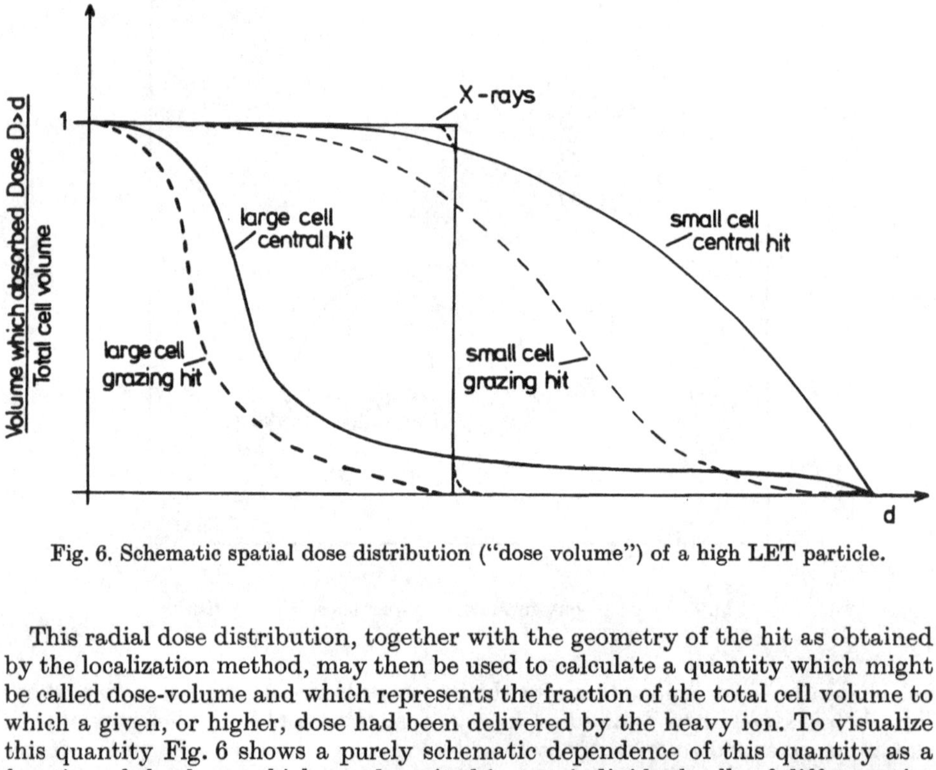

Fig. 6. Schematic spatial dose distribution ("dose volume") of a high LET particle.

This radial dose distribution, together with the geometry of the hit as obtained by the localization method, may then be used to calculate a quantity which might be called dose-volume and which represents the fraction of the total cell volume to which a given, or higher, dose had been delivered by the heavy ion. To visualize this quantity Fig. 6 shows a purely schematic dependence of this quantity as a function of the doses which are deposited in two individual cells of different size by the same heavy ion. The solid curves can be correlated with hits into the central region of the cells, whereas the dashed curves should arise from grazing passage of the heavy ion. Detailed analysis of a sufficiently large sample of cells for various biological endpoints could then be correlated with the results obtained from X-ray irradiation of the same biological objects. The roughly step function for such an irradiation is also shown in Fig. 6. The smoothed edges are shown to indicate the effects of the stochastics of even "soft" radiations, which arise at small objects already at high doses.

References

[1] R. Pfohl, R. Kaiser, J. P. Massué and P. Cuer, Life Sciences and Space Research XII, **57** (1974).
[2] H. Bücker et al., Life Sciences and Space Research XIII, 161 (1975).
[3] G. W. Barendsen, Rad. Res. **18**, 106 (1963).
[4] W. Pohlit and M. Schäfer, Int. Atomic Energy Agency Vienna (1974) IAEA-SM-179/23 (p. 177).
[5] J. E. Donnellan and H. J. Morowitz, Rad. Res. **7**, 71 (1957).
[6] S. S. Easter and F. Hutchinson, Rad. Res. **15**, 333 (1961).
[7] R. E. Zirkle and W. Bloom, Science **117**, 487 (1953).
[8] T. R. Munro, Exp. Cell Res. **18**, 76 (1959).
[9] W. Pohlit, private communication.
[10] M. R. Raju et al., Rep. UCRL-20161 (1970). Lawrence Radiation Laboratory.
[11] A. M. Kellerer, Mikrodosimetrie, Grundlagen einer Theorie der Strahlungsqualität Gesellschaft für Strahlenforschung GSF-Bericht B1, Nov. 1968.
[12] A. Chatterjee, H. D. Maceabee and C. A. Tobias, Rad. Res. **54**, 5479 (1973).
[13] L. Muga and G. Griffith, Phys. Rev. B **8**, 4069 (1973).
[14] R. Katz and S. C. Sharma, Nucl. Instr. and Meth. **111**, 93 (1973).
[15] H. G. Paretzke, G. Leuthold, G. Burger and W. Jacobi, Fourth Symp. on Microdosimetry, Verbania 1973 (p. 123).

Life Sciences and Space Research XIV — Akademie-Verlag, Berlin 1976

LOCALIZATION AND TRACK EVALUATION OF HZE PARTICLES

H. Bücker, D. Hildebrand, G. Reitz and M. Schäfer

Arbeitsgruppe für biophysikalische Weltraumforschung an der Universität Frankfurt, Frankfurt/Main, FRG

A new method of track-etching was developed called "individual etching", allowing increased accuracy of the localization whereby a latent track is developed under microscopic control. The influence of the humidity of the detectors at the beginning of the etch process was investigated. This influence consists in an increase of bulk and track etching rate. The scattering in experimental data, which is mostly attributed to the inhomogeneity of the material, can be partially explained by this effect, especially the difference in isotopic resolution between accelerator experiments and cosmic-ray flights.

To study the radiation effects of HZE particles on biological objects we consider it necessary to reach a high accuracy in the determination of the region hit. This is necessary for the correlation of the penetrating particle and the biological effect. In the Biostack experiments an experimental arrangement is given allowing such a localization be made by using three types of track detectors: nuclear emulsions, silver chloride crystals and plastic detectors [1]. In this paper only plastic detectors will be discussed. Plastic detectors are organic and for this reason they have some unique advantages. Their density is nearly the same as tissue; their atomic composition includes only light elements like tissue; they respond to penetrating particles by forming radicals, crosslinks and cracked products. The change in behaviour towards chemical agents is commonly used for track etching to enlarge the track to microscopic visibility. Using plastic detectors it is possible to determine charge and energy from etch cones along the trajectory of the particle, even without knowing where the particle stops [2, 3]. Because plastics are nontoxic, they can be in fixed contact directly with the detector surface. The objects can even stay on the detector surface during the etching process, provided the biological layer is sealed against the etchant.

Up to now only overall etching in etching baths has been done to find the tracks. If the object is fixed between the detector sheets the spatial resolution is limited by the quality of extrapolation of the etch cones. This quality depends on the length, dip angle and the shape of the etch cone. For steep and long etch cones an accuracy in the range of 2 μm can be reached when only a single cone is used for the evaluation. Therefore a new method was developed called "individual etching" to increase the accuracy of the localization method, particular for flat and short tracks. This method was performed with cellulose nitrate as detector material.

The idea was to apply a drop of etching solution in the region of an etch cone, which was produced by pre-etching in the bath for a short time to make the track visible. One side of the detector was covered with a monolayer of *Bacillus subtilis* spores, which were embedded in PVA. Around the drop a plastilin ring was laid and a coverglass was pressed on the ring to seal this microchamber and to protect the spores against the etchant (Fig. 1, top). The construction of the chamber allows one to observe the track growth under the microscope. On reaching the biological layer the track was stopped by washing out the etchant with distilled

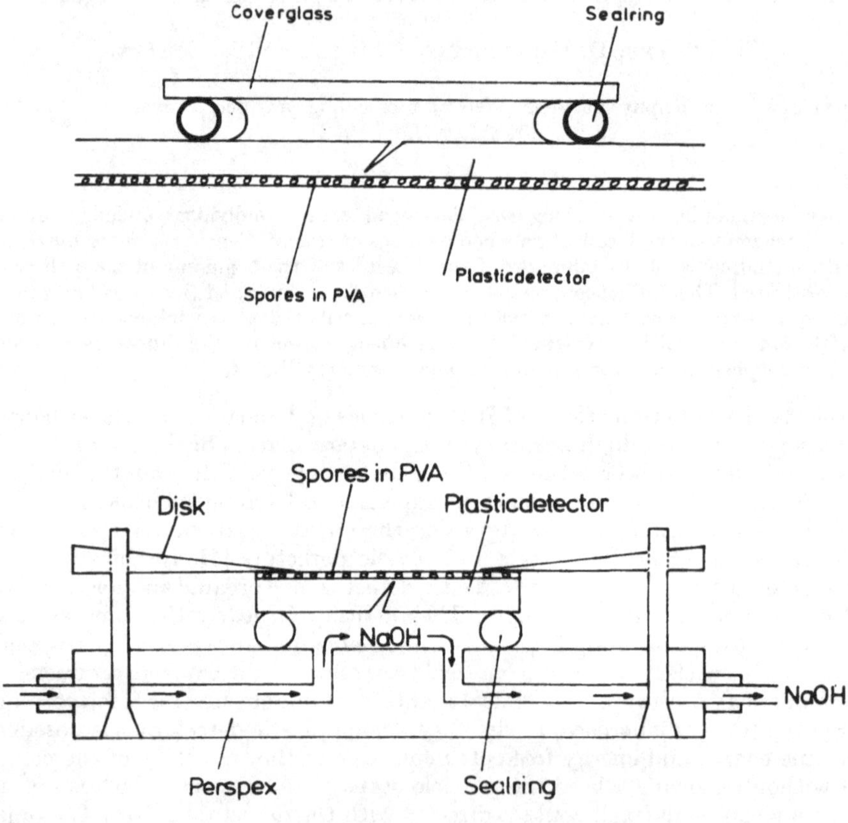

Fig. 1. Chambers for the "individual etching" method.

water. However, with this arrangement again only steep tracks can be etched satisfactorily: particularly for small tracks, which belong to particles with a low REL, the tip of the cone becomes round, which means that the ratio of the track etching rate to the bulk etching rate becomes unity. The velocity of this process depends on the temperature of the etchant. Among the other problems which arose the following were of major importance. First we had to deal with leakage through the seal ring, and secondly we get a great concentration of etching products in the drop. The last effect diminished the visibility of the tip and often proscribed microscopical observation of the cone. The observed differences in etching behaviour and the reduced visibility of the etch cone induced us to develop a new etching chamber, where etching is possible under microscopical observation and under nearly the same conditions as in the bath (Fig. 1 bottom). To protect the biological layer we used a viton seal ring. Inlet and outlet were chosen so that the foil surface was properly rinsed with the etchant. The etchant is in a separate reservoir which is moved through the system by a pump. The temperature of the etchant reservoir and the temperature of the microscope support are continuously variable up to 80 °C. By using this chamber the tip of the cone does not become round. The track etching rate is decreased, so that the etching time is always greater than in the etch bath. The only explanation we can give is the humidity content of the foil. Temperature effects can be excluded, also the influence of the light from the microscope. A further problem of the method is to stop the etching precisely at the moment when the track reaches the biological layer. Here we can choose two methods. The first is to etch the track two or three micrometres before the object is touched and then to stop it. In order to avoid the chance of damaging the biological objects a better approach is impossible with this method. A better method is to photograph the point of penetration on the opposite surface of the foil. Therefore first a photograph was taken of the region in which the extrapolated point of penetration was expected. Then the spores were removed by micromanipulation (this method was developed in our laboratory). Finally the track was etched until it penetrates the opposite side of the foil. In that moment a second photograph is made. By comparing the two photos the track can be correlated with the hit spore. The localization accuracy will be increased by each of these two methods. For steep tracks we obtained only a small improvement. For flat and small tracks the methods have a great advantage, because the former method of evaluation was affected by errors that were too large to enable a reliable determination of the hit regions to be made.

Another aspect of the new etching method is the observation of the formation of the etch cones and their growth along the latent track, also the attack of the etchant on the detector surface. We also observed layers of etching products which adhere on the detector surface and are so stable that they could be removed only by washing off with water. These layers are insoluble in water. The water content of the foils seems to be the cause of the difference in etching obtained in the microchamber and in the bath [4]. We have found that the humidity of the foils at the beginning of the etch process influenced their etching behaviour in a very unfortunate way. The findings are:

(i) Influence on the bulk-etching rate. Before etching the foils were laid in double distilled water for different times. There is an increase of the bulk-etching rate with the time of soaking (Fig. 2). The effect will vary with the temperature, being dependent on the ratio of the water diffusion velocity to the vertical track-etching rate. It is necessary to take this into account when making track radius measurements, for example. Published results on the bulk-etching rate in connection with etching techniques or investigation of etching mechanism should be revised in the light of this.

(ii) Influence on the track-etching rate. Foils from a Biostack exposed to carbon ions at Berkeley were used to measure cone length versus range diagrams. Again the foils were placed in double distilled water for different times before etching. The result is a significant increase of the track-etching rate with the time of soaking (Fig. 3 and 4).

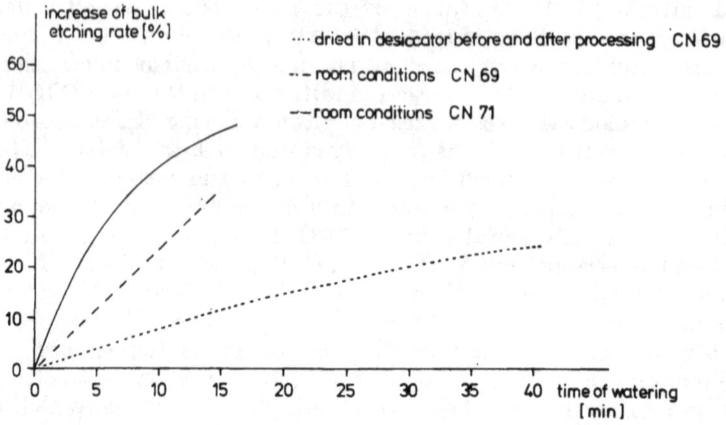

Fig. 2. Increase of bulk-etching rate with time of soaking.

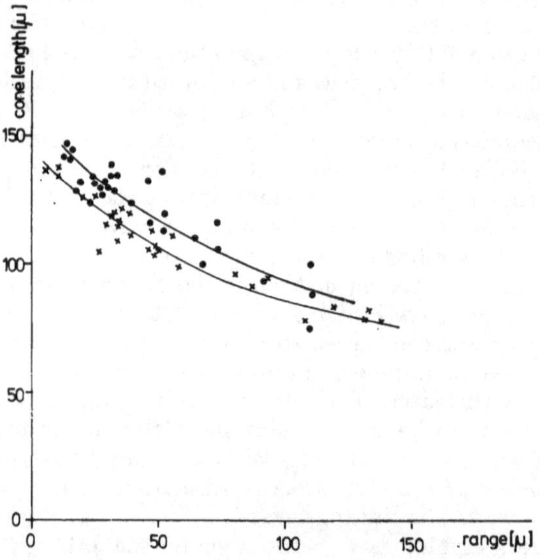

Fig. 3. Cone length versus range diagram for carbon ions.
• 15 min soaked in double distilled water; × not soaked.

What does this mean for etching foils taken from room conditions? The track-etching rate will be influenced by the local humidity and it will depend on the depth reached after the elapsed etching time. There will therefore be a dependence of the cone length on the dip angle (Fig. 5).

The scattering of experimental data, especially in cosmic ray measurements, which is mostly attributed to the inhomogeneity of the material can be partially explained by this effect, even the difference in isotopic resolution between accelerator

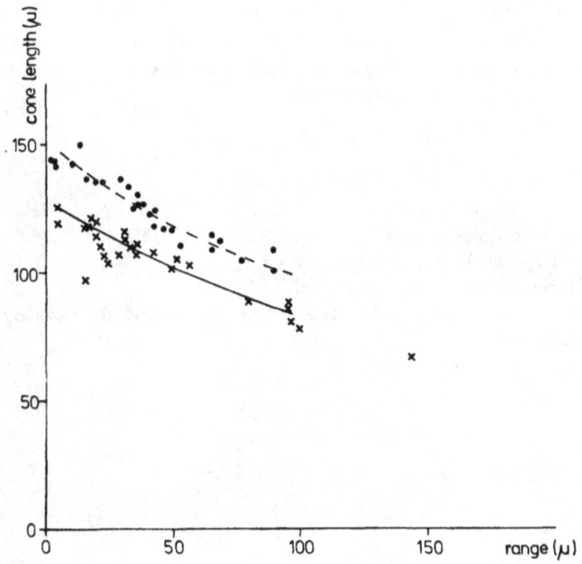

Fig. 4. Cone length versus range diagram for carbon ions.
--- 40 min soaked in double distilled water before etching; — not soaked.

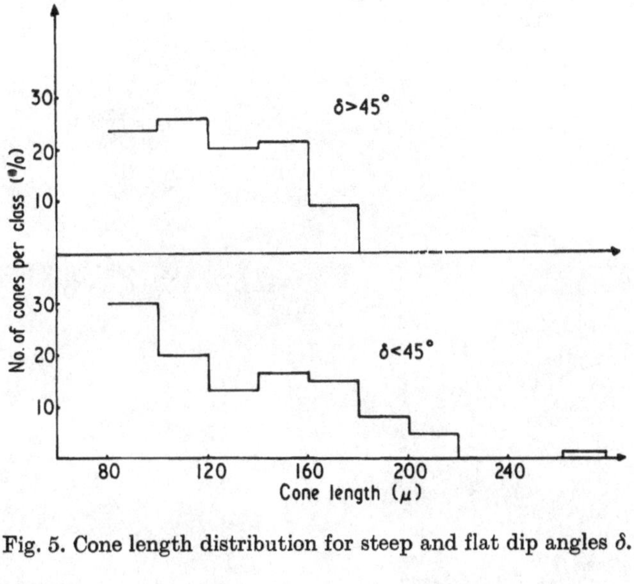

Fig. 5. Cone length distribution for steep and flat dip angles δ.

experiments and balloon flights. We also determined whether water changes the structure of the detector material by wetting foils and then drying them. Preliminary results show no difference between the track-etching rate in the two cases. When we wet foils before etching and then observe the track formation in the microchamber we can see an increase in the track-etching rate. This increase disappeared after some minutes etching. This may be explained by a humidity gradient. These observations show that the presence of water influences the etching process in an important way, and perhaps the presence of water is necessary for the etching process to occur.

References

[1] H. BÜCKER et al., Life Science and Space Research XII, (1974).
[2] D. HILDEBRAND, Diplom-Arbeit, University Frankfurt.
[3] R. BEAUJEAN, Dissertation, University Kiel.
[4] G. REITZ, D. HILDEBRAND and H. BÜCKER, paper presented at Working Party on Space Biophysics, Uppsala 1974.

Life Sciences and Space Research XIV — Akademie-Verlag, Berlin 1976

INFLUENCE OF HEAVY IONS ON THE TRANSFORMING ACTIVITY OF DNA

M. I. MINKOVA[a], N. I. RYZHOV[b] and T. P. PANTEV[a]

[a] Medical Academy, Institute of Roentgenology and Radiobiology, Sofia, Bulgaria
[b] Institute of Medico-Biological Problems, Moscow, USSR

Changes in functional activity following treatment of DNA with heavy ions were analysed by transformation assay. Biological response to exposure to charged particles, ^{10}B, ^{12}C, and ^{22}Ne, was evaluated by the extent of inactivation of ability of DNA to transfer the genetic marker of tryptophane independence. The possibility of protecting the biological activity of DNA was studied using a conventional protector, cysteamine hydrochloride, or the preparation cytriphos, synthesized at the Institute of Radiobiology and Radiation Hygiene, Sofia.

Dried samples of donor DNA isolated from a *Bacillus subtilis* prototroph strain were exposed to doses of 50 krad to 1 Mrad. For the three types of charged particles used, dose response of the inactivation process was defined by an exponential function. Comparison of response curves for the three types of radiation showed the effect to be dependent on LET, being highest for ^{10}B and lowest for ^{22}Ne. In all three cases, addition of a protective substance failed to produce any change in the biological effect observed.

1. Introduction

Prolonged manned space flights involve a serious hazard to astronauts from galactic cosmic radiation (GCR) since estimations indicate that in interplanetary space GCR dose rates may occasionally approach values of as much as 200 to 250 mrem per day. According to calculations, about 90% of absorbed dose is contributed by heavy ions [1]. To ensure radiation safety in such flights a knowledge is needed of the biological response to energetic light and heavy ions as well as of the possibility of modifying the damage inflicted by them. The biological effectiveness of heavy ions has been studied on various test models [2—6]. However, few studies have been devoted to changes induced in DNA. Bearing in mind that DNA plays an essential part in the ultimate effect of radiation damage, we have attempted to shed light on some aspects of high LET radiation effects on biological activity of the DNA molecule.

2. Materials and Methods

A *Bacillus subtilis* transforming system with prototroph strain 23 as donor and BS 168 ind⁻ strain as recipient was used. The method for isolation of biologically active DNA and the transformation assay procedure are described in [7].

The basic test by which changes in DNA functional activity were measured was genetic transfer of the tryptophan-independence marker.

The ions used were those of ^{10}B (LET = 1200 MeV g^{-1} cm^2), ^{12}C (LET = 4800 MeV g^{-1} cm^2) and ^{22}Ne (LET = 6000 MeV g^{-1} cm^2) accelerated on cyclotron UMZ-300 of the Laboratory of Nuclear Reactions at Dubna, to yield a surface dose rate of 1 krad s^{-1}. Since ions have a low penetrating ability, DNA was irradiated in the dgy state at 15 μm film thickness. To modify biological response to heavy ions exposure, additions were made to DNA solutions of S-containing protectors, namely cysteamine hydrochloride (MEA), 150 μg ml^{-1} or cytriphos (CATP), 750 μg ml^{-1}.

3. Results and Discussion

Exposure of DNA to heavy ions clearly tended to reduce its ability to transfer genetic information with increase in radiation dose. For any of the three accelerated particles, dose-response kinetics were found to be fitted by a straight line

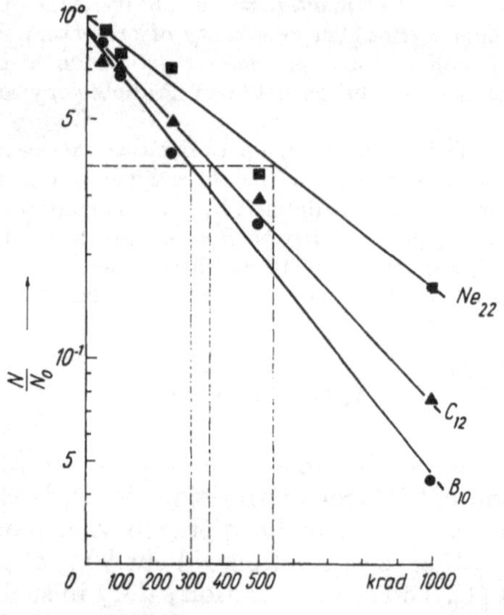

Fig. 1. Transforming DNA inactivation curves.

on a semilogarithmic plot (Fig. 1). Mathematical data treatment yielded the following equation for the relationship studied: $\log y = \log a + D \log b$, indicating that the inactivation process followed an exponential curve within the dose range of 5 krad to 1 Mrad.

Radiation type intercomparison of dose-response regression lines showed that the effectiveness in reducing the ability of DNA to transfer genetic information was most marked for ^{10}B and least for ^{22}Ne. D_{37} dose values estimated by the equation amounted to 304 krad for ^{10}B, 359 krad for ^{12}C and 543 krad for ^{22}Ne. Differences in effectiveness were statistically significant in the dose range studied.

Based on the experimental results obtained, calculations were made of the two parameters, formal target size α and inactivation cross section σ. As may be

Table 1

Values of the Parameters D_{37}, Target Size and Cross Section for the Types of Radiation used

Treatment	dE/dx (MeV g^{-1} cm^2)	Radiat. energy (MeV/nucleon)	D_{37} krad	Coeff. α (10^{-6} rad^{-1})	Inactivation cross section (10^{-11} cm^2)	$G_{inactiv.}$
^{10}B	1200	8.7	304	3.3	6.3	0.20
^{10}B + MEA	1200	8.7	309	3.2	6.2	0.19
^{10}B + CATP	1200	8.7	292	3.4	6.5	0.20
^{12}C	4800	2.0	359	2.8	21.5	0.17
^{12}C + MEA	4800	2.0	386	2.6	20.0	0.15
^{12}C + CATP	4800	2.0	381	2.6	20.0	0.16
^{22}Ne	6000	6.7	543	1.8	17.3	0.11
^{22}Ne + MEA	6000	6.7	486	2.1	20.2	0.12
^{22}Ne+CATP	6000	6.7	541	1.8	17.3	0.11

seen from the data in Table 1, the effective size of the target, in this case the tryptophan marker, falls as the LET is raised, i.e. the probability of biological inactivation of DNA decreases with increasing LET.

A somewhat different relation was obtained for inactivation cross section versus LET. Inactivation cross section rose parallel to LET up to a certain limiting value beyond which values decreased (Table 1). For instance with ^{22}N ion exposures inactivation cross sections are below those with exposures to ^{12}C ions, even though the latter have a smaller LET than ^{22}Ne ions. Table 1 also shows, however, that inactivation cross section in the transformation test changed in direct proportion to energy per nucleon.

These findings lead us to conclude that the inactivation cross section relating to reduction of functional ability of DNA depends, as observed for other biological tests [8], not only on the LET of particles but also on their energy per nucleon. With LET rising from 1200 to 6000 MeV g^{-1} cm^2, G values for biological inactivation of DNA were reduced from 0.2 to 0.1. A comparison of these figures with those cited in the literature [9] indicated that radiation-induced decline in DNA transforming ability is most probably to be traced back to breaks in the polynucleotide chain of the molecule. As for protection experiments, it appears from the determined end-points that an S-containing protector of high dose reduction factor for sparsely ionizing radiation, such as cysteamine hydrochloride or cytriphos, will not modify damage by charged particles in the effect studied here. Failure in the present experiments to observe effectiveness of compounds that are established protective drugs with respect to other biological models could be accounted for in terms of the mechanisms of their action. It is known that sulphur-containing protectors exert their action primarily by radical scavenging, disulphide-bound formation or migration of energy from target to protector. Since in our case irradiation was preceded by drying of solutions containing DNA and protector, it may be presumed that this form of the materials prevented the occurrence of these processes underlying the anti-radiation effect.

References

[1] Yu. Grigor'ev, in: Proc. 4th Symp. on Cosmic Biology and Medicine, Prague 1971.
[2] Yu. Grigor'ev, in: Materials Symp. on Cosmic Biology and Medicine, Budapest 1970 (p. 190).
[3] N. I. Ryzhov et al., in: Materials Symp. on Cosmic Biology and Medicine Berlin 1973.
[4] T. Alper, J. Moore and D. Bewly, Radiat. Res. **32**, 277 (1967).
[5] T. Brustad, Radiat. Res. suppl., **2**, 65 (1960).
[6] C. Tobias and P. Todd, in: Semiannual Reports Biology and Medicine, Donner Laboratory and Donner Pavilion, Berkeley, 1964 (p. 25).
[7] M. Minkova, C.R. Acad. bulgare Sci. **26**, 125 (1973).
[8] E. A. Krassavin, Dissertation, Moscow 1971.
[9] H. Dertinger and H. Jung, Molecular Radiation Biology, Atomizdat, Moscow 1973.

Life Sciences and Space Research XIV — Akademie-Verlag, Berlin 1976

STUDY OF BASIC ELECTROSTATIC RADIATION SHIELD CHARACTERISTICS ON BOARD THE COSMOS 605 SATELLITE

E. E. KOVALEV and T. YA. RIABOVA

Institute of Medico-Biological Problems, Moscow, USSR

Direct measurement of conduction currents in the electrostatic radiation shield model carried out on board the Cosmos 605 satellite yielded additional information on the dose, weight and size characteristics and helped to evaluate the power requirements for a real electrostatic radiation shield. It was found that the electrostatic shield power requirements (at an electric field strength of 10^7 V m^{-1} and total shielded surface of 10^2 m^2) did not exceed 10 watts, i.e. it was at least one order of magnitude lower than had been assumed from data obtained in ground-based studies.

These results show the feasibility of protective electric fields of the required strength, with the space vacuum around the vehicle being used as an insulating medium.

Studies of basic characteristics of electrostatic shielding were carried out aboard the Cosmos 605 satellite in order to investigate the possibility of utilizing electric fields to protect space vehicles against charged particles in space. The principal objective of the experiment was to find out whether it was possible to create and maintain around the space vehicle protective electric fields with strengths of up to 10^7 V m^{-1} using the space vacuum as the insulating medium. Considerable difficulties were expected from possible adverse effects of magnetospheric plasma ions, space vehicle surface outgassing, etc. upon the efficiency of the natural vacuum insulation. To answer this question data had to be obtained on the vacuum gap conductivity with electric fields of $4 \times 10^6 - 7 \times 10^7$ V m^{-1} inside the electrostatic shield model.

The electrostatic radiation shield model used in the studies included an electrode system and a high-voltage generator placed outside the satellite, as well as current meters, microdischarge counters, a time-programming device, etc. The model operation and electrode movements were controlled from the ground. The high-voltage electrode system (using space vacuum as the insulating medium) was shielded from the solar ultraviolet radiation, either direct, scattered or reflected from the earth's atmosphere.

In the experiment aboard Cosmos 605 the following major results were obtained: The constant component of the conduction current in the high-voltage vacuum gap did not exceed 10^{-7} amp m^{-2} even at 1.5×10^7 V m^{-1} electric field strength. The pulse component of the conduction current was manifested only at the maximum voltage, represented by spontaneous solitary pulses of dark current of 3—5 pulses min^{-1} and $0.5 - 1 \times 10^{-7}$ amp m^{-2} mean current density. The mean microdischarge current density did not exceed 10^{-7} amp m^{-2}.

Comparison of the Cosmos 605 experimental data with those obtained in ground-based studies has shown that no increase in the conduction current in the electrostatic radiation shield model was produced by an unfavourable — as far as the electric strength is concerned — space environment at altitudes of 221—424 km. These results apparently show that at the above altitudes the space vacuum can be used as an efficient insulation despite the presence of

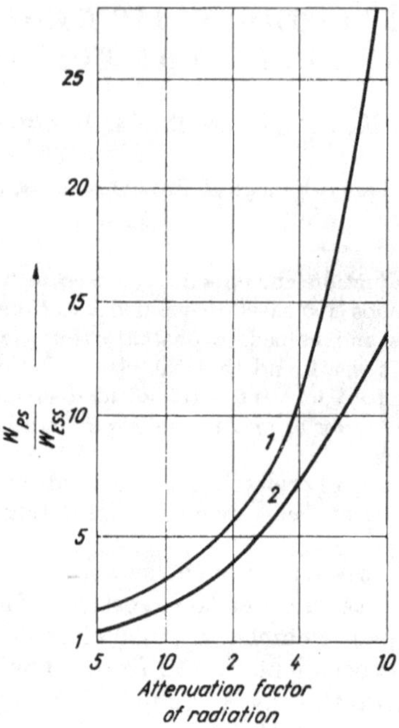

Fig. 1. Ratio between the weights of the passive W_{PS} and electrostatic W_{ESS} radiation shields as a function of the radiation attenuation factor. Curve 1, data from the Cosmos 605 experiment; curve 2, ground-based data.

charged and neutral particles and outgassing from vehicle surfaces. Direct measurements of conduction currents allowed additional information to be acquired on the dose, weight and size characteristics and evaluation of the power requirements for a real electrostatic radiation shield [1]. It was found that the electrostatic shield power requirements (at the electric field strength of 10^7 V m^{-1}, operating voltage of 10^5 V and total shielded surface of 10^2 m^2) did not exceed 10 W, i.e. was at least one order of magnitude lower than had been assumed according to data obtained in ground-based studies [2].

Fig. 1 illustrates the dependence of the ratio between the weight of the passive shield and that of the electrostatic shield on the radiation attenuation factor based on data from Cosmos 605 [1]. Calculations from ground-based data [2] are given for comparison. It can be seen that at high attenuation factors the calculated weight ratio between passive and electrostatic radiation shields was underrated at least by one half.

The results this experiment have caused reconsideration of previous views on the expected difficulties associated with the creation of protective electric fields and utilization of environmental vacuum as an insulating medium. An efficient electrostatic shielding against cosmic radiation, with minimum weight requirements, can be developed by present day high-voltage technology.

References

[1] K. A. TRUKHANOV et al., An Active Protection of Space Vehicles, Atomizdat, Moscow 1970.
[2] E. E. KOVALEV et al., XXIV Int. Astronaut. Congr. Abstracts, Baku, October 1973 (pp. 234—237).

Gravitational Biology

Life Sciences and Space Research XIV — Akademie-Verlag, Berlin 1976

AMINO ACID SPECTRUM OF HUMAN BLOOD PLASMA DURING SPACE FLIGHT AND IN ANTIORTHOSTATIC HYPOKINESIA

A. S. Ushakov and T. F. Vlasova

Institute of Medico-Biological Problems, Moscow, USSR

The paper summarizes results of experiments on the influence of space flight and antiorthostatic hypokinesia on the amino acids spectrum of human blood plasma. Our findings give evidence for: (i) A specific norm of the content of free amino acids in plasma during training and (ii) consistent changes of plasma aminograms during space flight related to its duration and to individual features of cosmonauts. The content of free amino acids in the plasma of bed-rested subjects varied phasically and tended to increase practically at every experimental stage. The paper discusses the findings and possible mechanisms of the detected changes.

Space flight effects on the human body include consistent changes in nitrogen metabolism. Various changes in protein metabolism previously found during hypokinetic syndrome [1, 2, 5, 7, 12, 13] are indicative of consistent changes which involve a decline in the synthesis of protein and in utilization of nutritive nitrogen, a change in the activity of enzymes involved in protein metabolism, a change in the protein content in muscles and their functional activity, development of negative nitrogen balance, etc. Nitrogen metabolism in the body can be assessed by amino acid metabolism. The most informative study is the measurement of the free pool of amino acids in the plasma that are metabolically related to tissue amino acids. The published literature contain few reports on the amino acid metabolism during diminished motor activity [3, 6, 8, 10, 11] and no information on this factor in space flight. Such information could contribute to the proper understanding of the mechanisms that obtain under these conditions.

We carried out an investigation of the content of free amino acids in the plasma of cosmonauts both during the pre-flight training and post-flight recovery as well as of test subjects during antiorthostatic hypokinesia. The amino acid composition of the plasma was investigated by ion-exchange chromatography in an automatic analyser. The results obtained during examinations of cosmonauts pre-flight were summarized and the average values of the content of free amino acids in their plasma were determined (Table 1). This was necessary because the data so far obtained were mainly derived from clinical observations, and this includes large scatter of the amino acid concentrations which lie between wide physiological limits. These data usually ignore the subject's state, age, diet, occupation and other factors, which makes their objective assessment difficult.

An analysis of the amino acid content in the plasma of the Soyuz 12, 13 and 16 cosmonauts has shown that the concentrations of free amino acids undergo

Table 1

Concentration of Free Amino Acids in Blood Plasma of Adult Subjects (mg%, mean and s.d., physiological normal)
Authors reporting results and number (n) of subjects. A blank indicates an undetectable level

Amino acid	Jagat 1968	Weller 1970	Swendseid 1966	Felig 1969	Adibi 1968	Semenov 1971	Our data	Normal of special subjects
	$n = 10$	$n = 20$	$n = 6$	$n = 5$	$n = 25$		$n = 80$	$n = 30$
Isoleucine	0.97	1.60 ± 0.16	0.77 ± 0.06	0.73 ± 0.07	0.85 ± 0.13	0.69 — 1.28	1.05 ± 0.05	0.77 ± 0.13
Leucine	2.00	2.81 ± 0.26	1.38 ± 0.09	1.43 ± 0.17	1.75 ± 0.21	1.42 — 2.30	2.03 ± 0.08	1.48 ± 0.10
Lysine	4.10	3.39 ± 0.44	1.54 ± 0.09	2.63 ± 0.12	3.06 ± 0.36	2.52 — 3.02	3.22 ± 0.15	2.24 ± 0.13
Valine	2.67	4.30 ± 0.47	2.13 ± 0.09	2.40 ± 0.24	2.71 ± 0.39	2.37 — 3.71	2.54 ± 0.06	2.33 ± 0.10
Threonine	1.87	2.41 ± 0.24	1.43 ± 0.14	1.72 ± 0.13	2.08 ± 0.25	1.21 — 1.72	1.91 ± 0.09	2.56 ± 0.13
Tryptophan	0.37	—	—	0.83 ± 0.04	0.82 ± 0.09	1.11	1.53 ± 0.04	0.95 ± 0.08
Methionine	0.38	0.48 ± 0.07	0.39 ± 0.02	0.35 ± 0.02	0.42 ± 0.05	0.33 — 0.43	0.54 ± 0.04	0.26 ± 0.01
Phenylalanine	1.32	1.39 ± 0.12	0.84 ± 0.06	0.78 ± 0.03	1.06 ± 0.10	0.69 — 0.95	1.13 ± 0.06	0.73 ± 0.04
Tyrosine	1.09	1.29 ± 0.12	1.02 ± 0.10	—	1.16 ± 0.09	0.82 — 1.45	1.04 ± 0.06	0.80 ± 0.05
Cystine	0.80	—	2.60 ± 0.21	2.45 ± 0.05	2.21 ± 0.13	1.08 — 1.30	0.91 ± 0.10	0.73 ± 0.04
Aspartic acid	0.16	0.44 ± 0.08	—	—	0.17 ± 0.01	0.01 — 0.07	0.07 ± 0.05	0.11 ± 0.01
Glutamic acid	1.47	2.55 ± 0.72	1.27 ± 0.37	—	1.21 ± 0.11	0.43 — 1.15	3.97 ± 0.26	2.01 ± 0.10
Serine	—	1.92 ± 0.18	1.28 ± 0.13	1.35 ± 0.09	1.60 ± 0.27	1.01 — 1.25	2.02 ± 0.09	1.34 ± 0.07
Proline	2.00	3.73 ± 0.80	1.80 ± 0.19	2.20 ± 0.03	2.29 ± 0.29	2.01 — 3.34	2.76 ± 0.08	2.28 ± 0.07
Glycine	2.14	1.34 ± 0.28	2.62 ± 0.28	1.98 ± 0.43	1.85 ± 0.32	1.34 — 1.73	2.32 ± 0.09	1.32 ± 0.09
Alanine	2.98	4.00 ± 0.74	2.98 ± 1.04	2.61 ± 0.29	3.23 ± 0.90	3.01 — 3.73	3.80 ± 0.14	2.44 ± 0.10
Histidine	1.58	1.55 ± 0.10	—	1.18 ± 0.19	1.84 ± 0.19	0.79 — 1.48	1.64 ± 0.08	0.88 ± 0.06
Arginine	1.49	1.55 ± 0.12	—	1.17 ± 0.10	1.78 ± 0.12	1.22 — 1.93	1.07 ± 0.08	1.04 ± 0.08

consistent changes (Table 2). A short-term flight of two days (Soyuz 12, Makarov, Lazarev) produced no significant changes in the amino acid content as compared with the physiological normals we have established, except for lowered glutamic acid. Longer-term flights of six and eight days (Soyuz 16, Rukavishnikov, Filipchenko; Soyuz 13, Lebedev, Klimuk) caused a marked decrease in the concentration of leucine, valine, tyrosine, glutamic acid, serine and proline and sometimes cystine.

A study of the amino acid content in the plasma of the cosmonauts who performed a 30-day flight supported this regularity. These cosmonauts showed a noticeable decline in the content of free amino acids, particularly of methionine and cystine, and a radical decrease of arginine. The amino acid composition in the plasma of the cosmonauts a week post-flight did not differ from the pre-flight values. This gives evidence that the changes in the free amino acid content induced by space flight effects were reversible.

It has been recently recognized [1, 4, 5, 12, 13] that hypokinesia and antiorthostatic hypokinesia in particular is an adequate model of weightlessness, simulating its effects on the human body. Therefore, it is of interest to compare the data obtained during bed rest and real space flight.

The results of a 49-day hypokinetic experiment (at an angle of inclination of —4°) in which eight test subjects took part differed from the data obtained in space flight. The hypokinetic exposure increased the concentration of most free amino acids in the plasma (Table 3). The content of proline, glycine and aspartic acid remained unaltered throughout the experiment. The level of all the other amino acids varied in a different way: (i) the concentration of the following amino acids increased during the experiment: isoleucine ($P < 0.02$), leucine ($P < 0.05$), phenylalanine ($P < 0.02$), valine ($P < 0.001$), methionine ($P < 0.05$), serine ($P < 0.001$), glutamic acid ($P < 0.001$), alanine ($P < 0.01$), tyrosine ($P < 0.05$); (ii) the concentration of the following amino acids increased during the first hypokinetic days and returned to the pre-test level by the 30th day: histidine, arginine, cystine; (iii) the concentration of the only acid-lysine increased at the onset of the experiment ($P < 0.02$) and decreased below the pre-test level beginning with the 30th day. The findings on the increase in the concentration of free amino acids during hypokinetic exposures are supported by other authors [6, 8, 3].

The variations in the content of free amino acids of the test subjects exposed to antiorthostatic hypokinesia seem to be induced by a reduction of the rate of protein synthesis and predominance of catabolic processes in the metabolism. Of particular interest are the data on alanine and glutamic acid whose concentrations increased during the experiment and continued to increase after it. This increase may be associated with an enhanced entry of these amino acids from the muscles and lymph tissues due to a decline in the rate of protein synthesis as well as with the alteration of amino acid metabolism to a new level. The above changes in the concentrations of free amino acids in the plasma were reversible and did not differ from the pre-test values two months after the experiment.

The differences in the content of free amino acids in the plasma during real space flight and hypokinetic exposure can be attributed to the different environments. During the hypokinetic experiment the major factor is diminished motor activity and concomitant changes in the musculo-skeletal, cardiovascular and metabolic systems. In space flight the major factor is weightlessness combined with acceleration, emotional stress, artificial atmosphere, etc. This difference determines the pattern and selectivity of the metabolic changes.

Table 2

Concentration of Free Amino Acids in Blood Plasma of Soviet Cosmonauts (mg %)

Amino acid	Normal of test subjects $n = 30$	"Soyuz-12" (2 days)		"Soyuz-13" (6 days)		"Soyuz-13" (8 days)		"Soyuz-17" (30 days)	
		Makarov	Lazarev	Rukavishnikov	Filipchenko	Lebedev	Klimuk	Grechko	Gubarev
Isoleucine	0.77 ± 0.13	0.68	0.64	0.60	0.50	0.49	0.53	0.39	0.29
Leucine	1.48 ± 0.10	1.41	1.39	1.00	1.09	0.89	0.92	0.70	0.55
Lysine	2.24 ± 0.13	2.57	2.44	1.75	1.83	—	—	2.24	1.88
Valine	2.33 ± 0.10	3.00	3.06	1.33	1.59	1.27	1.36	0.94	0.83
Threonine	2.56 ± 0.13	5.07 + serine	4.69 + serine	3.07	3.20	2.21	2.48	2.36 + serine	2.91 + serine
Tryptophan	0.95 ± 0.08	—	—	1.00	1.40	—	—	0.74	0.60
Methionine	0.26 ± 0.01	0.30	0.28	0.16	0.12	0.22	0.22	0.08	0.07
Phenylalanine	0.73 ± 0.04	0.70	0.69	0.52	0.72	0.49	0.41	0.53	0.52
Tyrosine	0.80 ± 0.05	0.79	0.84	0.56	0.73	0.51	0.43	0.49	0.49
Cystine	0.73 ± 0.04	0.55	0.43	0.63	0.68	0.29	0.29	0.17	0.16
Aspartic acid	0.11 ± 0.01	0.08	0.06	0.06	0.05	0.11	0.05	0.06	0.04
Glutamic acid	2.01 ± 0.10	1.01	1.00	1.78	1.67	1.54	1.76	0.57	0.75
Serine	1.34 ± 0.07	+ threonine	+ threonine	1.01	1.12	0.67	0.60	+ threonine	+ threonine
Proline	2.28 ± 0.07	2.48	2.12	1.33	1.03	1.28	0.73	1.18	1.45
Glycine	1.32 ± 0.09	1.26	1.23	0.74	0.86	0.87	1.05	0.76	0.65
Alanine	2.44 ± 0.10	3.02	2.97	1.49	2.33	1.43	1.99	1.26	1.55
Histidine	0.88 ± 0.06	1.22	1.18	1.02	0.97	—	—	1.03	0.98
Arginine	1.04 ± 0.08	1.33	1.33	1.21	1.14	—	—	trace	trace

Table 3

Concentration of Free Amino Acids in Blood Plasma of Subjects (mg %, mean and s.d. *m* antiorthostatic hypokinesia)

P values indicate significance of the difference between the experimental and the reference periods

Amino acid	Reference period 5th day	Experimental period									
		12th day		21st day		30th day		39th day		48th day	
	$M \pm m$	$M \pm m$	P	$M \pm m$	P	$M \pm m$	P	$M \pm m$	P	$M \pm m$	P
Isoleucine	0.86 ± 0.08	1.02 ± 0.04	> 0.1	1.14 ± 0.06	< 0.01	1.19 ± 0.08	< 0.05	1.24 ± 0.07	< 0.01	1.13 ± 0.05	< 0.02
Leucine	1.83 ± 0.13	2.26 ± 0.11	< 0.01	2.39 ± 0.12	< 0.02	2.63 ± 0.11	< 0.02	2.56 ± 0.14	< 0.01	2.56 ± 0.27	< 0.05
Lysine	2.96 ± 0.09	3.64 ± 0.08	< 0.001	2.72 ± 0.12	> 0.1	2.41 ± 0.12	< 0.01	2.26 ± 0.20	< 0.05	2.29 ± 0.18	< 0.01
Valine	2.55 ± 0.07	3.37 ± 0.21	< 0.01	3.70 ± 0.09	< 0.001	3.59 ± 0.18	< 0.001	3.46 ± 0.16	< 0.001	3.27 ± 0.10	< 0.001
Tryptophan	1.23 ± 0.06	1.57 ± 0.14	> 0.1	1.52 ± 0.10	> 0.05	1.60 ± 0.12	< 0.05	0.90 ± 0.12	< 0.05	1.49 ± 0.22	> 0.1
Methionine	0.40 ± 0.05	0.49 ± 0.03	> 0.1	0.52 ± 0.02	< 0.05	0.54 ± 0.02	< 0.05	0.61 ± 0.03	< 0.01	0.59 ± 0.02	< 0.05
Phenylalanine	1.09 ± 0.11	1.26 ± 0.08	> 0.1	1.39 ± 0.05	> 0.05	1.20 ± 0.14	> 0.1	1.57 ± 0.08	< 0.01	1.65 ± 0.13	< 0.02
Tyrosine	0.95 ± 0.09	1.27 ± 0.04	< 0.05	1.31 ± 0.06	< 0.02	1.40 ± 0.12	< 0.05	1.34 ± 0.12	< 0.05	1.38 ± 0.13	< 0.05
Cystine	0.45 ± 0.05	1.19 ± 0.13	< 0.001	0.58 ± 0.02	> 0.05	0.62 ± 0.03	> 0.1	0.48 ± 0.05	> 0.1	0.56 ± 0.09	> 0.1
Aspartic acid	0.17 ± 0.02	0.13 ± 0.01	> 0.1	0.23 ± 0.02	> 0.05	0.25 ± 0.03	> 0.05	0.22 ± 0.05	> 0.05	0.24 ± 0.09	> 0.1
Glutamic acid	2.35 ± 0.12	2.72 ± 0.18	> 0.1	4.46 ± 0.12	< 0.001	4.73 ± 0.12	< 0.001	3.87 ± 0.24	< 0.001	3.43 ± 0.43	< 0.05
Serine	1.52 ± 0.08	2.43 ± 0.08	< 0.001	2.75 ± 0.14	< 0.001	2.26 ± 0.10	< 0.05	3.40 ± 0.19	< 0.001	3.40 ± 0.23	< 0.01
Proline	2.66 ± 0.11	2.86 ± 0.21	> 0.1	2.46 ± 0.12	> 0.1	2.62 ± 0.09	> 0.1	2.76 ± 0.07	> 0.1	2.68 ± 0.09	> 0.1
Glycine	2.32 ± 0.12	2.08 ± 0.26	> 0.1	2.34 ± 0.14	> 0.1	2.60 ± 0.14	> 0.05	2.38 ± 0.10	> 0.1	2.60 ± 0.17	> 0.05
Alanine	3.49 ± 0.09	3.75 ± 0.18	> 0.1	3.74 ± 0.12	> 0.1	3.15 ± 0.16	> 0.1	4.78 ± 0.21	< 0.001	4.20 ± 0.19	< 0.01
Histidine	1.23 ± 0.07	1.88 ± 0.11	< 0.002	1.72 ± 0.05	< 0.001	1.38 ± 0.14	> 0.1	1.46 ± 0.17	> 0.1	1.27 ± 0.05	> 0.1
Arginine	1.75 ± 0.12	2.11 ± 0.13	> 0.05	2.20 ± 0.09	< 0.02	1.77 ± 0.11	> 0.1	1.34 ± 0.19	> 0.1	1.54 ± 0.04	> 0.1

References

[1] Y. S. BALAKHOVSKY and Y. V. NATOCHIN, in: Problems of Space Biology, Moscow 1973 (pp. 22, 123).
[2] E. I. VOROBYOV, A. D. EGOROV, L. I. KAKURIN et al., Kosm. Biolog. i Med. 6, 26 (1970).
[3] T. F. VLASOVA and S. K. SHISHKINA, Proc. Conf. of Young Scientists, Leningrad 1971 (p. 81).
[4] A. M. GENIN and L. I. KAKURIN, Kosm. Biolog. i Med. 4, 26 (1972).
[5] L. I. KAKURIN, Doctorate dissertation, Moscow 1968.
[6] N. S. PETROV, in: Proc. Dept. of Phys. Culture and Sports **30**, 128 (1962).
[7] V. V. PARIN, T. N. KRUPINA, G. P. MIKHAILOVSKY et al., Kosm. Biolog. i Med. **5**, 59 (1970).
[8] N. P. RASSOLOVA, V. A. ZAKHARCHENKO and E. M. KOGAN, Space Biology and Aerospace Med. **1**, 93 (1972).
[9] I. V. FEDOROV, V. I. VINOGRADOV, Y. I. MILOV et al., Space Biology and Med. **1**, 53 (1967).
[10] I. V. FEDOROV, Space Biology and Med. **5**, 35 (1973).
[11] I. V. FEDOROV, Space Biology and Aerospace Med. **1**, 3 (1975).
[12] C. A. BERRY, Weightlessness Bioastronautics Data Book, NASA, Washington 1973 (p. 349).
[13] G. D. WHEDON, L. LUTWAK, P. C. RAMBAUT et al., Skylab 1/2 Prelim. Biomed. Rep., NASA 1973 (p. 162).

Life Sciences and Space Research XIV — Akademie-Verlag, Berlin 1976

SPACE FLIGHT EFFECT UPON THE BIOENERGETICS OF THE SKELETAL MUSCLES IN RATS

E. S. MAILYAN and E. A. KOVALENKO

Institute for Medico-Biological Problems, Moscow, USSR

Studies of tissue bioenergetics are very important to gain insight into the effects of weightlessness on the animal and human body. A reduction of the muscular load in zero gravity should affect reactions of biological oxidation because their pattern and level are related to energy requirements of the body. After the 22-day flight of the Cosmos 605 biosatellite tissue respiration and oxidative phosphorylation (P/O coefficient) in skeletal muscles of rats was investigated. The investigations showed that on the 2nd post-flight day oxidation assimilation by the muscle tissue was reduced compared with the control. In addition, the intensity of phosphorylation which determines the rate of ATP formation was decreased. These changes may reduce significantly the energy potential of the muscle cell and seem to form the basis for the functional inadequacy of skeletal muscles which can be found in men during the first post-flight days.

At later stages (the 26th post-flight day) oxygen consumption and phosphorylation rate increased and reached the control level. Therefore, readaptation to the earth gravity is followed by adaptive changes in the processes responsible for the accumulation and utilization of the energy of biological oxidation.

1. Introduction

There are theoretical reasons for believing that energy expenditure involved in locomotion and work must decrease significantly in weightlessness [15]. Nevertheless, specific studies on humans and animals carried out under weightlessness have yielded controversial results. It has been shown that short-term weightlessness during real and simulated aerospace flights results in increase of the oxygen uptake and energy expenditure [11—13]. It is believed that the metabolic activity of man under weightlessness may be greater than that at $1g$ [8]. It is also known that with an increase of exposure to weightlessness the metabolic activity tends to decrease: this follows from the comparison of the Mercury and Gemini 5 data [2—5]. Several authors support the idea that under prolonged weightlessness energy expenditures are likely to reduce [9, 10]. Apollo 14 crew members during locomotion and work on the surface of the moon showed decreased levels of metabolic activity [6, 7]. Finally, clear-cut reduction of gas exchange and metabolic activity was demonstrated during bed-rest experiments and on a decreased gravity simulator [1, 14, 18, 21].

In summing up the above-mentioned literature it should be noted that objective evaluation and differentiation of the effects of weightlessness require certain corrections due to the following factors. It is known that weightlessness effects are accompanied, at least during the first flight days, by emotional stress, disturbance of motor coordination and the influence of inertial moments of movement during complicated work. Apparently, all these factors may primarily result in a significant increase of gas exchange and metabolic activity. Hence, the effects of weightlessness itself may become manifest in the later flight stages, after primary reactions of the body accompanied by high energy expenditure have diminished.

There is another important factor. In flight cosmonauts perform, as a rule, physical exercises which may counterbalance and conceal the effects of weightlessness. It can therefore be postulated that animal studies during prolonged space flights are the only possible model which can clearly demonstrate the role of zero gravity upon metabolic activity.

This paper presents the results of studying tissue respiration and oxidative phosphorylation (P/O ratio) in the skeletal muscles of rats flown in the Cosmos 605 biosatellite for 22 days.

2. Methods and Materials

Examinations were performed on the 2nd and 26th post-flight days. Parameters of tissue respiration were determined in three experimental series: (i) flight experiment, rats flown for 22 days; (ii) synchronous ground experiment, animals kept in a mock-up of the space vehicle for 22 days; (iii) control experiment, animals selected from the vivarium population kept on a different food ration.

Homogenates and suspension of mitochondria isolated from the posterior femoral muscles were used. Tissue respiration was measured in the Warburg apparatus: oxygen uptake (ΔO), assimilation of inorganic phosphates (ΔP) and phosphorylation ratio (P/O). Tissue preparations were incubated for 15 min at 26 °C and 0.02 M succinic acid was used as the substrate. Inorganic phosphates were assayed by the method of Lowry and Lopes modified by Skulachev [20]. Protein in the mitochondrial suspension was measured according to Lowry et al. [16].

3. Results

Experimental results of examinations of homogenates of skeletal muscles are given in Table 1 and in Fig. 1. They show parameters of oxidative phosphorylation obtained on the 2nd and 26th days after termination of flight and synchronous experiments and in the control group. On the basis of these data it can be concluded that both the respiratory and phosphorylative activities (ΔO and ΔP) in the homogenates decreased significantly (40—50%) on the 2nd post-flight day compared with the synchronous and control experiments. The P/O ratio remains unchanged; hence, the coupling of the processes is not disturbed despite their inhibition. In contrast, on the 26the post-flight day no abnormalities in the oxygen uptake and assimilation of inorganic phosphates of flight rats were noted compared with synchronous and control experiments.

Examinations of mitochondria isolated from the same muscles (Fig. 2) confirmed the above consistency. It should be noted that the P/O ratio obtained in the control group was slightly lower. The respiratory and phosphorylative activities showed a significant decline on the 2nd post-flight day (40%) and no changes on the 26th day compared with both synchronous and control experiments.

Table 1

Rate of Oxidative Phosphorylation in Homogenates of Skeletal Muscles of Rats Post-flight (ΔO and ΔP — μA/100 mg tissue/hour)

Experimental series	ΔO	ΔP	P/O
Control	7.30 ± 0.29	8.07 ± 0.59	1.11 ± 0.22 $n = 18$
Synchronous ground experiment on 2nd day	5.46 ± 0.56 $P_1 < 0.01$	5.71 ± 0.68 $P_1 < 0.05$	1.06 ± 0.12 $n = 7$
In-flight experiment on 2nd day	3.37 ± 0.36 $P_1 < 0.001$ $P_2 < 0.05$	3.34 ± 0.28 $P_1 < 0.01$ $P_2 < 0.05$	1.02 ± 0.10 $n = 4$
Synchronous ground experiment on 26th day	6.11 ± 0.56	8.63 ± 0.91	1.43 ± 0.11 $n = 6$
In-flight experiment on 26th day	7.59 ± 0.31	10.60 ± 0.96	1.40 ± 0.11 $n = 5$

Note: P_1 = significant difference from the control
P_2 = significant difference from the synchronous experiment

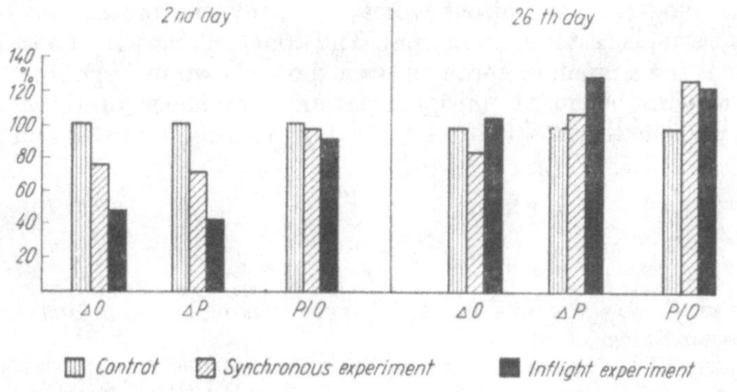

Fig. 1. Post-flight changes in the parameters of oxidative phosphorylation of homogenates of skeletal muscles in rats (as percentage of the controls).

4. Discussion

The post-flight period is characterized by a phasic pattern of changes in the tissue respiration of skeletal muscles. Oxygen uptake and phosphate assimilation in muscles were decreased on the 2nd post-flight day but later recovered. This change in the oxidative processes may greatly decrease the energy potential of

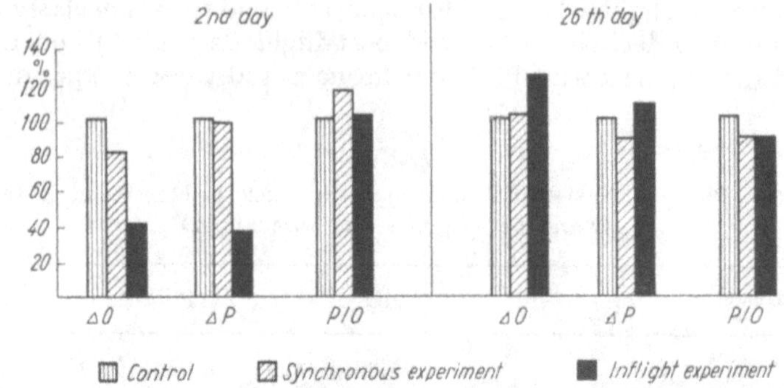

Fig. 2. Post-flight changes in the parameters of oxidative phosphorylation of mitochondria of skeletal muscles of rats (as percentage of the controls).

the cells and is very likely to underlie the functional inadequacy of skeleta muscles which is found in cosmonauts during the first days post-flight [6, 7, 19]. It is difficult to say whether this decreased level of the bioenergetic processes meets the requirements of the animal body in weightlessness, but this level is inadequate on return to the earth. Here the general mechanism of adaptation probably comes into force.

We suppose that the marked decrease of the respiratory and phosphorylative activities in muscles during the first post-flight days is the factor responsible for the reduced synthesis of high-energy compounds and this results in stimulation of metabolic processes (resulting in recovery) on return to $1g$. This signal probably enhances the biogenesis of mitochondria, whereby the production of ATP per unit of mass tissue is restored to normal. The above changes in the tissue respiration of muscles have much in common with those shown in hypokinetic rats [17]. The significant inhibition of phosphorylation, and hence of the formation of high-energy phosphates, is a feature typical of both hypokinetic and flown rats.

References

[1] A. S. BARER, N. K. GNOEVAYA and A. N. SEREBRYAKOV, in: Space Biol. and Aerospace Med., Moscow-Kaluga, **1**, 55 (1972).
[2] C. A. BERRY, XVII Cong. Int. Astronaut. Fed., Madrid 1966 (p. 43).
[3] C. A. BERRY, XVII Cong. Int. Astronaut. Fed., Madrid 1966 (p. 43).
[4] C. A. BERRY, in: Gemini Summary Conf., NASA-SP-138, Manned Spacecraft Center, Houston, Texas, Feb. 1967.
[5] C. A. BERRY, Aerospace Med., **40**, 245 (1969).
[6] C. A. BERRY, XXII Int. Astronaut. Cong., Brussels, Belgium 1971.
[7] C. A. BERRY, 4th Int. Symp. on Basic Environmental Problems of Man in Space, Yerevan 1971 (p. 61).
[8] D. H. CALLOWAY, in: Foundations of Space Biol. and Med. Vol. III, Part II 1975, (Ch. 1).
[9] V. N. CHERNOV, V. I. YAKOVLEV, in: Earth's Artificial Satellites, **1**, 80 (1958).
[10] A. M. GENIN, G. I. VORONIN and A. G. FOMIN, 2nd Int. Man-in-Space Symp., Paris 1965.
[11] P. K. ISAKOV, E. M. YUGANOV and I. I. KASYAN, Izv. Akad. Nauk SSSR ser. biol. **6**, 913 (1964).

[12] I. I. Kasyan and G. F. Makarov, et al., Izv. Akad. Nauk SSSR, ser. biol., **1**, 20 (1969).
[13] I. I. Kasyan, G. F. Makarov and V. I. Sokolkov, Izv. Akad. Nauk SSSR, ser. biol. **5**, 673 (1971).
[14] B. S. Katkovsky, Kosm. Biol. i Med. **5**, 67 (1967).
[15] E. A. Kovalenko, in: Weightlessness (Biomedical Studies), Nauka, Moscow 1974 (p. 237).
[16] O. H. Lowry, R. J. Rosenbronk, A. L. Farr and K. J. Randall, J. Biol. Chem. **193**, 265 (1951).
[17] E. S. Mailyan, L. N. Grinberg and E. A. Kovalenko, in: Adaptation to the Muscular Activity and Hypokinesia, Novosibirsk, 1970 (p. 111).
[18] M. I. Mikhasev, V. I. Sokolkov and M. A. Tikhonov, in: Problems of Space Biology, Nauka, **13**, 71 (1969).
[19] Yu. G. Nefyodov, L. I. Kakurin and A. D. Egorov, in: Space Biology and Aerospace Medicine, Moscow-Kaluga 1972 (p. 87).
[20] V. P. Skulachev, Relationship of Oxidation and Phosphorylation in the Respiratory Chain, Nauka, Moscow 1962.
[21] E. C. Wortz and E. J. Prescott, Aerospace Med. **37**, 1217 (1966).

Life Sciences and Space Research XIV — Akademie-Verlag, Berlin 1976

BODY COMPOSITION CHANGES IN MEN AND WOMEN AFTER 2–3 WEEKS OF BED REST

N. Pace[a], A. M. Kodama[a], D. C. Price[b], B. W. Grunbaum[a], D. F. Rahlmann[a] and B. D. Newsom[c]

[a] White Mountain Research Station, University of California, Berkeley, Calif., USA
[b] Department of Radiology, University of California, San Francisco, Calif., USA
[c] NASA Ames Research Center, Moffett Field, Calif., USA

Several parameters of body composition were measured in eight men before and after 14 days of continuous recumbency, and in eight women before and after 17 days of recumbency. The parameters measured included body weight, body water, body potassium, plasma volume, and plasma protein concentrations. From these, values were derived for body fat content, lean body mass, body cell mass, and circulating plasma proteins. In general, the men and women responded similarly to continuous recumbency. Characteristically, there was significant reduction of plasma volume and body potassium in both groups. The women showed a significant reduction in circulating plasma protein, entirely in the albumin fraction; a similar change was observed in the men. The women, but not the men, showed a significant increase in circulating fibrinogen. Both men and women lost body cell mass, while body fat content remained the same or tended to increase slightly. It is expected that similar changes would occur in weightlessness. It is further concluded that women should tolerate the weightlessness of space flight physiologically as well as men.

1. Introduction

During the past two years, a series of studies on the effects of prolonged recumbency in normal subjects has been carried out at the Human Research Facility of the NASA Ames Research Center. In one of these studies, the effects of 14 days of bed rest were examined in eight men ranging in age from 19 to 26 years, mean age 21 years [1]. In another, the effects of 17 days of bed rest were examined in eight women ranging in age from 23 to 35 years, mean age 29 years [2]. This paper compares the changes observed in several body composition parameters for the men and women of these two studies. The parameters measured were: body weight, body potassium, body water, plasma volume, and plasma protein concentrations.

2. Procedures and Methods

The general conditions under which the two groups were studied were closely similar. The subjects entered the facility, and for the first several days remained ambulatory to permit acquisition of baseline data under standardized conditions before the start of the continuous bed-rest period. During the latter period, eating up on one elbow, excreting, and bathing all were accomplished horizontally. A nutritionally balanced diet was provided throughout the study on an *ad libitum* basis, although the subjects were kept apprised daily of their body weight.

Shortly before the bed-rest period was started, and on the last day of prolonged recumbency, maintaining the horizontal position, the subjects were transported from Ames to Berkeley for the measurements reported here. Body potassium content was obtained by measurement of the gamma radiation emitted by the natural radioisotope ^{40}K in a 2Π whole-body liquid scintillation counter. Body water content and plasma volume were measured as the dilutional volume of known doses of heavy water, D_2O, and the blue dye, T-1824. Plasma was separated from a heparinized blood sample, and total protein concentration was measured colorimetrically by the biuret reaction. A plasma protein electrophoretogram was prepared on cellulose acetate, and the major fractions were determined densitometrically.

Statistical comparison of the parametric values obtained before and after bed rest in each of the two groups were made by the paired t-test. A probability of 0.05, or lower, was used as the criterion that two means were significantly different from each other.

3. Results

The mean and standard deviation for each of the primary parameters measured in the study are presented in Table 1. The bed-rest values which were statistically significantly different from the values measured before the start of bed rest are indicated by an asterisk. It may be seen that body weight did not change significantly in either group as a consequence of bed-rest experience. Body potassium content, on the other hand, exhibited a highly significant decrease in both groups by the end of the bed-rest period. Mean body water content was decreased by about 1 liter in both groups at the end of the bed-rest period, but the difference

Table 1

Mean and Standard Deviation of the Individual Values for Various Parameters measured in 8 Men and 8 Women before and after 2—3 weeks of Continuous Bed Rest

Parameter	Men		Women	
	Before bed rest	Bed-rest Day 14	Before bed rest	Bed-rest Day 17
Body weight (kg)	74.9 ± 9.0	74.7 ± 9.2	57.6 ± 5.6	56.4 ± 6.1
Body potassium (eq)	4.11 ± 0.23	3.91 ± 0.25*	2.50 ± 0.15	2.42 ± 0.16*
Body water (liters)	51.1 ± 4.0	50.2 ± 4.5	31.6 ± 2.5	30.6 ± 3.5
Plasma volume (liters)	3.85 ± 0.62	3.34 ± 0.51*	2.62 ± 0.23	2.29 ± 0.21*
Plasma protein (g/100 ml)	6.48 ± 0.40	7.00 ± 0.49*	6.57 ± 0.44	6.65 ± 0.26
Plasma albumin (g/100 ml)	3.88 ± 0.22	4.08 ± 0.32	3.86 ± 0.23	3.46 ± 0.09*
Plasma globulins (g/100 ml)	2.08 ± 0.26	2.34 ± 0.31*	2.24 ± 0.35	2.48 ± 0.19*
Plasma fibrinogen (g/100 ml)	0.52 ± 0.09	0.57 ± 0.14	0.47 ± 0.05	0.72 ± 0.06*

* $P < 0.05$ that the mean values before and after bed rest are the same.

did not achieve statistical significance. However, an appreciable, and highly significant, decrease in plasma volume was found in both groups, coupled with significant changes in concentration of the plasma proteins.

4. Discussion

Normalized mean values for four of the primary body composition parameters measured in men and women after 2—3 weeks of bed rest are shown in Fig. 1. For each individual in the study, the value measured at the end of the bed-rest

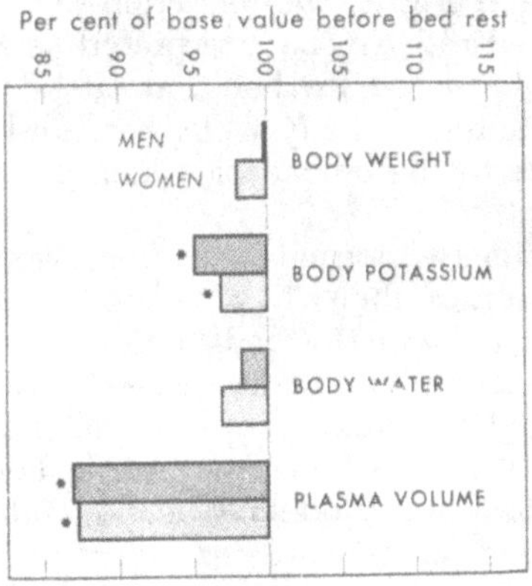

Fig. 1. Mean values of 4 major body components measured in 8 men after 14 days of bed rest and in 8 women after 17 days of bed rest, expressed as percentages of the corresponding values measured before the start of bed rest. The asterisks indicate significant differences at the $P < 0.05$ level.

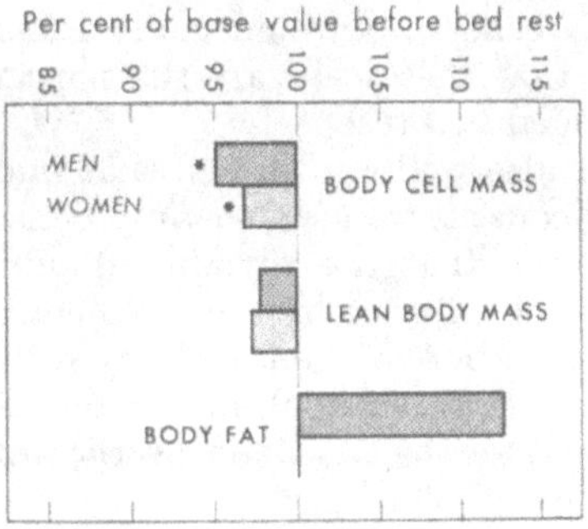

Fig. 2. Derived values for 3 major body components in 8 men and 8 women after 14 and 17 days of bed rest, respectively, expressed as percentages of the values before bed rest. Asterisks—as in Fig. 1.

period was expressed as a percentage of the base value measured for that individual before the bed-rest period started.

As mentioned earlier, a statistically significant change in body weight was not demonstrable in either group. It should be noted that a few of the women subjects made a deliberate effort to lose weight by restricting their caloric intake during the period of the study, with some success. However, the effect on the group mean was not significant.

On the other hand, both the men and the women lost significant quantities of total body potassium, amounting to about 5% for the men and 3% for the women. Also, there was indication of a tendency toward a net loss of body water. While the loss of body water did not attain statistical significance when the men and the women were considered separately, it is of interest that if the data for the two groups are pooled, the decrease in body water becomes significant. It appears, therefore, that a real loss of body water occurs in bed rest.

This conclusion is concordant with the well-documented finding [3, 4], substantiated in the present study, of a consistent loss of 10—15% of the plasma volume of the body in prolonged bed rest, probably through the Gauer-Henry reflex [5]. It is of particular importance to note the close correspondence in relative degree of plasma volume loss between the men and the women, because of the major role postulated for the hypovolemia of bed rest in the development of orthostatic intolerance.

Because more than 95% of the body potassium is contained intracellularly, and because the intracellular potassium concentration remains constant under all but extreme conditions, the body potassium content is an excellent measure of the protoplasmic, or active, cell mass of the body [6]. The body cell mass can be estimated to contain 0.1 equivalent of potassium per kilogram; therefore, simple multiplication of the body potassium content in equivalents by 10 yields the body cell mass in kilograms. Thus, as shown in Fig. 2, the reduction of 3—5% in body potassium content which occurs in bed rest can be interpreted as a loss of 3—5% in the body cell mass. It should be noted further that the skeletal muscle cells account normally for about 80% of the body cell mass, and it is likely that these cells are the most responsive to the unloading which occurs in prolonged bed rest.

It is also well established that under conditions of normal hydration, the body water constitutes a reasonably constant fraction of about 73% of the lean body mass, and that the variable amount of stored fat in the body acts as a gross diluent of water content expressed as a percentage of total body mass [7]. A good approximation of the lean body mass in kg can be obtained simply as the quotient of the body water content in liters divided by 0.73. The fat content of the body in kg can then be obtained as the difference between total body mass and lean body mass. The results of these computations for the present study are also shown in Fig. 2, where it may be seen that neither lean body mass nor body fat content showed a statistically significant change. However, there was a tendency toward a loss of lean body mass in both groups and a tendency toward body fat deposition among the men, who were not so assiduous as some of the women subjects in deliberately restricting caloric intake during the period of bed rest.

Turning now to the summary of the changes observed in normalized plasma protein concentrations, shown in Fig. 3, a number of differences may be noted between the men and the women. A statistically significant increase in total protein concentration occurred among the men, but not among the women. Plasma albumin concentration appeared to change in opposite directions in the two groups, and was significantly decreased among the women. Plasma globulin concentration was significantly increased in both groups; and plasma fibrinogen level was markedly increased among the women, but not among the men.

In the light of the substantial change in plasma volume in both groups, however, it would be more useful to compare the total quantities of circulating plasma protein rather than the plasma protein concentrations. The total quantity of circulating protein can be obtained as the product of the concentration in the plasma and the total volume of plasma. The result of treating the present data in this fashion is shown in Fig. 4, where it may be seen that most of the differences between the men and women in response to bed rest disappear. Furthermore, it is evident that there is a net drop in total circulating plasma protein, which is the result of a drop in circulating plasma albumin while the quantity of circulating globulins remains unchanged. This would appear to point to an active role for plasma albumin as a regulator of the osmotic pressure of the blood, responding to the changes in blood water content which are a consequence of the Gauer-Henry reflex. Search for a possible linking mechanism might prove fruitful.

While the men and women responded in similar fashion with respect to plasma albumin and globulins, it may be seen in Figs. 3 and 4 that both the concentration and the total circulating quantity of plasma fibrinogen were substantially elevated among the women. Because of the clinical implication of a possibly greater tendency toward intravascular clotting with elevated levels of plasma fibrinogen, this finding seems worthy of further investigation, with respect to its significance

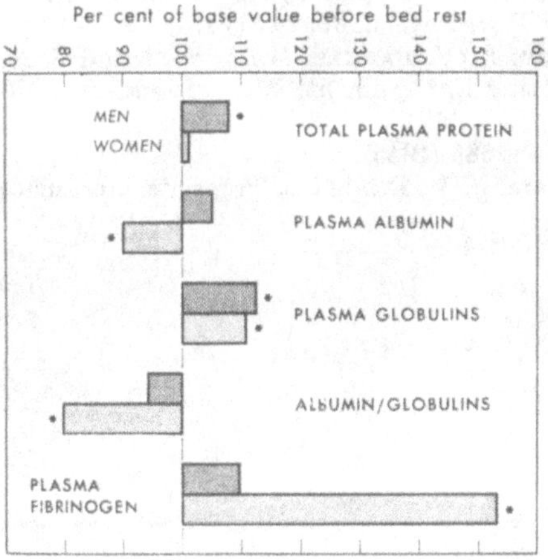

Fig. 3. Mean plasma protein concentrations measured in 8 men and 8 women after 14 and 17 days of bed rest, respectively, expressed as percentages of the values before bed rest. Asterisks—as in Fig. 1.

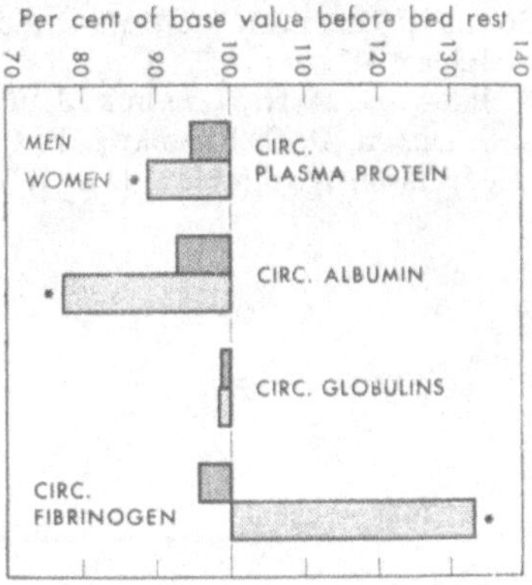

Fig. 4. Mean values for circulating plasma proteins in 8 men and 8 women after 14 and 17 days of bed rest, respectively, expressed as percentages of the values before bed rest. Asterisks—as in Fig. 1.

both for women during prolonged bed rest and for women exposed to prolonged periods of weightlessness.

On the whole, however, in terms of the parameters examined in the present study, it appears that women respond physiologically to prolonged recumbency in much the same manner, both qualitatively and quantitatively, as do men. There is growing evidence [8] that while bed rest does not duplicate weightlessness in all respects, it can serve as a useful model for evaluation of many of the physiological changes to be expected in long-term space flight. On these grounds, it may be predicted that metabolically women should tolerate the weightlessness associated with space flight as well as men, a prediction strengthened by the heroic pioneering space flight of Valentina Tereshkova in Vostok 6 in 1963.

Acknowledgment

The work described in this paper was supported in large part under NASA Grant NGR 05-003-470.

References

[1] N. PACE, B. W. GRUNBAUM, A. M. KODAMA, D. F. RAHLMANN and B. D. NEWSOM, Life Sciences and Space Research XIII, 41 (1975).
[2] N. PACE, B. W. GRUNBAUM, A. M. KODAMA, D. C. PRICE and B. D. NEWSOM, Preprints, Aerospace Med. Assn., Washington, D.C. 1975 (p. 143).
[3] H. L. TAYLOR, L. ERICKSON, A. HENSCHEL and A. KEYS, Am. J. Physiol. **144**, 277 (1945).
[4] W. VAN BEAUMONT, J. E. GREENLEAF and L. JUHOS, J. Appl. Physiol. **33**, 55 (1972).
[5] O. H. GAUER, J. P. HENRY and C. BEHN, Ann. Rev. Physiol. **32**, 547 (1970).
[6] F. D. MOORE, K. H. OLESEN, J. D. MCMURREY, H. V. PARKER, M. R. BALL and C. M. BOYDEN, The Body Cell Mass and its Supporting Environment, W. B. Saunders, Philadelphia 1963.
[7] N. PACE and E. N. RATHBUN, J. Biol. Chem. **158**, 685 (1945).
[8] C. S. LEACH, P. C. RAMBAUT, P. C. JOHNSON and J. V. DANELLIS, Preprints, Aerospace Med. Assn., Washington, D.C. 1975 (p. 147).

Life Sciences and Space Research XIV — Akademie-Verlag, Berlin 1976

ANTIORTHOSTATIC TEST AS A MODEL TO STUDY ANTIGRAVITY MECHANISMS OF THE CARDIOVASCULAR SYSTEM

KH. KH. YARULLIN, T. D. VASILYEVA and D. A. ALEKSEEV

Institute of Biomedical Problems, Moscow, USSR

The paper presents rheographic investigations of regional haemodynamics (brain, lungs, liver and limbs) during antiorthostatic exposures of varying intensity (−15, −30, −45°, times of exposure 20, 40 and 60 min). Our findings show that the pattern and time of the function of compensatory mechanisms preventing excessive vascular complicance under the influence of the hydrostatic blood column depend on the value and length of antiorthostasis, because prolonged venous congestion results not only in congestive circulatory hypoxia but also in arterial hypoxia due to compensatory limitation of arterial inflow.

Almost all cosmonauts experienced rush of blood to the head during insertion into orbit which was similar to the sensation developing in the head-down position. This necessitates the search for and the development of provocative tests which may help to simulate haemodynamic changes occurring in weightlessness and to assess compensatory-adaptive capabilities of the vascular system, particularly of the brain and the lungs. The antiorthostatic test is a provocative test of this type [1−3].

The present communication gives the results of rheographic studies of regional haemodynamics (brain, liver, lungs and limbs) in 18 healthy male test subjects, aged 22 to 38. Each test subject was exposed to eight modifications of antiorthostatic tilt tests which differed in the length and value of antiorthostasis: (1), angle of tilt −15°, time of exposure 20 min; (2) −15°, 40 min; (3) −30°, 20 min; (4) −30°, 40 min; (5) −30°, 60 min; (6) −45°, 20 min; (7) −45°, 40 min; (8) −45°, 60 min; altogether 144 experiments were carried out.

In our investigations we used four-channel transistorized rheographs based on the alternating current demodulator bridge to record simultaneously rheoencephalograms (REG) in the fronto-mastoidal and bimastoidal leads, and rheograms of the right lung or liver, forearm and leg. Rheograms and rheoplethysmograms of the brain and legs were recorded simultaneously from the same electrodes in some test subjects.

The state of blood-filling, elasticity and tone of cerebral, pulmonary and peripheral vessels was estimated on the basis of quantitative analysis of basic rheographic parameters. The relative value of pulse blood-filling of the regions tested was determined using the maximum value of rheographic waves measured as ohm fractions. The ratio $\alpha/T\%$ of the anacrotic phase of the curve α to the cardiac cycle time T was also measured. The outline analysis of rheograms facilitated the determination of the dicrotic index (the ratio of the amplitude at the incision height to the maximal amplitude of the rheogram) indicating mainly the state of the arteriolar tone [4—7]. The diastolic index (the ratio of the amplitude at the dicrotic wave peak to the maximal amplitude of the REG) showing the state of the venous outflow and tone was also determined [7, 8]. The increase in the diastolic pressure (on the average by 14.2 mmHg, $P < 0.05$) and the decrease of the pulse arterial pressure (at an unchanged maximal pressure) in the brachial artery can result from the pressor systemic reaction developing in response to blood redistribution in vessels of the upper body and to increase of blood-filling of the right heart [9].

Variations in the REG parameters, increase of $\alpha/T\%$, maximal amplitude and appearance of venous wave (Fig. 1) gave evidence that with an increase of the

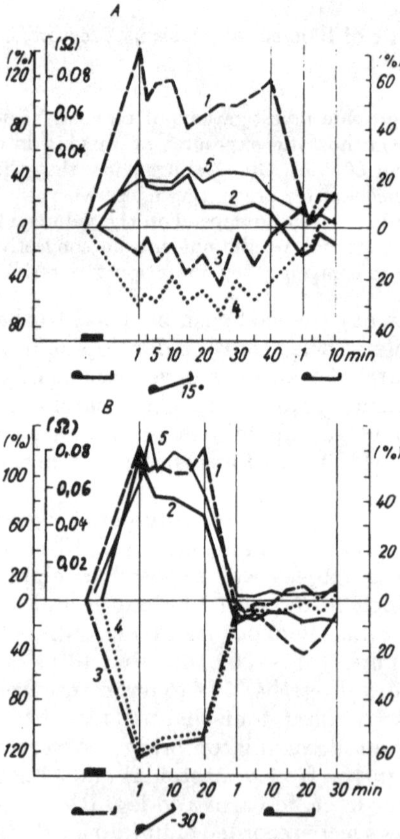

Fig. 1. Variations in the mean values of fronto-mastoidal rheography during a 40 min exposure to $-15°$ (A) and a 20 min exposure to $-30°$ (B).

Abscissa: time: (A), 1 to 40, (B), 1 to 20 min of the exposure, (A), 1 to 10, (B), 1 to 30 min of the exposure, (A), 1 to 10, (B), 1 to 30 min of the after effect. Ordinates: left outer, percentage of the initial level (maximum amplitude of the REG wave); left inner, ohm (amplitude of the venous wave); right, percentage of the initial level ($\alpha/T\%$, dicrotic and diastolic indexes). 1, $\alpha/T\%$; 2, maximum amplitude of REG wave; 3, dicrotic index; 4, diastolic index; 5, amplitude of the venous wave.

angle of the tilt table the pulse blood-filling of cranial vessels and intracranial venous congestion increased. If during the first minute of the $-15°$ exposure the maximal amplitude of the fronto-mastoidal REG (Fig. 1a) increased on the average from 0.173 ± 0.053 to 0.253 ± 0.028 ohm (by 46.6%, $P < 0.05$), then during the -30 and $-45°$ exposure it grew to 0.320 ± 0.019 and to 0.368 ± 0.028 ohm (by 109 and 131%, respectively, $P < 0.001$). The mean value of the amplitude of the venous wave appearing on the first antiorthostatic minute at $-15°$ was 0.025 ± 0.009 ohm ($P < 0.01$) and at $-30°$ and $-45°$ was 0.064 ± 0.012 and 0.120 ± 0.025 ohm ($P \leqq 0.01$). All this, as follows from the pronounced decline of dicrotic and diastolic indexes, was accompanied by a decrease of the arteriolar and venous tone in most test subjects which was greater for greater values of exposure. If during the first minute of the $-15°$ exposure the mean values of the dicrotic and diastolic indexes were less than the reference data by 11.8 and 30.7% ($P < 0.05$), then during the -30 and $-45°$ exposure they were less by 53—62% ($P < 0.001$). At $-30°$ and $-45°$ rheographic signs of arteriolar and venous hypotension and intracranial congestion were noted in practically all test subjects and at $-15°$ only in half of them (Fig. 1).

As follows from the changes in the right rheoplethysmoencephalograms recorded synchronously with fronto-mastoidal REG from the same electrodes, the total blood-filling of the right hemisphere increased with increase of the inclination angle (Fig. 2). The increase of rheoplethysmography at $-30°$ noticeably exceeded its level at $-15°$ (Fig. 2; II) and reached maximum at $-45°$ (Fig. 2, III).

All the test subjects well tolerated the exposure to $-15°$ and reported only slight rush of blood to the head, whereas during an exposure to $-30°$ and $-45°$ they felt a distinct rush of blood to the head, high temperature and full-headedness, pulsation in the temples and neck, nasal congestion, cold feet and legs. By the 10—15th min these changes were slightly reduced. Objectively reddening of the face and the neck, hyperemic sclera and engorgement of neck veins were noted. At 20 min exposures to -30 and $-45°$ these subjective sensations and objective signs of the head hyperemia increased and decreased alternately, i.e. were of phasic pattern.

During a 20 min exposure to $-30°$ changes in the pulse blood-filling and tone of brain vessels were similar and moderate (Fig. 1b). The increase in the blood-filling and decrease in the tone of the cerebral vessels were produced not only by the hydrostatic effect of a large quantity of blood but also by the enlargement of cerebral vessels due to the excitation of baroreceptors of the carotid sinus and other reflexogenic zones associated with the pressure increase in the carotid artery, dural sinuses, aorta, right heart and pulmonary vessels.

After the first minutes of the -30 and $-45°$ exposure increasing venous congestion, especially the pressure rise in the intracranial venous sinuses due to the involvement of the venous-arterial reflex, resulted in the compensatory reflex limitation of the influx of the arterial blood to the cranial cavity [10] due to the elevation of the tone of brain vessels that decreased significantly at the beginning of the antiorthostatic exposure. In other words, a significant increase of the intracranial pressure was followed by the increased capacity of the walls of cerebral vessels to withstand the increased intravascular hydrostatic pressure. This was suggested by the variations of REG parameters (Fig. 1b): pulse blood-filling (maximal amplitude) was increased during the first minute then decreased progressively, whereas the arteriolar and venous tone (dicrotic and diastolic indexes) decreased significantly at the beginning of the exposure and then

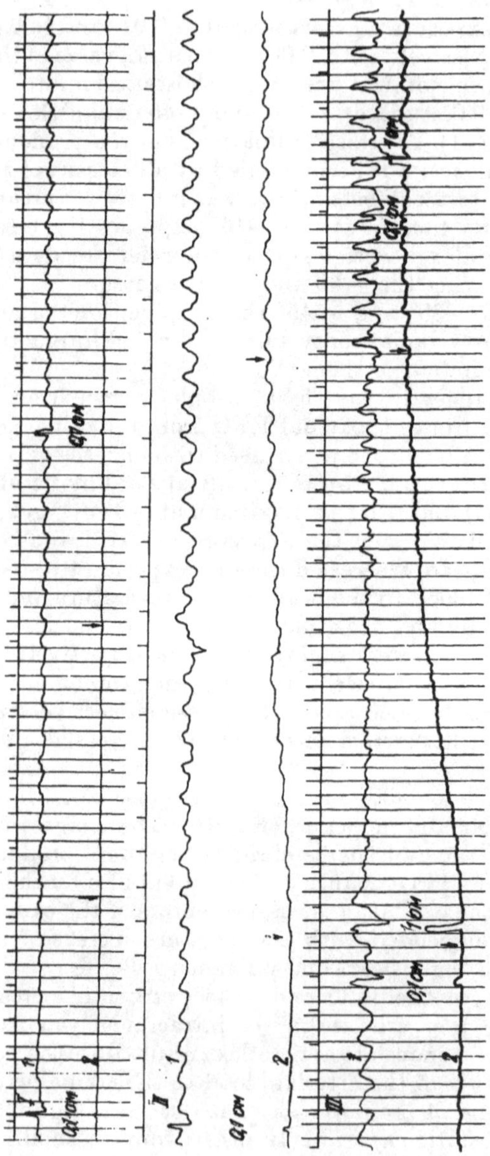

Fig. 2. Changes in fronto-mastoidal rheoencephalography (1) and rheoplethysmoencephalography (2) recorded from the same electrodes during the head tilt down by $-15°$ (I), $-30°$ (II) and $-45°$ (III) in the test subjects c, aged 38 (I, II) and x, aged 22 (III).

Time scale, sec. Arrows indicate the beginning and end of the tilt.

increased slightly. The values of $\alpha/T\%$ and amplitude of the venous wave increased at the beginning of the exposure and decreased beginning with the 2nd–5th min (Fig. 1b). This indicated a decline of the intracranial venous congestion which was pronounced during the first minutes of the exposure. These data show that the REG changes, beginning from the 3rd–5th min of the exposure, demonstrate the adaptation-compensatory tolerance of the cerebral vascular system to the negative gravity effect.

At 40 and 60 min exposures to −45° and −30° venous and arterial hypotension was more pronounced than at a 20 min exposure. During the 20 min exposure to −30° and −45° the dicrotic and diastolic indexes tended to increase from the 3rd–5th minute, i.e. compensatory increase of the arteriolar and venous tone was observed. During the 40 and 60 min tests these compensatory mechanisms overstrained; this was accompanied by a further decrease of the mean values of these indexes induced by congestive cerebral hypoxia which developed during the 15–20 min exposure.

Simultaneous rheoencephalography and electroencephalography suggest that in our 40 and 60 min exposures to −30° and −45° cerebral hypoxia was indicated not only by the decrease of the amplitude of the REG systolic wave but also by the appearance (after 15–20 min) of slow bioelectric activity. Cerebral hypoxia is associated with disturbances in the vasomotor regulation and autoregulation of cerebral circulation, particularly with metabolic changes and acid-base equilibrium [11–13] induced by metabolic acidosis which brings about vasoparesis. The time of the reduction of the increased cerebral blood-filling and enhancement of the vascular tone (period of active compensation) is changed with the phase of vasodilatation and repeated increase of venous congestion. This was typical of our tests at −30° and −45° for 40 and 60 min. The compensatory decrease of the arterial and venous congestion of the brain from the 3rd–5th min of the 20 min exposure to −30° was more uniform than during an exposure to −45°.

At −45° (Fig. 3b) all test subjects showed a marked increase of the pulse blood-filling of the lungs (on the average by 14%, $P < 0.05$) against vascular hypotension (decline of the dicrotic index by 43.3%, $P < 0.05$), whereas at −30° (Fig. 3a) this pulse blood-filling increased to a lesser extent and the venous tone increased in one third of the subjects. At −30° the mean amplitude of the venous wave rose to 0.036 ± 0.008 ohm ($P < 0.05$) and at −45° grew to 0.055 $\pm$ 0.011 ohm ($P < 0.05$).

The increase of the blood-filling in the liver during an exposure to −30° (by 31%, $P < 0.05$) which took place together with a marked increase in the tone of liver vessels of large size (on the average by 36%, $P < 0.002$) and arterioles and veins (on the average by 23%, $P < 0.05$) indicate compensatory blood-filling aimed at the maintenance of normal circulation in the liver itself and at the reduction of the venous inflow. Less intense or no increase of the blood-filling of the liver noted at −45° may be due to redistribution of blood from the liver to the lungs and the head as a result of a significant gradient of hydrostatic pressure. Since these haemodynamic changes at −45° developed against the decrease of the arteriolar and venous tone (by 42 and 25%, respectively, $P < 0.05$) of liver vessels, it can be postulated that this antiorthostasis was too large, it overstrained compensatory mechanisms and broke autoregulation.

As shown above, a 20-min antiorthostatic test at −30° reveals most convincingly (better than at −15° and −45°) the adaptive-compensatory potentialities of the cardiovascular system, especially cerebral and pulmonary vessels, to the antiorthostatic exposure and proves to be an adequate test in studying its antigravity mechanisms. This test carried out before and after flight may help to measure the effectiveness of antigravity mechanisms of the cardiovascular system which provide adequate blood supply of the brain under the conditions of altered gravity.

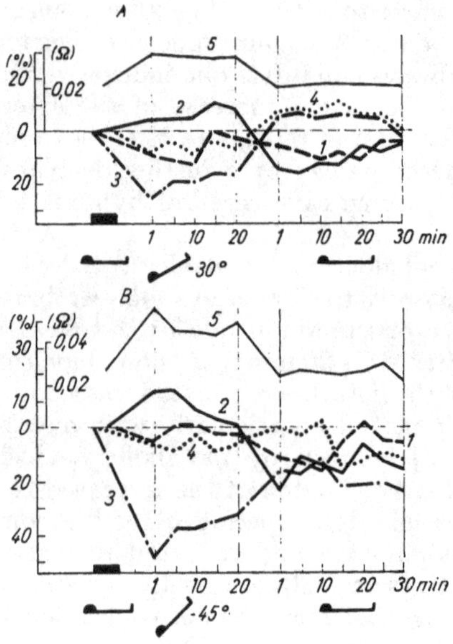

Fig. 3. Changes in mean values of the rheogram of the right lung during a 20 min exposure to −30° (A) and to −45° (B).

Ordinates: outer, percentage of the initial level (maximum amplitude of the REG wave, $\alpha/T\%$, dicrotic and diastolic indexes); inner, ohm (amplitude of the venous wave). See details in Fig. 1.

References

[1] D. Denison, I. Ernsting and D. S. Fryer, J. Physiol. **172**, 2, 49 (1964).
[2] Kh. Kh. Yarullin, Clinical Rheoencephalography, Medizina, Leningrad 1967.
[3] M. A. Abrikosova, Fiziol. Zh. SSSR **54**, 1039 (1968).
[4] E. Donzelot, J. A. Milovanovich and A. Meyer-Heine, Arch. Malad. Coeur. **11**, 1013 (1950).
[5] H. Chlebus, Amer. Heart J. **64**, 22 (1962).
[6] G. J. Eninya, Klin. Med. **45**, 5, 93 (1967).
[7] Kh. Kh. Yarullin, Klinicheskaya Neirofisiologia, Nauka, Leningrad 1972 (p. 545).
[8] A. Meyer-Heine, S. Giblin and G. Desertenne, Arch. Malad. Coeur. Vaiss. **42**, 337 (1949).
[9] G. P. Konradi, Regulation of the Vascular Tone, Nauka, Leningrad 1973.
[10] M. I. Khologenko, Disorders of Venous Circulation in the Brain, Medgiz, Moscow 1963.
[11] N. A. Lassen, Acta Anaesth. Scand. 15 Suppl. **45**, 78 (1971).
[12] D. H. Ingvar, Acta Anaesth. Scand. 15 Suppl. **45**, 110 (1971).
[13] S. Mellander, Angiologica 8, 187 (1971).

Life Sciences and Space Research XIV — Akademie-Verlag, Berlin 1976

METABOLIC PROCESSES IN HYPOKINETIC AND REHABILITATED MEN

V. P Bychkov and M. V. Markaryan

Institute of Biomedical Problems, Moscow, USSR

The present paper summarizes the experimental data on influence of antiorthostatic hypokinesia (49 days) and rationalized nutrition during the rehabilitation period (30 days) on metabolic processes. On the basis of these data a well balanced diet is recommended for the rehabilitation period which follows a long-term bed rest.

In space flight weightlessness is one of the major factors continuously influencing crew members. An experimental model in which hypokinesia is combined with recumbency is often used to simulate weightlessness effects. Studies show that the hypokinetic state gives rise to substantial changes in metabolism [1, 2, 6, 9, 10, 12]. These investigations revealed the main trends in the changes of the internal medium of the human body which can be characterized as a prevalence of catabolic processes over anabolism. Changes in the internal medium of the body which develop under the influence of prolonged hypokinesia form an important problem that badly needs solution. This first requires the development of effective methods for their prevention, and the acceleration of readaptation processes.

The purpose of the present investigation was to study the effect of 49-day antiorthostatic hypokinesia on the protein and lipid metabolism, glycaemic control and assimilation of the main nutrients, and to examine the possibility of correcting metabolic changes arising from hypokinesia by a rational regime of nutrition.

Experiments were carried out on eight healthy male test subjects, aged 26 to 40. The subjects were kept in the recumbent position during bed rest. The foot of the bed was lifted at an angle of 4°. The test subjects were allowed to turn and lie on their sides and stomach. Meals were taken and all hygienic procedures were carried out in the recumbent position.

The 49-day test period was preceded by a 12-day control period and followed by a 30-day rehabilitation period. The nutrition was strictly controlled and included special diets made from natural foods. The chemical composition and caloric value of the diets are given in Table 1. The experimental diets differed in the content of main ingredients and in their calorific value. During the rehabilitation period the test subjects were divided into two groups: one group (control) consisting of three persons was kept on the control diet and another group (experimental) that included five persons was kept on the diet in which the composition and the ratio of food ingredients were specially selected. The nutritive rehabilitation measures were carried out by feeding animal protein with a known ratio of amino acids (beef, cottage cheese), poly-unsaturated fatty acids (sunflower seed oil), and mineral salts (potassium, calcium, phosphorus in cottage cheese and dried stoneless apricots).

Table 1

Chemical Composition and Calorific Value of the Experimental Diets

Diet	Proteins (g)	Fats (g)	Carbohydrates (g)	Ash (g)	Calorific value (kcal)
Reference	117.0	83.3	373.0	14.1	2874
Experimental	102.0	77.5	256.4	12.8	2190
Rehabilitation					
Control group	122	108.7	434.0	15.0	3290
Experimental group	144	93.4	402.0	21.0	3170

In 24-hour urine samples total nitrogen, urea, amine nitrogen, ammonia, uric acid and creatinine were measured. In the blood serum total protein and its fractions, cholesterol, total lipids, α- and β-lipoproteins were assayed. The glycaemic control was determined with respect to the sugar concentration in the blood after excess administration, i.e. during provocative tests for glucose tolerance. The sugar concentration was measured when fasting and in six blood samples withdrawn during three hours after provocation. At the end of each test urine was analysed for sugar.

Investigations of balance were carried out and assimilation of the basic food substances was determined.

Our investigations demonstrated that under the influence of prolonged antiorthostatic hypokinesia the concentration of albumins decreased and, therefore, the content of plasma proteins declined. On the 30th experimental day the albumin concentration fell significantly from 5.41 to 4.89 g% and on the 48th day it increased to 5.18 g%. The fall in the concentration of plasma proteins, especially albumins in bed rested subjects, has been reported also by others [9].

The renal excretion of end products of nitrogen metabolism — ammonia, creatinine and uric acid — underwent no noticeable changes compared with the control. The urine excretion of total nitrogen tended to increase at the expense of amine nitrogen. Both the decrease in the concentration of plasma proteins and the increase of the renal excretion of amine nitrogen can be attributed to the predominance of catabolism over anabolism and subsequent reduction of the protein synthesis [2, 3, 8]. The level of urinary excretion of nitrogen was correlated with weight losses of the test subjects.

It was shown that nitrogen balance in all the test subjects was negative throughout the bed rest experiment. It is important to note that the negative balance was most pronounced during the early stage of bed rest (9th—11th days), averaging −2.69; later the value decreased and became −1.21 by the 45th—46th day.

The experimental procedure did not affect seriously the content of total lipids and cholesterol in the serum. The study of lipid fractions revealed a decrease in the relative content of α-lipoproteins.

In response to the sugar test on the 31st day of bed rest three test subjects displayed a marked hyperglycaemic effect. Two test subjects showed a level which can be described as so-called "high and delayed glycaemic curves" with an increased sugar content in the blood after fasting (125 and 155 mg%); two hours after the sugar consumption the glycaemic level had not returned to the normal. The third test subject showed two peaks in the sugar content on the 30th and 90th minutes of the test (a two-peak curve) and 1% sugar in the urine. On the 42nd day a decrease of glucose tolerance was found in three other test subjects. The sugar content in the blood after fasting and three hours following sugar consumption varied from 140 to 165 mg%. Our findings are in good agreement with the results obtained by other researchers who also noted an elevated excretion of end products of protein metabolism and negative nitrogen balance during bed rest [4, 11, 12]. These changes were, however, more marked than in our experiment. This can be explained by the fact that the test subjects were kept on diets which adequately met the requirements of the hypokinetic body and had a well balanced composition of amino acids. The disorders in glycaemic control and lipid metabolism have also been mentioned in the literature [5, 7]. The decrease in the protein synthesis must reduce the utilization of energy producing materials (carbohydrates, lipids). Under conditions of reduced energy expenditure (hypokinesia) this must decrease the utilization of these products by cells and tissues and, therefore, may increase their content in the blood. Metabolic changes can be induced not only by hypokinesia but also by the concomitant emotional stress.

An attempt to correct the metabolic changes by a rational regime of nutrition in five test subjects that was made during the rehabilitation period resulted in a positive nitrogen balance (+3.06), increase of the albumin fraction of the serum proteins and more rapid compensatory retention of nitrogen. For instance, in the experimental group the value of the excreted total nitrogen was within the limits of the reference value beginning with the 4th day of the rehabilitation period (14.0—16.6 g versus 15.0 g); in the control group the value was markedly lower than the reference value of 10.4—13.5 g throughout the rehabilitation period. Nevertheless, disorders in the glycaemic control due to bed rest were not eliminated. During the rehabilitation period both the experimental and control groups showed an increase of the relative content of α-lipoproteins compared with the pre-test level.

Throughout the entire experiment the utilization of proteins, fats and carbohydrates was high, amounting in the experimental and control groups to 90, 93; 95, 98; 96, 98% respectively.

The results obtained allowed the following conclusions and recommendations to be made:

1. 49-day hypokinesia with a negative orthostatis of 4° produced changes in the protein, lipid, carbohydrate metabolism and weight losses; this was indicative of a reduction of the anabolic processes in the body under these conditions.

2. Rehabilitation by dietary measures (increase of proteins, polyunsaturated fatty acids and some mineral elements in the diet) tended to give rise to a more rapid recovery of some parameters of protein metabolism (albumins of the blood serum and total nitrogen of the urine).

3. It is recommended that in the recovery period following prolonged hypokinesia the diets used in the rehabilitation period be given (a tentative menu is attached).

4. The recommended composition of the diet must be tested in combination with other rehabilitation measures (physical exercises and drugs).

A tentative diet for one day

Breakfast

1. Meat and cabbage
2. Cottage cheese with sugar
3. Coffee with sugar
4. Wheat bread
5. Dried stoneless apricots

Lunch

1. Herrings and onions
2. Potato soup and meat
3. Meat and vegetables
4. Rye bread
5. Grape juice
6. Dried stoneless apricots

Supper

1. Vermicello and egg
2. Cheese
3. Wheat bread
4. Dried plums and sugar
5. Dried stoneless apricots
6. Tea and sugar

The tentative diet presents names of foad products that were given to the test subjects on a strietly measured basis. The content of basic nutrients in an average daily diet is shown in Table 1.

References

[1] V. P. Bychkov et al., Proc. Vth Tsiolkovsky lectures, Moscow 1971 (p. 132).
[2] J. E. Deitrick et al., Amer. J. Med. 4, 3 (1948).
[3] I. V. Fedorov et al., Kosm. Biolog. i Med. No. 3, 18 (1970).
[4] I. V. Fedorov and L. A. Grishanina, Kosm. Biolog. i Med. No. 3, 43 (1967).
[5] F. R. Lecocq et al., Hypogravic and Hypodynamic Environments, NASA 8P-269, 1971 (p. 289).
[6] R. L. Lipman, Metabolism 19, 980 (1970).
[7] T. M. Lobova, Kosm. Biolog. i Med. No. 5, 32 (1973).
[8] V. V. Portugalov et al., Kosm. Biolog. i Med. No. 6, 18 (1967).
[9] M. S. Seregin, I. G., Popov et al., in: Problems of Space Biology 13, 78 (1969).
[10] Y. K. Syzrantsev et al., in: Problems of Space Biology 7, 342 (1967).
[11] H. S. Taylor et al., J. Appl. Physiol. 2, 223 (1949).
[12] F. Schonheyder, N. C. Heilskov and K. Olesen, Scand. J. Clin. Lab. Invest. 6, 178 (1954).

Life Sciences and Space Research XIV — Akademie-Verlag, Berlin 1976

EFFECT OF ANTIORTHOSTATIC BED REST ON THE HUMAN BODY

T. N. KRUPINA, B. M. FYODOROV, L. M. FILATOVA, N. I. TSYGANOVA and E. I. MATSNEV

Institute of Biomedical Problems, Moscow, USSR

Head-down (−4°) bedrest brings about a more rapid development of changes in haemodynamics, fluid displacement and nervous tone than traditional recumbent bed rest. Changes occurring during head-down hypokinesia have much in common with the changes typical of real weightlessness. They are characterized by phasic changes in the vascular tone. Under these conditions changes in the tone and blood-filling of the brain are important for the functional state of the vestibular apparatus.

The results of the Soviet and American manned space missions suggest that the processes involved in man's adaptation to weightlessness are complex. Ample theoretical and factual data have been accumulated to show that many physiological effects which develop in ground-based experiments with diminished motor activity are very similar to those occurring in real flights. Some authors attribute many effects to haemodynamic changes. In order to investigate haemodynamic effects, the experimental model of diminished motor activity has recently been supplemented with a qualitatively new element — antiorthostatic bed rest, i.e. negative tilt of the body at an angle of −4°. This position causes deterioration circulation in the upper body and the head. The model was used to reproduce one of the most important aspects of the acute period of adaptation to weightlessness, and to study further development of physiological reactions which occur in space flight.

We carried out examination on nine healthy test subjects, aged 22—36, who were exposed to antiorthostatic (−4°) bed rest for 49 days. Many changes found in different systems of the human body during antiorthostatic bed rest have a cyclic pattern. An acute stage, a stage of relative adaptation and a stage of readaptation can be distinguished in the bed rest experiment.

Changes in the nervous system during the acute stage were very marked: during the first week of bed rest venous congestion in the pulmonary circulation, increase of arterial flow to the head and decrease of outflow from the head were found.

At the same time the test subjects reported a sensation of blood rushing to the head and face together with severe headaches. On the second and third days they complained of unpleasant feelings of chilled legs and toes, and hyperhydrosis. Many subjects noted nasal congestion and impaired nasal breathing. Rhinoscopically marked and stable swelling of the nasal concha was found. These symptoms were more pronounced on the side on which the test subjects preferred to lie. Beginning with the second and third days of bed rest nasal bieathing became easier in most subjects but remained difficult till the end of bed rest in a few subjects. Rhinopneumometric examinations confirmed objectively these disorders of intranasal circulation.

From the first days changes in the vestibular function were found: they included various illusory sensations and mild vertigo upon abrupt movements of the eye. During the caloric test an inhibition of the nystagmic reaction was noted; this was expressed in an increase of the latent period and decrease of the length, amplitude and frequency of the nystagmus. The test was followed by inadequately vivid autonomic and sensory reactions.

During the first days after termination of bed rest almost all the test subjects showed a deterioration of postural equilibrium and an elevation of vestibular reactivity to adequate stimuli.

The discrepancy between the results of caloric tests of the labyrinthine function (inhibition of the nystagmic reaction) and of studies with the use of adequate "stop-stimuli" (sharp excitation of the post-rotatory nystagmus) suggests not only changes in the peripheral compartment of the vestibular analyser but also changes of the function of the central vestibular formations at the diencephalic level. Thus, haemodynamic changes are not the only factor responsible for changes in the nervous system.

It is well known that a decrease in the tone of skeletal muscles and a reduction of the load on the muscular system also bring about changes in the functional state of the central nervous system, including higher autonomic centres of the hypothalamus. Among the changes typical of the state of the nervous system dysfunction of the autonomic compartment should be first mentioned. This includes a reduction of the adaptation of the cardiovascular system to physical exercises and postural effects, as well as a number of objective signs: changes in the dermographic pattern, perversion of autonomic reflexes, changes in the reflexes, thermotopometric changes, disturbed co-ordination.

Another typical manifestation of the hypokinetic syndrome is increasing asthenization. Asthenization is not only a consequence of hypokinesia but also a factor causing further abnormalities in the emotional state of test subjects, decline of their work capacity, disturbances in their emotional reactivity, etc.

It was demonstrated that tests which required mental work produced cardiovascular changes in many subjects. For instance, some subjects developed a good number of extrasystoles when solving an arithmetic problem. They also showed an increase of heart rate and arterial pressure. Many subjects stated that during bed rest mental tests were the most demanding tasks. At later hypokinetic stages the T voltage of the ECG decreased. In this situation the emotional stress did not produce increase of T waves as occurs in the normal but induced further decrease.

During the bed rest experiment we also examined the immunobiological reactivity of the body. We found that humoral parameters of immunity (for instance, the content of properdine in the blood) decreased. It is very important to note that hypokinesia-induced decrease in immunobiological reactivity persisted for a long time after the experiment (up to 30 days). The reduction of immunological reactivity may contribute to the activation of inflammatory processes. This is suggested by an increase of the content of sialic acids, haptoglobin and globulin fraction of protein, and by the appearance of C-reactive protein in the blood during bed rest.

Many investigations of different authors have shown that muscle inactivity for various causes (bed rest, confinement) gives rise to an increase of the frequency of atherosclerosis and various thromboembolic complications. As shown experimentally and clinically, the major role in the development of thrombosis and atherosclerosis is played by disturbances of lipid metabolism and the blood-coagulating system.

During 49 days of bed rest a 15—20% increase in cholesterol and α-lipoproteins was found as compared with the pre-test data in our subjects. There was an increase in the clot-forming properties of the blood during bed rest. This was indicated by a statistically significant increase of plasma tolerance to the effect of heparin, shortening of calcification time and activation of the prothrombin complex. With prolongation of bed rest the coagulatory activity of the blood continued to increase. On the 30—40th day an increase in the content of fibrinogen, VIII factor (antihaemophilous globulin) and XIII factor (fibrino-stabilizing factor) was noted. The combination of a change in the cholesterol metabolism and an increase in the coagulatory activity of the blood under aggravating conditions may bring about thrombotic complications.

Life Sciences and Space Research XIV — Akademie-Verlag, Berlin 1976

DECONDITIONING DURING PROLONGED IMMERSION AND POSSIBLE COUNTERMEASURES

E. B. Shulzhenko, I. F. Vil-Vilyams, M. A. Khudyakova and A. I. Grigoryev

Institute of Medico-Biological Problems, Moscow, USSR

Test subjects covered with a waterproof highly elastic cloth were exposed to 13-day water immersion up to the neck. They were divided into two groups. The first (control) group consisting of six persons was exposed to immersion alone and the second (experimental) group was exposed daily to accelerations of 0.6—2 G_z for 60—90 min during the last 6 days of immersion. Before and after immersion all the test subjects were exposed to +3 G_z for 5 min which served as a provocative test. These experiments give evidence that the use of dry immersion allows experimentation during prolonged immersion without concomitant complications. Variations in the physiological parameters (cardiovascular system, fluid-electrolyte balance, blood-coagulatory system) are indicative of the preventive effect of periodic accelerations during 13-day immersion.

1. Introduction

It is recognized that weightlessness brings about functional changes in the physiological systems of the body which can be characterized as deconditioning of the body. The main pathogenic mechanisms contributing to this state are: diminished load on the musculo-skeletal system, elimination of the hydrostatic component of the pressure of the blood and its subsequent redistribution, and change in afferentation due to zero gravity [1—3]. Methods for simulating the effects of weightlessness and possible countermeasures against them are therefore important. One of the methods simulating functional changes in the body that are typical of weightlessness is water immersion [4—8].

It is also known that periodic centrifugation is recommended as a countermeasure against the deconditioning which develops in weightlessness [9, 10]. In this connection it was decided to evaluate changes in physiological functions during prolonged immersion and to assess body tolerance to post-immersion acceleration. A modified method of prolonged immersion, the so-called "dry immersion", has been developed: this method eliminates bodily contacts with the immersion medium which cause problems, especially skin maceration.

2. Methods

Experiments on 14 healthy male test subjects were carried out during a 13-day period of "immersion". The test subjects were divided into two groups, one of which (control) included six persons who were exposed to immersion alone and the other (experimental) group consisted of eight persons who were exposed to immersion and acceleration of 0.6—$2G_z$ for 60—90 min during the last six days of immersion. In all experiments before and after immersion the test subjects were exposed to head-to-feet acceleration of $+3G_z$ for 5 min as a provocative test. The experiments were carried out in a centrifuge with a radius of 7.25 m. When the test subjects reported visual disorders (greyout, blackout), centrifugation was terminated before the scheduled time.

Prolonged immersion was performed in a specially designed bath 200 cm × 100 cm × 100 cm for two subjects (Fig. 1). The body of the test subject was

Fig. 1. General view of the bath with test subjects immersed up to the neck.

separated from the water environment by elastic waterproof cloth fixed to the outer perimeter of the bath. The test subjects were kept in the medium up to the neck. The cloth surface area was much larger than the water surface area. Therefore, folds of the waterproof cloth closed in on the mid-body line of the subject, covering him loosely. No couches were used in the immersion bath as a support. The water temperature was automatically maintained at 33—34°. The air temperature in the immersion room was 22—24°.

During the experiments the following recordings and calculations were made: electrocardiography in the Neb leads, arterial pressure by Korotkoff sounds, sphygmography of the carotid and femoral arteries, and phonocardiography to measure haemodynamic parameters. The amount of water consumed and urine excreted was measured on a daily basis. The test subjects drank water *ad libitum* but the amount was carefully measured. The sodium and potassium content in every urine sample was determined by flame photometry. The coagulatory properties of the blood were estimated by the prothrombin concentration and fibrinolytic activity.

3. Results and Discussion

The method of dry immersion and the use of highly elastic and waterproof cloth allowed long-duration investigations which induced no concomitant problems that typically result from human body contacts with the liquid environment.

The experimental data showed that cardiovascular parameters varied sinusoidally, remaining within physiological limits. The dynamics of these parameters

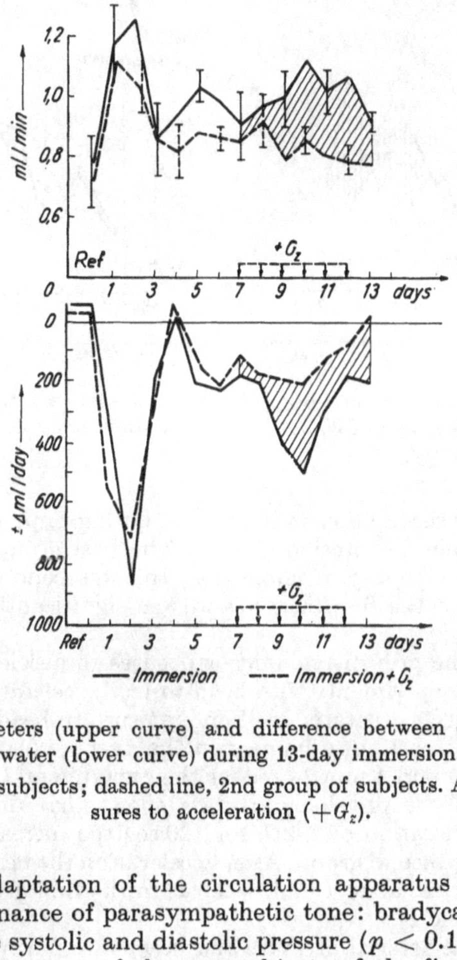

Fig. 2. Diuretic parameters (upper curve) and difference between consumed and excreted water (lower curve) during 13-day immersion.

Solid line, 1st group of subjects; dashed line, 2nd group of subjects. Arrows show daily exposures to acceleration ($+G_z$).

at rest indicated adaptation of the circulation apparatus to immersion which showed the predominance of parasympathetic tone: bradycardia, decrease of the T voltage, fall of the systolic and diastolic pressure ($p < 0.1$—0.3). No important differences in the responses of the test subjects of the first and second groups were noted. Fluid-electrolyte metabolism exhibited a significant ($p < 0.05$) increase in the rate of urine excretion (Fig. 2). These data indicated that the diuretic pattern differed markedly between the two groups during the second half of immersion. For instance, during the first half of immersion total diuresis in the first and second groups was larger than the amount of fluid consumed by 1887 ml and 1956 ml respectively; during the second half of immersion the difference was 1650 ml and 647 ml respectively. Therefore the test subjects who were daily exposed to acceleration (the experimental group) showed a 1000 ml greater loss of fluid than the control group.

The renal excretion of sodium increased significantly ($p < 0.05$) in both groups during the first two days (Fig. 3). During the second half of immersion the sodium excretion increased again in the first group, reaching the level of the first day of immersion. During the same period sodium excretion in the second group did not increase and remained close to the reference level. The difference between the two groups was statistically significant ($p < 0.05$).

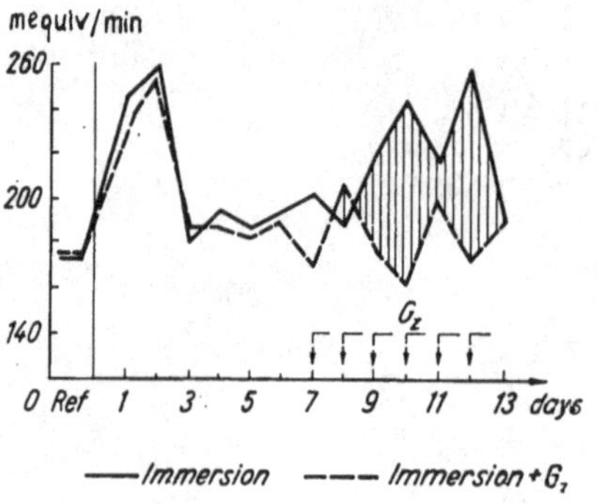

Fig. 3. Renal excretion of sodium during 13-day immersion. Key as in Fig. 2.

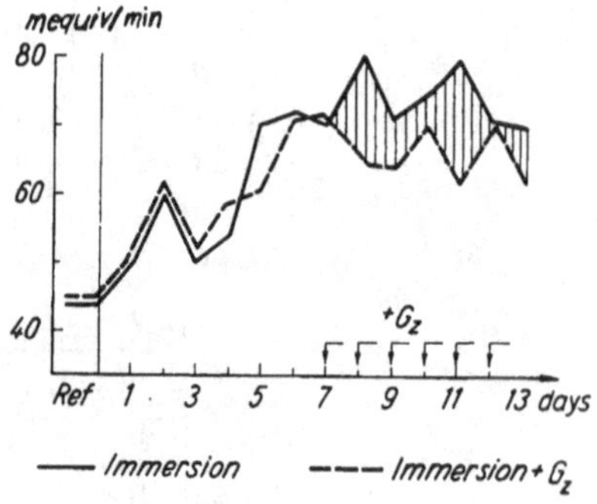

Fig. 4. Renal excretion of potassium during 13-day immersion. Key as in Fig. 2.

Beginning with the second day of immersion both groups showed a progressive increase in the potassium excretion (Fig. 4). The first group displayed maximal excretion on the 10—11th day of immersion. In the second group the potassium excretion stabilized on the 6—7th day and was significantly lower than in the first group ($p < 0.02$).

The efficiency of the preventive daily exposures to acceleration also confirms the data obtained in experiments with head-to-feet acceleration. Prior to immersion all the test subjects tolerated well an exposure to head-to-feet acceleration of $+3G_z$ for 5 min. After 13-day immersion the average plateau time was 3 min 46 sec for the controls and 4 min 14 sec for the experimental group ($p < 0.05$).

An examination of the physiological data shows that during a post-immersion exposure to acceleration of $+3G_z$ for 120 sec the increase in heart rate was less pronounced in the second group. After acceleration the heart rate in the second group returned to normal after 1 min, whereas in the first group tachycardia was noted for 10 min (Table 1).

A comparison of the arterial pressure and heart rate during the first minute of an exposure to $+3G_z$ before and after immersion also indicates the positive effect of regular centrifugation (Fig. 5). An exposure to acceleration prior to immersion brings about an increase of heart rate and arterial pressure in both groups at ratios that indicate an adequate level of circulatory adaptation. An exposure to acceleration which followed 13-day immersion induced marked tachycardia, decreased diastolic pressure and slightly increased systolic pressure in the first

Table 1

Differences in Absolute Values of the Heart Rate (beats/min) during an Exposure to Acceleration ($+3G_z$ for 5 min) before and after Immersion for the First Group (immersion alone) and the Second Group (immersion combined with acceleration)

Group	Reference	During acceleration after						After effect after	
		10 s	60 s	120 s	180 s	240 s	300 s	1 min	10min
I	+10	+17	+25	+36	+30	+27	—	+15	+16
II	+3	+6	+17	+20	+31	+32	—	0	+1

group. This suggested a decline in the compensatory capabilities of the circulation system. Regular centrifugation during the immersion regime of the second group helped to maintain an adequate ratio of arterial pressure and heart rate at $+3G_z$. Stroke volume, cardiac output and total peripheral resistance in the second group returned to the initial level by the 40—50th min. During the same period of time the first group subjects showed a 48% decline in the stroke volume, a 28% decrease in the cardiac output, and a 56% increase in the total resistance of the vascular bed.

An examination of the coagulatory capacity of the blood also demonstrated a positive effect of regular centrifugation. During an exposure to acceleration prior to and on the 2nd day of immersion both groups showed similar changes indicating an increase of coagulability. At the end of the second half of immersion there were significant changes in the concentration of prothrombin and in fibrinolytic activity

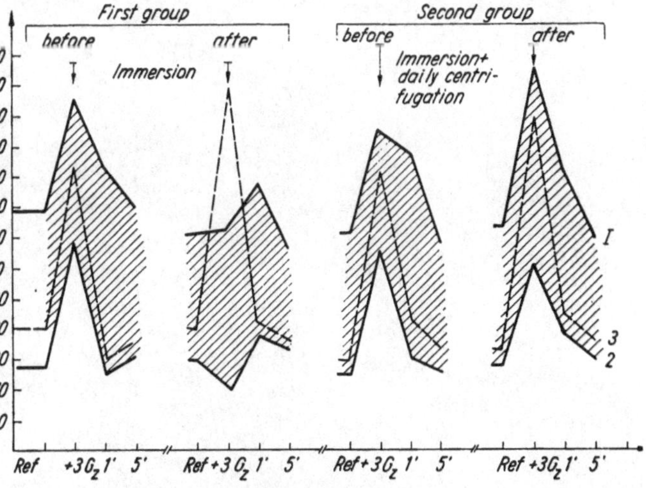

Fig. 5. Ratio of arterial pressure and heart rate in the first and second group subjects before and after immersion. 1st group: parameters at 1st min during exposure to $+3G_z$ and at 1st and 5th min of after effect; 2nd group: 1, systolic pressure; 2, diastolic pressure; 3, heart rate.

between the groups. The first group showed an increase in prothrombin and a decrease in fibrinolytic activity. The second group test subjects showed the opposite effects. It is possible that periodic acceleration reduces prothrombin and stimulates the anticoagulatory system.

References

[1] V. V. PARIN, R. M. BAYEVSKY, M. D. YEMELYANOV and I. M. KHAZEN, Essays on Space Physiology, Medizin, Moscow 1967 (in Russian).
[2] O. G. GAZENKO, A. A. GURYIAN and V. V. PARIN, in: Problems of Space Biology, **4**, 22 (1967).
[3] C. A. BERRY, Aerosp. Med. **40**, 245 (1969).
[4] O. G. GAZENKO, A. M. GENIN and V. B. MALKIN, Nauka i Zhizn, **4**, 92 (1972).
[5] A. M. GENIN and I. A. SOROKIN, Problems of Space Biology **13**, 9 (1969).
[6] D. E. GRAVELINE, B. BALKE, R. E. MCKENZIE and B. HARTMAN, Aerosp. Med. **32**, 387 (1961).
[7] A. GRAYBIEL and B. CLARK, Aerosp. Med. **32**, 181 (1961).
[8] O. H. GAUER, Proc. Symp. on Hypogravic and Hypodynamic Environments, Indiana 1969, Indiana University, A-3652, 1970 (p. 383).
[9] W. I. WHITE, Status Rep., Douglas Aircraft Co., Santa Monica, Calif. 1965.
[10] T. E. PIEMME, A. S. HYDE, M. MCCALLY and G. POTOR, Aerosp. Med. **37**, 16 (1966).

Life Sciences and Space Research XIV — Akademie-Verlag, Berlin 1976

STUDY OF SPACE PERCEPTION FUNCTIONING DURING SIMULATION OF CERTAIN SPACE FLIGHT FACTORS

I. Y. Yakovleva, B. B. Bokhov and L. N. Kornilova

Institute for Biological Problems, Moscow, USSR

The results are reported of spatial perception function studies in 130 healthy males of 17—35 years of age and in 33 people with complete or partial inhibition of the function of the labyrinth (deaf-mutes). The comparative magnitudes of gravitational vertical perception impairments were studied during vestibular stimulation (Coriolis and linear acceleration), as well as during clino-orthostatic and antiorthostatic hypokinesia. Hypokinesia to a certain extent permitted the simulation of the blood redistribution in weightlessness. The ability of the subjects to determine the subjective visual vertical was used as the test criterion.

The experiments have shown that the magnitude of the observed changes in the human sensory area depends on the physical properties of the vestibular stimulus, on the angle of inclination of the head of the bed during hypokinesia and on the duration of bed rest, but not necessarily on the level of vestibular tolerance. Possible mechanisms of impairment of perception in the space flight environment are discussed. Examinations of spatial perception function are useful during selection procedures for astronauts and airmen.

Several American and Soviet astronauts reported instances of illusions of changed body position and spatial disorientation during earth orbital and lunar flights. Mechanisms of spatial disorientation in space flight are not yet fully understood. There are now several hypotheses that interpret disorientation in space flight either (i) as an effect of stimulation of gravitational receptors by weightlessness, or (ii) as a consequence of an impairment of the interaction between the otolith and the cupular system in weightlessness, or else (iii) as a result of a changed interaction between the afferent systems of the body providing spatial orientation. However, as well as the possible development of spatial disturbances as a classic sensory conflict produced by unusual functioning of the vestibular system and its associated analysers, it is probable that disorientation in space flight can also be caused by haemodynamic shifts.

A comparative study was therefore carried out to examine changes in spatial perception during simulation of certain space flight factors, namely, vestibular exposure and clino-orthostatic and antiorthostatic hypokinesia. Assessment of spatial perception function was performed by a technique based on comparison of subjective spatial coordinates with the gravitational ones. For this purpose a device "Vertical" (Fig. 1) was designed [1, 2]. The procedure and methods of operating this device were similar to those used in A. Graybiel's laboratory. Subjects were examined in three positions: sitting head upright and lying on the right or left side with head maintained at 90°. Spatial perception studies were conducted at rest (base-line study), during experimental simulation of space flight factors and during the rehabilitation period. Studies were made on 130 healthy males 17—35 years of age, 33 people with complete or partial loss of labyrinth function (deaf-mutes) and 37 control subjects. In the first series of experiments the effects of vestibular stimulation on the spatial perception function was studied, as well as the relationship between spatial perception function indices and vestibular-vegetative and vestibular-sensory responses. A test with cumulative exposure to

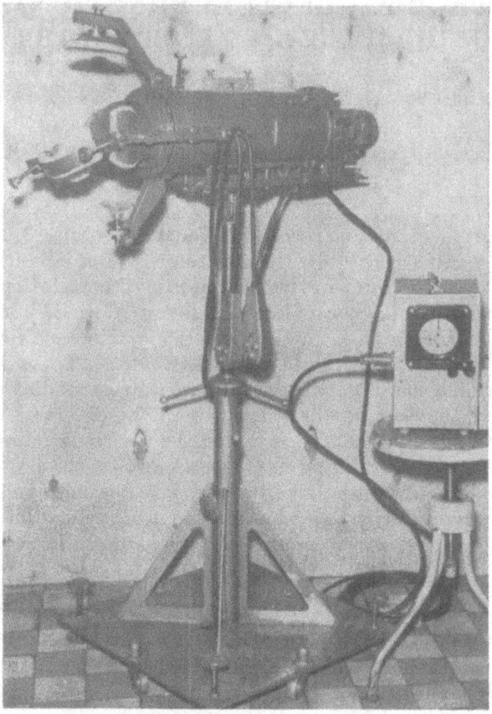

Fig. 1. General view of the device for studying function of orientation relative to gravitational vertical.

Coriolis acceleration [3], and a test with cumulative exposure to linear accelerations by swinging on a four-bar Khilov swing, were used as vestibular stimuli. Polyeffector monitoring was used to record physiological responses. During base-line examination of vestibular tolerance the subjects were divided into a "tolerant" group (high tolerance level) and a "non-tolerant" group (medium and low tolerance level). An analysis of experimental data (Fig. 2) showed that in both groups, regardless of the base-line level of vestibular tolerance, an increment in perception error was present after both the Coriolis acceleration test and the Khilov swing test. Analysis of the relationship between spatial perception function indices and clinically recorded indices of vestibular exposure tolerance demonstrated that the group tolerant to swinging showed statistically significant superiority in the base-line accuracy of spatial coordinate determination over that of the "non-tolerant" group. The same superiority was maintained after experimental test with linear accelerations.

During base-line study using Coriolis acceleration test there was practically no difference in the accuracy of determination of spatial coordinates between the two groups. The experimental study with the same test resulted in a significant increment in error shown only by the "non-tolerant" group. Analysis of relationship between indices of vestibular-vegetative tolerance and those of perception function (base-line and experimental) revealed correlation only for cumulative linear acceleration exposures. In cases of repeated exposure to Coriolis acceleration

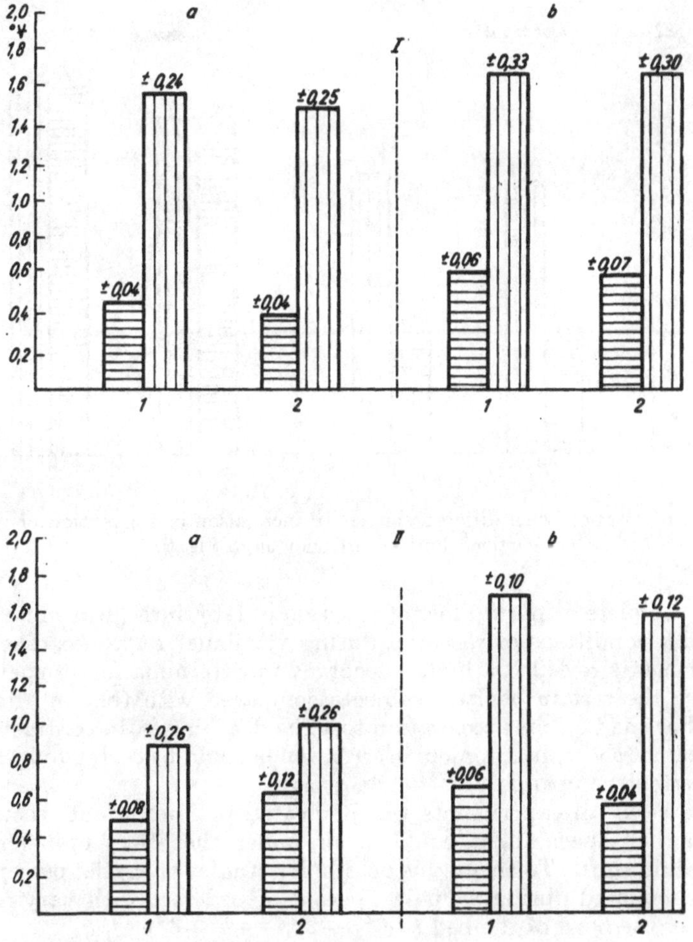

Fig. 2. Effect of vestibular tests on spatial perception as a function of vestibular-vegetative tolerance: I, test with the 4-bar Khilov swings; II, test with cumulative Coriolis accelerations: a, "tolerant" group; b, "non-tolerant" group; 1, determination of the vertical; 2, determination of the horizontal. The graph shows angles between objective and subjective verticals (horizontals). Transverse shading, values obtained before bed rest; vertical shading, values obtained after bed rest.

test no such correlation was detected. There was also no correlation between indices of specific vestibular-sensory response (latent period of the illusion of "swinging") and spatial perception function.

Thus the first series of experiments has shown that linear and Coriolis accelerations of the type that can occur in space flight produce sensory response represented by an incremental error in perception of spatial coordinates. Linear accelerations evoke greater response.

The objective of the second series of experiments was to determine the role of vestibular receptors in the development of spatial disturbances. For that purpose a comparative study of spatial coordinate perception by healthy individuals and

Fig. 3. Effect of rotation with different angles of inclination on perception of gravitational vertical (horizontal). Key as in Fig. 2.

those with complete or partial loss of function of labyrinth (deaf-mutes) was performed under conditions of rest and during vestibular exposure. The results of these experiments revealed a higher accuracy in determination of spatial coordinates before the test in healthy subjects compared with that of "deaf-mutes" (Fig. 3). After the Coriolis acceleration test the deaf-mutes, in contrast to normal subjects, did not show an increment in error which could be correlated with impairment of vestibular apparatus receptor activity.

The third series of experiments was dedicated to finding out whether or not disorientation in space flight could occur under the effect of haemodynamic changes or fluid shifts. To check this possibility studies of spatial perception function were conducted during a 30-day period of bed rest with varying angle of inclination of the head of the bed (−6°, −2°, −4°, +6°).

In the group of subjects with bed inclination of −4° physical exercises and electric stimulation of abdominal and leg muscles were used as preventive measures. In groups with antiorthostatic exposure without preventive measures a sharp increment in error was observed in the first two days of bed rest (Fig. 4). The intensity of response depended on the extent of antiorthostatic stress, i.e. subjects with −6° inclination of bed showed the greatest error. Toward the middle of the experiment the error decreased but showed a trend to another increase nearer the end. No such changes were detected in the group with orthostatic hypokinesia or in control.

The first two days of the experiment were marked by objective as well as subjective symptoms. All subjects in antiorthostatic position experienced a sensation of fullness in the head and some subjects reported an illusion of being in the head-down position (misjudging the angle of inclination) after they closed their

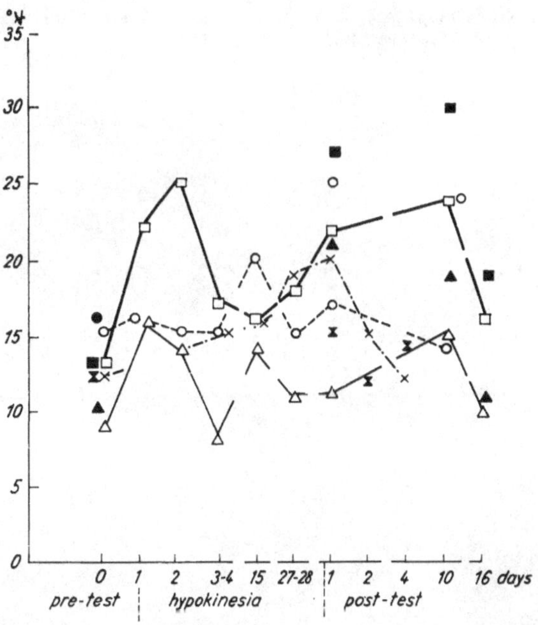

Fig. 4. Changes in perception of subjective visual vertical during antiorthostatic and orthostatic hypokinesia (mean values).

The light squares, circles, crosses and triangles represent the mean values obtained from subjects lying on the right side, the dark ones those from subjects lying on the left side. Thick line, subjects exposed to hypokinesia with $-6°$ inclination from the horizontal; dashed line, $+6°$ inclination; chain line, $-4°$ inclination; thin line, $-2°$ inclination. The base-line and rehabilitation period data were obtained in the horizontal position (90° from the vertical).

eyes. During the rehabilitation period either consistent (in the $-6°$ group) or single increments in error were observed. A specific effect of hypokinesia was expressed in response asymmetry in lateral positions. Within 4—10 days after the end of bed rest the spatial perception and response asymmetry data returned to normal. In the antiorthostatic groups where preventive measures had been used (exercises, electric muscle stimulation) subjects showed smaller spatial perception changes and no response asymmetry.

In conclusion, general haemodynamic disorders following blood and fluid redistribution can lead to the development of illusory responses probably for two reasons: (i) an increased afferentation from compartments with high blood and fluid content, leading to a sensory conflict between the vestibular and tactile-kinesthetic perception of gravitational vertical; and (ii) development of frontal asymmetry of skeletal muscle toxicity leading to an asymmetry of vertical perception error.

References

[1] B. B. Bokhov, I. Y. Yakovleva, L. N. Kornilova et al., Kosm. Biol. i Med. No. 6, 51 (1973).
[2] B. B. Bokhov, L. N. Kornilova, I. Y. Yakovleva, Kosm. Biol. i Med. No. 1,51 (1975)
[3] I. I. Bryanov et al., J. Military Med. No. 11,54 (1963).

Life Sciences and Space Research XIV — Akademie-Verlag, Berlin 1976

CARDIAC OUTPUT DURING PHYSICAL EXERCISES FOLLOWING REAL AND SIMULATED SPACE FLIGHT

B. S. KATKOVSKY and YU. D. POMYOTOV

Institute of Medico-Biological Problems, Moscow, USSR

During moderate physical work (after 5 min) the cardiac stroke volume in the cosmonauts who made 2—8-day flights in the Soyuz spacecraft was lower and the pulse rate was higher than the pre-flight level. By the end of the 49-day bed-rest experiment the stroke volume during physical exercises (after 5 min) was lower than at the 5th min of the pre-test period in both the supine and sitting positions. The cosmonauts performed physical exercises in the sitting position. Therefore, it can be assumed that the major factor changing cardiac output may be the effect of the earth's gravity against the background of decreased orthostatic tolerance. Results of simulated experiments give evidence that in both cases one of the major factors responsible for changes in the cardiac output was a decline in the contractile capacity of the myocardium.

Decreased tolerance of man to physical exercises has been consistently recorded after space missions [1, 2, 9, 16, 19]. Similar reduction in exercise tolerance was also found in ground-based experiments simulating weightlessness [5, 8, 10, 11, 14, 15, 17, 21, 28]. There are essentially two factors common to real space flights and ground-based experiments simulating them: (i) diminished motor activity (hypokinesia) which leads to deconditioning of the mechanisms responsible for exercise tolerance, and (ii) adaptation of cosmonauts and test subjects to altered haemodynamics due to the lack of hydrostatic pressure of fluids (in weightlessness) or its decrease (in simulated flights). Due to this, during readaptation to 1*g* the capacity of maintaining a vertical posture deteriorates.

The role of each of the above factors needs further study because until very recently pre- and post-flight responses to physical exercises were carried out in the upright position of the body. This is also true of the bed rest and water immersion experiments to simulate weightlessness. After real and simulated space flights orthostatic tolerance was noticeably lowered [1, 6, 7, 12, 13, 19, 23, 24, 26]. This could significantly affect the responses of cosmonauts and test subjects to physical exercises carried out in the upright position.

The purpose of the present investigation was (i) to assess the space flight effects on the adaptation of the main haemodynamic parameters (cardiac output and stroke volume) to moderate physical work and (ii) to identify experimentally the factors responsible for the decline in work capacity after real or simulated weightlessness.

The four Soyuz 12 and 13 crew members were examined three weeks pre-launch and 2—3 days post-flight. The work performed by a sitting cosmonaut on a

Table 1

Haemodynamic Parameters in Soyuz 12 and 13 Crew Members during Work Capacity Tests

Space flight Cosmonaut	Resting in recumbent position						Physical work sitting					
	Cardiac output ($l\ min^{-1}$)		Stroke volume (ml)		Heart rate (beats/min)		Cardiac output ($l\ min^{-1}$)		Stroke volume (ml)		Heart rate (beats/min)	
	Before	After	Before	After	Before	After	Before	After	Before	After	Before	After
Soyuz 12												
V. G. Lazarev	4.85	4.39	78	60	62	73	20.0	18.4	175	144	114	128
O. G. Makarov	5.37	5.53	94	89	57	62	16.2	14.7	155	128	104	115
Soyuz 13												
P. I. Klimuk	7.05	6.98	93	89	76	78	18.2	15.4	138	109	132	142
V. V. Lebedev	5.92	7.26	76	97	78	75	19.0	20.0	130	121	146	165

bicycle ergometer at the pedalling rate of 60 ± 5 rev min^{-1} was assigned individually on a basis of 400 kg m min^{-1} per m^2 of body surface area. The work done pre- and post-flight was identical.

Cardiac output was measured by recording the CO_2 content in the lung-bag system during rebreathing which was compared at rest and immediately after the 5th minute of work. During this procedure the cosmonaut continued to pedal at the same rate (60 ± 5 rev min^{-1}) and at the same power (400 kg m min^{-1} m^{-2}).

Cardiac output was derived from the Fick equation:

$$\dot{Q} = \frac{\dot{V}\,CO_2}{C_{\bar{v}}\,CO_2 - C_{\bar{a}}\,CO_2},$$

where $\dot{V}\,CO_2$ is the CO_2 production (ml min^{-1}) in the steady state (at the 5th min) of physical exercises; $\dot{Q}$ is the cardiac output (l min^{-1}); $C_{\bar{a}}\,CO_2$ is the CO_2 content in arterial blood (vol.%) at the same time (this equals the CO_2 content in the alveolar gas) which was recorded exhalation after exhalation in a low-inertial infrared gas analyser ("Capnograph"), and $C_{\bar{v}}\,CO_2$ is the CO_2 concentration in the mixed venous blood (vol.%) which was estimated by CO_2 rebreathing on the basis of the plateau in the lung-bag system [25]; in the absence of the plateau the concentration was assessed on the basis of the exponential CO_2 rise during rebreathing [22].

It is important to note a consistent decrease in the stroke volume of the cosmonauts who made 2-day and 8-day flights during the bicycle ergometer exercises. For instance, post-flight V. G. Lazarev and O. G. Makarov who made a 2-day flight showed a 18% and 16.5% decrease compared with the pre-flight test; P. I. Klimuk and V. V. Lebedev who made an 8-day flight showed a 21% and 6.6% decrease (Table 1). Cardiac output also tended to decline in the three cosmonauts and only in V. V. Lebedev did it grow by 5% despite the decrease in stroke volume; the rise was evidently due to the increase of the heart rate from 146 to 165 beats/min. No strict correlation between the haemodynamic parameters and the flight time has been found. It should be noted that P. I. Klimuk showed the greatest reduction of the stroke volume (21%). Changes in the cardiac output at rest before provocative tests were insignificant. However in three cosmonauts in the recumbent position a decrease of stroke volume was noted, and in the fourth cosmonaut an increase of the parameter was found (Table 1). Since the cosmonauts performed physical exercises in the sitting position, one of the major factors responsible for the decrease of stroke volume could have been insufficient influx of blood to the right heart due to deconditioning of antigravity mechanisms affecting the haemodynamics. In order to clarify the contribution of the hydrostatic pressure to the reduction of stroke volume during physical exercises, antiorthostatic bed rest experiments (with the foot end of the bed lifted at an angle of 4°) were carried out. This model induced sensations similar to those that cosmonauts experienced during the first hours and days of weightlessness. They were never noted during recumbent bed rest experiments (3, 4, 18).

Prior to and at the end of the 49-day bed-rest experiment twelve test subjects performed physical exercises of 600 kg m min^{-1} in the sitting and supine position. The basic haemodynamic parameters were measured at the same periods of time and using the same equipment as those of the Soyuz 12 and 13 crew members.

At rest the cardiac output and stroke volume did not significantly differ in the supine or sitting position either at the beginning or at the end of the experiment; the heart rate was significantly higher in the sitting than in the supine position (Table 2).

Table 2

Haemodynamic Parameters of 12 Test Subjects at Rest and in the Steady State (5th min) of Physical Exercises performed in the Supine and Sitting Positions

Parameter	Body position			
	Supine		Sitting	
	Before	After	Before	After
At rest				
Cardiac output (l min^{-1})	5.6 ± 0.3	6.0 ± 0.4	5.3 ± 0.5	6.2 ± 0.5
Stroke volume (ml)	80 ± 4.8	73 ± 6.2	67 ± 6.3	62 ± 6.6
Heart rate (beats/min)	70 ± 3.1	83 ± 3.1*	79 ± 3.6	99 ± 4.5*
Physical exercises				
Cardiac output (l min^{-1})	13.6 ± 0.8	11.5 ± 0.5	13.9 ± 0.7	12.9 ± 0.5
Stroke volume (ml)	114 ± 7.3	86 ± 6.4*	114 ± 5.7	83 ± 5.2*
Heart rate (beats/min)	118 ± 3.9	135 ± 5.3*	122 ± 4.8	157 ± 5.7*

* Significant difference $P < 0.05$

For physical exercises performed before bed rest no statistical difference between the supine and sitting positions was noted in any of the parameters. There was a statistically significant decrease of stroke volume in both body positions for comparisons before and after bed rest (Table 2). The heart rate increased (most markedly in the sitting position), cardiac output decreased in the supine position but remained unaltered in the sitting position.

Blomquist et al. [20] estimated changes in cardiac output induced by 30 days of bed rest by means of dye dilution. The results they found in five test subjects were very close to our findings, although we used the CO_2 rebreathing method. For instance, at a work load of 600 kg m min^{-1} in the supine position the test subjects of Blomquist et al. showed an increase of the heart rate from 129 to 154 beats/min (19.4%), a decrease of the cardiac output from 14.4 to 12.4 l min^{-1} (13.9%) and a reduction of the stroke volume from 116 to 88 ml (24.1%) before and after bed rest. In our experiment the changes were 14.4%, 15.5% and 24.6% respectively.

Similar data were obtained by the American investigators [27] during pre- and post-flight examinations of Skylab 4 crew members. The American researchers attribute these changes to decline of the post-flight tolerance to the upright position due to inadequacy of the mechanisms responsible for the venous return to the heart. Our study on the Soviet cosmonauts supports this idea. It should be noted that the flights were short-term and that physical exercises were performed in the sitting position. Nevertheless, bed-rest experiments have demonstrated that by the end of the experiment the stroke volume also decreased during physical exercises performed in the supine position. Our results give evidence that decrease in myocardial contractility is one of the factors responsible for the changes in the cardiac output and its constituents.

References

[1] Yu. M. Volynkin and P. V. Vasilyev, Problems of Space Biol. **6**, 53 (1967).
[2] E. I. Vorobyov, Yu. G. Nefyodov and L. I. Kakurin et al., Kosm. Biol. i Med. No. **2**, 65 (1970).
[3] A. D. Voskresensky, B. B. Egorov and I. D. Pestov et al., Kosm. Biol. i Med. No. **4**, 28 (1972).
[4] A. M. Genin and L. I. Kakurin, Kosm. Biol. i Med. No. **4**, 26 (1972).
[5] V. S. Georgiyevsky and L. I. Kakurin et al., in: Oxygen Profile and its control in the Body, Kiev 1966 (p. 225).
[6] V. S. Georgiyevsky, V. A. Gornago and L. Ya. Divina et al., Kosm. Biol. i Med. No. **6**, 61 (1973).
[7] N. N. Gurovsky, L. I. Kakurin and A. D. Egorov et al., Symp. Space Biol. and Med., Intercosmos, Budapest 1970 (p. 1).
[8] B. B. Egorov, M. M. Mirrakhimov and V. I. Korolkov et al., in: Space Biology and Aerospace Medicine, Moscow-Kaluga **1**, 116 (1972).
[9] L. I. Kakurin and B. S. Katkovsky, in: Physiology of Humans and Animals: Scientific Achievements 1963, Moscow 1966 (p. 6).
[10] L. I. Kakurin, B. S. Katkovsky and V. S. Georgiyevsky et al., Voprosy Kurortologii, fiziotherapii i lechebnoi fizkultury, **1**, 19 (1970).
[11] L. I. Kakurin, Yu. N. Purakhin and V. S. Georgiyevsky et al., Voprosy pitania, **3**, 7 (1971).
[12] L. I. Kakurin, V. A. Gornago and B. S. Katkovsky et al., in: Space Biology and Aerospace Medicine, Moscow-Kaluga, **1**, 75 (1972).
[13] V. V. Kalinichenko, V. A. Gornago and G. V. Machinsky et al., Kosm. Biol. i Med. No. **6**, 68 (1970).
[14] B. S. Katkovsky, I. A. Pilyavskyand G. A. Smirnova, Kosm. Biol. i Med. No. **2**, 49 (1969).
[15] B. S. Katkovsky and V. S. Georgiyevsky et al., Voprosy pitania, **4**, 55 (1971).
[16] B. S. Katkovsky, G. V. Machinsky and V. A. Gornago et al., in: Space Biology and Aerospace Medicine, Moscow-Kaluga, **1**, 75 (1972).
[17] B. S. Katkovsky, G. V. Machinsky and P. S. Toman et al., Kosm. Biol. i Med. No. **4**, 43 (1974).
[18] Yu. D. Pomyotov and B. S. Katkovsky, Kosm. Biol. i Med. No. **4**, 39 (1972).
[19] C. A. Berry, in: Bioastronautics Data Book, NASA, Washington, D.C. 1973 (p. 349).
[20] C. G. Blomquist, J. H. Mitchell and B. Saltin, Proc. Symp. on Hypogravic and Hypodynamic Environments, Indiana Univ. **A-3652**, 197 (1970).
[21] D. Cardus, Aerospace Med. **37**, 993 (1966).
[22] J. G. Defares, J. Appl. Physiol. **13**, 159 (1958).
[23] L. F. Dietlein, Proc. Symp. on Hypogravic and Hypodynamic Environments, Indiana Univ. **A-3652**, 1 (1970).
[24] D. E. Graveline et al., Aerospace Med. **32**, 387 (1961).
[25] H. L. Jones et al., Clin. Sci. **32**, 311 (1967).
[26] L. E. Lamb et al., Aerospace Med. **35**, 420 (1964).
[27] E. L. Michel and J. A. Rummel et al., in: Proc. Skylab Life Sciences Symp. Vol. 2, 1974.
[28] B. Saltin et al., Circulation, **38**, No. 5, Suppl. 7 (1968).

Life Sciences and Space Research XIV — Akademie-Verlag, Berlin 1976

RENAL OSMOREGULATORY FUNCTION DURING SIMULATED SPACE FLIGHT

G. I. KOZYREVSKAYA, A. I. GRIGORIEV and YU. V. NATOCHIN

Institute for Medico-Biological Problems, Moscow, USSR

Experiments are described on the dynamics of the renal excretion of water and electrolytes, osmotically active substances and osmotically free water during bed rest. In short-term experiments (up to 10—14 days) renal changes are mainly associated with haemodynamic disturbances and fluid redistribution, whereas in longer duration experiments (up to 50 days) they are brought about by metabolic changes.

The Soviet and American space programmes give special attention to the study of fluid-electrolyte balance and renal function in flight and post flight. This can be attributed to the fact that man's adaptation to space flight effects involves changes in the functional state of almost every physiological system. These changes are associated with disturbances of hydro-ionic equilibrium.

Our examinations of Soyuz cosmonauts have shown that with an increase of the length of space missions new, qualitatively different changes in the renal excretion of water and electrolytes and in the control of fluid-electrolyte balance develop [1, 2].

Although we have extensive information on variations in clinical and biochemical parameters of the fluid-electrolyte balance and renal function during space flight, ground-based simulation of typical physiological states is still very important for the solution of many problems of space medicine.

The present communication summarizes the results of several experiments in which 50 young healthy male test subjects were kept in bed for 10—50 days. The dicts were carefully weighed daily. The control diet was dried to measure water content and burnt to study the mineral content. The subjects were given bread and drinking water *ad libitum* and the consumed products were strictly measured.

Every day the diuresis, renal excretion of sodium, potassium and calcium as well as the urine osmotic concentration were measured. The same parameters were recorded regularly in the blood serum. The rate of glomerular filtration was assayed using the endogenous creatinine clearance. The circulating blood volume was measured by means of ^{125}I-albumin. The cardiac output was estimated by the method of rebreathing CO_2 and the stroke volume was calculated. In addition, the functional test with a water load which made 2% of body weight was carried out. Data were collected from the subjects during the experimental period as well as from six subjects who were used as controls and were kept for 30 days in the clinic and allowed normal activity.

From the very first hours of bed rest the subjects showed distinct symptoms of thirst. Thus, 75% of test subjects displayed a significant ($p < 0.05$) decrease of water consumption averaging 22%. At the same time, their diuresis increased. For instance, during the first day of bed rest most subjects showed a 57% diuretic increase; some subjects exhibited an unchanged diuresis with a decreased amount of water consumption; this resulted in a reduced difference between water con-

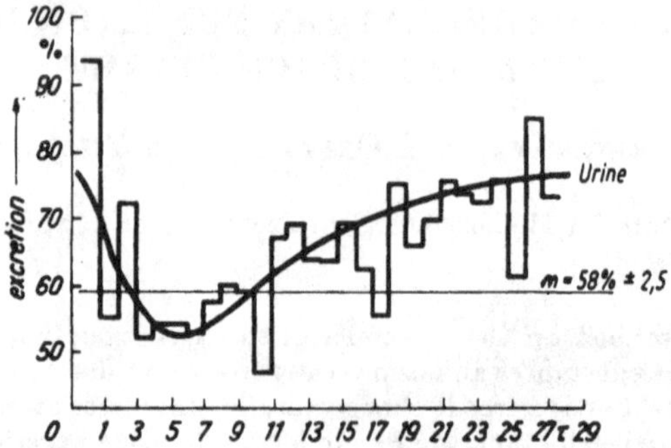

Fig. 1. Renal excretion of water during bed rest (%).

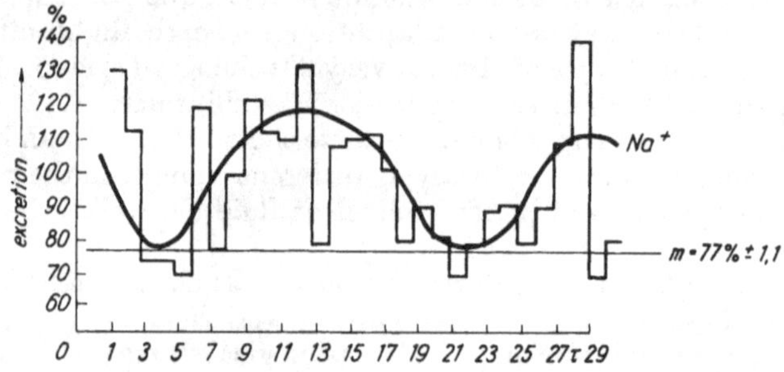

Fig. 2. Renal excretion of sodium during bed rest (%).

sumed and urine excreted. Therefore, in the first day of bed rest an absolute and relative increase in diuresis took place.

Enhanced renal excretion of water and salts was always followed by an increase of the sodium content in the urine. At an early stage of bed rest the water requirements decreased and the excretion of water and salts in the urine decreased (Fig. 1, 2).

As a result, the test subjects showed weight losses; this was very noticeable in the experiments in which the calorific value of the diet equalled that of the basal metabolism. It is very probable that this loss was associated with increased water losses; the circulating blood volume was decreased by 400—700 ml. In some cases the discrepancy between the weight deficit by the second day of bed rest and the decrease of the circulating blood volume suggested that some blood was pooled in the body.

An increase in diuresis and excretion of sodium may develop for several reasons; change in the total and renal haemodynamics, enhancement of the load on the glomerular system of the kidneys, or shift in the renal reabsorption.

In some experimental runs the cardiac stroke volume was measured. As a rule, in the first 24 hours of bed rest this index increased for 4—6 hours and returned to the pre-test level by the end of the day [3]. The rate of glomerular filtration was increased by 17—32% and returned to normal by the end of the fourth day. The renal blood flow measured by means of diodone remained raised (8—11%) at that time. At present it is recognized that the polyuric reaction which controls the fluid volume in the body is induced by changes in the glomerular filtration rate. However, the major factor responsible for the control of the diuretic rate is a change in the tubular reabsorption of water and salts. Gauer et al. [5] have presented convincing data on the role of vasopressin in the control of fluid volumes in the body. The decline in the renal response to vasopressin was found in most test subjects during the first day of bed rest. They showed a significant decrease of the reabsorption of osmotically free water which was very marked during the first hours of the experiment. Besides this, excretion of osmotically active substances (first of all sodium) increased. However, in some experiments reabsorption of osmotically free water did not decrease but even increased. High renal excretion of sodium, increased osmolar clearance and elevated sodium concentration in the urine were noted (Table 1). These symptoms were more typical of osmotic diuresis than of the reduced renal response to vasopressin. In one of his investigations Gauer [4] revealed an increase of osmotic clearance and reabsorption of osmotically free water in an immersed subject with reduced hydration.

In our experiments well trained test subjects took part. During the reference period they led an active mode of life and trained regularly. Their diets were rich in water and salts. Small variations in the body weight, osmotic concentration and electrolyte composition of the blood as well as in the functional test with water load, have convinced us that the amount of water consumed made up completely for its excretion. The circulating blood volume of the test subjects was 4 ± 0.35 l.

It appears that in these test subjects during the first day of bed rest in which there was a drastic decrease in motor activity, the changes of total and renal haemodynamics were the most important in the development of adequate homeostasis. In this case an increased load of water and salts on the glomerular apparatus led to an enhancement of the maximum transport function of the tubules; this finally resulted in an accelerated renal excretion of water and salts against the background of incomplete inhibition of ADH.

Thus, the first acute period of adaptation of the human body to altered gravity is characterized by distinct haemodynamic changes. Later essentially all parameters of fluid-electrolyte balance show cyclic changes: increased renal excretion of water and salts is followed by their increased reabsorption in the tubules. It should, however, be noted that the renal excretion of water and salts at different stages of the experiment varied. For instance, in the first 2—3 weeks of bed rest the increase and decrease of the excretion of water and sodium occur in parallel. By the 20th day diuresis increases, whereas sodium concentration in the urine and its renal excretion decrease. By the 25—30th day the sodium concentration in the urine continues to decrease and its excretion increases. This is followed by a reduction of the reabsorption of osmotically free water in the distal tubules.

At present it is still difficult to give an adequate interpretation of the above

Table 1

Values of the Renal Osmoregulatory Function Tests

Group	Control			Experiment					
				First day			Second day		
	C_{osm}	$T^{C}_{H_2O}$	$U_{osm}V$ (ml h^{-1})	C_{osm}	$T^{C}_{H_2O}$	$U_{osm}V$ (ml h^{-1})	C_{osm}	$T^{C}_{H_2O}$	$U_{osm}V$ (ml h^{-1})
1	2.43 ± 0.1	1.6 ± 0.059	46 ± 1.82	3.23 ± 0.2	2.23 ± 0.118	53 ± 4.2	2.44 ± 0.138	1.59 ± 0.1	44 ± 2.8
2	2.57 ± 0.1	1.68 ± 0.045	45 ± 1.95	4.45 ± 0.15	2.54 ± 0.067	80 ± 5.1	2.62 ± 0.03	1.67 ± 0.017	47 ± 1.6
3	2.6 ± 0.06	1.7 ± 0.046	46 ± 1	3.6 ± 0.26	2.22 ± 0.16	63 ± 4.25	3.07 ± 0.138	1.74 ± 0.03	56 ± 3.3

1, bed rest +6° (with the head kept at an angle of +6° relative to the body)
2, bed rest −2° } the head down position
3, bed rest −6° }

phenomena. It is unclear in which segments of the tubules and under the influence of which hormones the ion transport changes. Apparently, the change in the reabsorption of osmotically free water indicates that distal segments of renal tubules are involved in this process. It is important to note marked excretion of calcium by the 30th day of bed rest. Increase in the calcium excretion and in its concentration in the tubular liquid may bring about a decrease of water reabsorption in the presence of ADH and a decline of sodium reabsorption (Fig. 3).

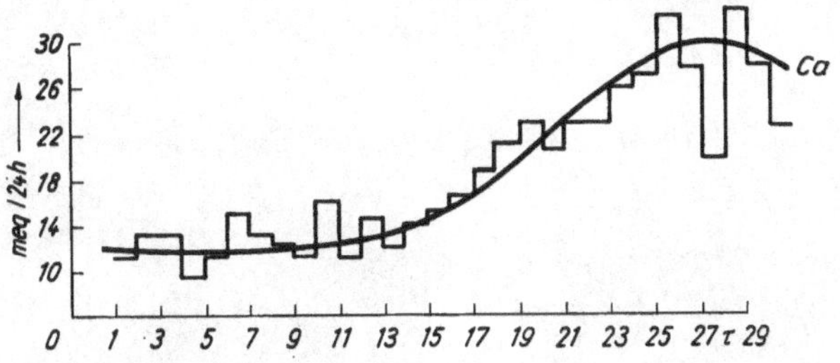

Fig. 3. Renal excretion of calcium during bed rest, mequiv/24 hours.

References

[1] I. S. Balakhovsky, Yu. V. Natochin and G. I. Kozyrevskaya, Proc. IV Int. Man-in-Space Symp. Erevan 1971, Nauka, Moscow 1974 (p. 194).
[2] Yu. V. Natochin, G. I. Kozyrevskaya and A. I. Grigoriev, paper presented at V Int. Man-in-Space Symp., Washington 1973.
[3] B. S. Katkovsky, G. I. Kozyrevskaya, Yu. D. Pomotov and A. I. Grigoriev, paper presented at Soviet Symp. on Adaptation to Muscular Activity and Hypokinesia, Novosibirsk 1970, Nauka, Moscow 1970.
[4] O. Gauer, Proc. IV Int. Man-in-Space Symp. Erevan 1971, Nauka, Moscow 1974 (p. 176).
[5] O. Gauer, J. Henry and C. Behn, Ann. Rev. Physiol. 32, 547 (1970).

Life Sciences and Space Research XIV — Akademie-Verlag, Berlin 1976

CHARACTERISTICS OF POSTURAL SELF-REGULATION IN COMPLEX SPATIAL ENVIRONMENTS AND AFTER-EFFECTS OF WEIGHTLESSNESS

V. I. Myasnikov, O. P. Kozerenko and N. M. Rudomyotkin

Institute of Medico-Biological Problems, Moscow, USSR

The structure of the postural-spatial activity of parachuters, high-board divers and volunteers has been studied during postural tests on sports towers and that of cosmonauts before and after space flight. The experimental results give evidence that the stabilographic pattern shows several different types which correspond to the fixed, "searching" and pliable models of postural self-regulation, regardless of professional experience. During postural tests the cosmonauts exhibited deformation of the stabilographic pattern which lost its individual features and was converted into a stable curve which was represented by unmodulated fluctuations with a period of 0.15 sec that was similar to the fixed model. Psychophysiological mechanisms of the changes are discussed.

On the earth information on muscular tone and postural activity [10] of a healthy man is either not realized or preceived only partially when special efforts are made (e.g. in training of sportsmen) or are distorted in the consciousness of patients (e.g. phantom pains, metamorphopsia, etc.). During flights aboard aircraft or spacecraft situations due to change in gravity develop unavoidably and these require reorganization of normal postures and posture-coordination activity. The information which enters the consciousness on the re-organization of the posture-tonic apparatus may act as "noise", which can be accompanied by psychosensory and emotional changes and disturbances in planned activity. All this makes the postural-spatial activity an important object of study.

When searching for an experimental model, we selected regulation of the vertical pose because it reflects specifically the human interaction with other structural elements of the environment, and provides material for an analysis of the microstructure of this interaction as a pattern of displacements of the body-mass center over the plane of the support fixed stabilographically [14]. The experiments included postural tests to maintain the upright position with the eyes open and closed, the classical Romberg pose and balancing on a moving support. Postural self-regulation was studied in experiments using 18 m high parachute towers and 10 m high diving-board towers with a narrow support; healthy male volunteers, professional parachuters and high-board divers, aged 21—35, participated.

Observations on the behaviour of test subjects standing on the towers demonstrated constrained posture, noticeable tremor of the entire body, impossibility of moving within the limits of the platform, involuntary seizure of the safety belt, poor performance and occasionally neglect of the experimenter's instructions, and sometimes refusal to continue the experiment.

Among subjective sensations complaints of numb feet and legs, loss of sensitivity of the lower body, dissociated sensitivity, feelings of pushes in the back, falling, etc. should be mentioned; all this can be assessed as psychosensory changes within the limits of the functional change of the body scheme. These complaints were followed by the appropriate vegetative reactions: local and general hyperhydrosis, rapid changes in the vasomotor reflexes of the face, dissociation of skin temperature, increase in the pulse rate and respiration rate, pronounced variations of the skin-galvanic reflex. These trends were typical of both volunteers and sportsmen.

Stabilographic parameters assessed with reference to the integrated stabilographic pattern obtained in various experimental situations showed the following types of stabilographic parameters: (i) the first type was characterized by a stable picture irrespective of the position of the support and spatial conditions, minimum number of working zones and maximum localization of fluctuations of the body mass centre in one of them (Fig. 1, A); (ii) the second type was characterized by the maintenance of the main features of the picture without clear-cut variation during spatial changes without definite regularities in the distribution of fluctuations of the body mass centre about working zones (Fig. 1, B); (iii) the third type was characterized by variability of the picture in conformity with a change of the experimental situation (Fig. 1, C). According to modern concepts [1—6], the mechanism of regulation of the vertical posture, i.e. of the musculo-skeletal system, is maintained by a harmonious interaction of different cerebral systems responsible for the spatial analysis (afferent flows of kinesthetic, visual, vestibular sensations) or tuning of the musculo-skeletal system. It is interesting that the regulation of the postural activity of the first type in the above-mentioned situations is accompanied by well expressed behavioral and vegetative manifestations. In the second type of regulation of postural activity behavioral and vegetative reactions are indistinct; behavioral reactions may, in fact, be absent. The typical feature in the third type of regulation of the postural activity is elimination of the behavioral component or, figuratively speaking, phenomenon of absolute "spatial courage". All the types of postural models described were detected in every professional group independent of their occupational experience.

A space mission and its basic characteristic feature, zero gravity, demand special efforts (learning) for the organization of optimal functions. This is indicated by postural and motor discomfort which cosmonauts reported during adaptation to weightlessness. Feelings of discomfort included inconsistency of spontaneous muscular efforts suggested by the expected action and reduced accuracy of arbitrary movements (missing of the object); the absence of the normal (horizontal) direction and change of the pattern of body movements in space; necessity of mental control of regulation of muscle efforts and visual control of the results of movement; transient changes of self-control and self-perception in the form of spatial illusions (head-down position, back-tilt of the body, head, etc.) and spatial disorientation [8—11]. In other words, cosmonauts arrange their interaction with the environment, especially during adaptation to weightlessness, deliberately, actively, by means of trial and error in motor behaviour.

As is known [11—13], on return to earth cosmonauts reported a noticeable increase in the weight of objects, difficulties in movement and organization of musculo-coordination efforts, unwillingness to move and maintain the upright position. This manifested itself in an altered gait and statics and in neurological symptoms of ataxia (dyskinesia, asynergia) the extent of which was correlated

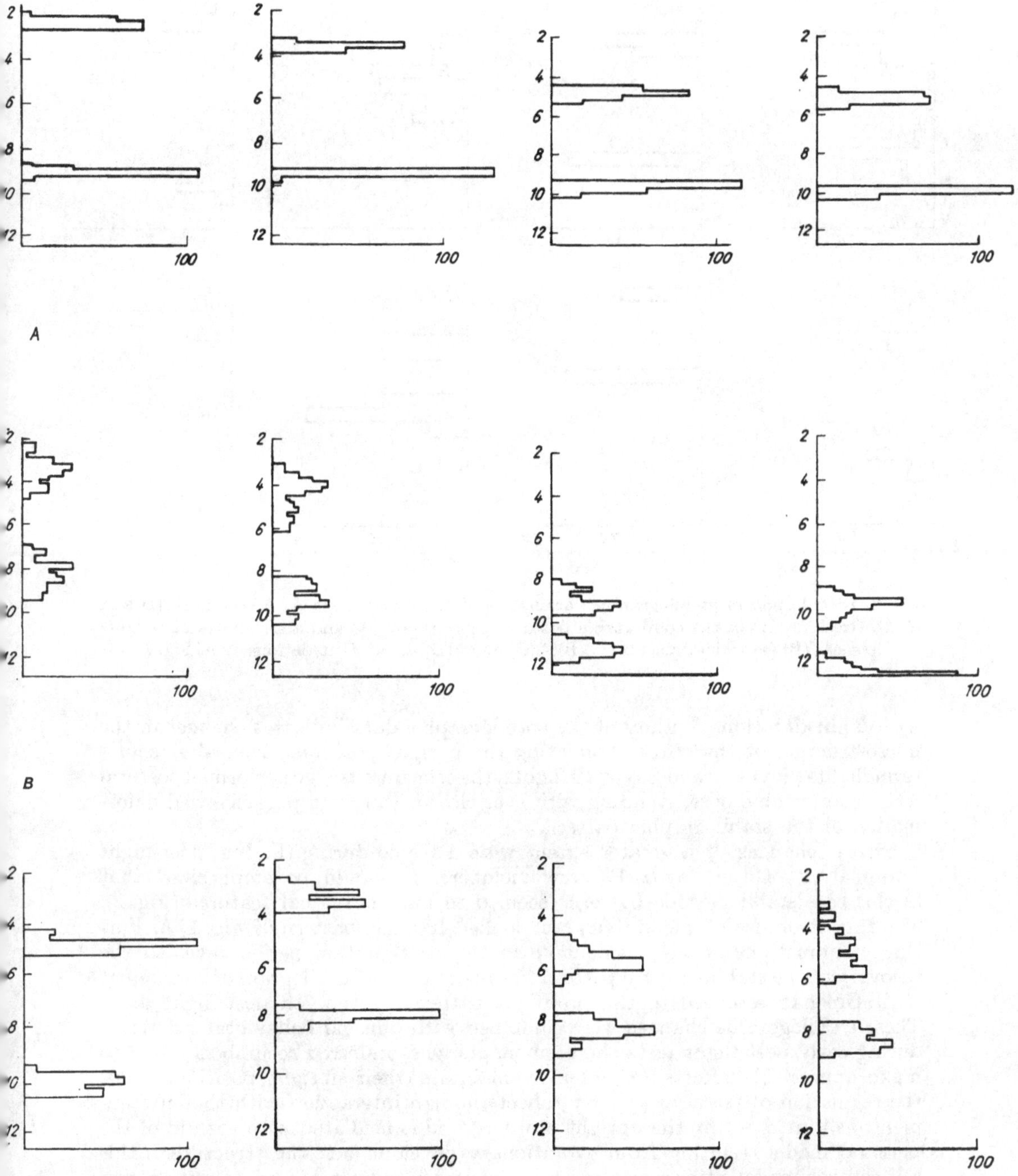

Fig. 1. Dynamics of the integrated stabilographic pattern obtained in various experimental situations.

Legend: The first (A), second (B) and third (C) types of regulation of the postural activity. Abscissa: number of shifts of the general centre of gravity of the body (GCGB) in sagittal plane (upper graph) and frontal plane (lower graph); ordinate: sequence number zones of the shifts GCGB.

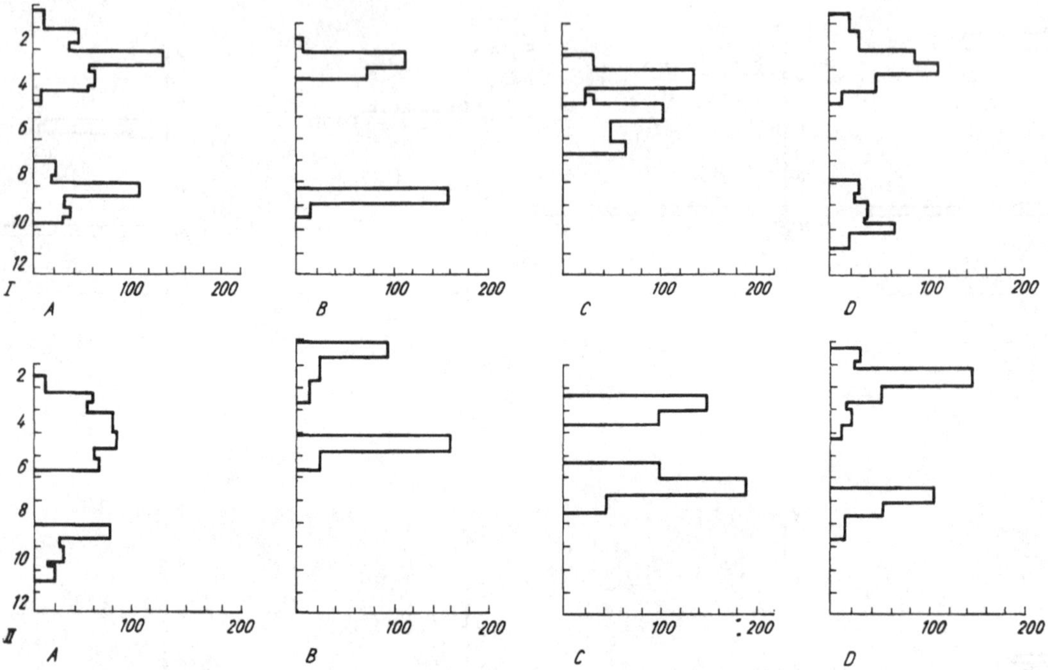

Fig. 2. Distribution of stabilographic parameters of the cosmonauts A. A. Gubarev (I) and G. M. Grechko (II) in the comfortable posture before flight (A) and after-effects of weightlessness (B) [examinations after 0(B), 3(C) and 7(D) days]. Conventions as in Fig. 1.

with flight duration. A study of the stabilographic data indicates changes in the microstructure of the efforts regulating the upright position. Even after short-term flights (Soyuz 12 and Soyuz 13 flights) the cosmonauts who performed postural tests (comfortable pose, standing with eyes closed, Romberg pose) showed deformation of the stabilographic pattern.

Similar changes of a greater extent were detected during the first post-flight examination of the Soyuz 17 crew members. It should be emphasized that in this case stabilographic patterns seemed to lose individual features (Fig. 2). The fine structure of the stabilogram looked like the pattern of Fig. 1, A. Further examinations of the cosmonauts in the readaptation period revealed the recovery of the stabilographic pattern (for instance, the stabilogram of cosmonaut G. M. Grechko returned to the pre-flight pattern by the 7th post-flight day). These stabilographic changes were combined with unusual behavioral reactions: during early post-flight tests the cosmonauts who preferred recumbency had to make appreciable efforts in order to maintain their upright position. Thus, the regulation of postural activity reflects human interaction with the environment (as illustrated by the upright position) and is mediated with the aid of the postural model resulting from evolutionary development; the structure of this self-control model utilizes information about the tuning of the executive segmentary mechanism of postural activity (the musculo-skeletal system) via modulating influences of all supraspinal levels of the cerebral organization.

References

[1] B. G. Ananyev, L. M. Vekker, B. F. Lomov and A. V. Yarmolenko, Tactile Perception in the Processes of Cognition and Labour, Izd-vo APN RSFSR, Moscow 1959.

[2] P. K. Anokhin, in: Biological Aspects of Cybernetics, Izd-vo AN SSSR, Moscow 1963,

[3] P. K. Anokhin, Principal Problems of the General Theory of Functional Systems, Izd-vo AN SSSR, Moscow 1971.

[4] N. A. Bernstein, On the Structure of Movements, Moscow 1947.

[5] N. A. Bernstein, Voprosy psychologii, **6** (1957).

[6] N. A. Bernstein, Essays on Physiology of Movement and Physiology of Activity, Medizina, Moscow 1966.

[7] E. I. Vorobyov, Yu. G. Nefyodov and L. I. Kakurin, Kosm. Biolog. i Med. No. **4**, 46, (1969).

[8] E. I. Vorobyov, L. I. Kakurin and A. D. Egorov, Kosm. Biolog. i Med. No. **6**, 26 (1970).

[9] L. S. Vygotsky, Development of Higher Psychic Functions, Izd-vo APN RSFSR, Moscow 1960.

[10] F. D. Gorbov, Diss. Doct. Med., Moscow 1962.

[11] N. N. Gurovsky, A. V. Eryomin, O. G. Gazenko, A. D. Egorov, I. I. Bryanov and A. M. Genin, Kosm. Biolog. i Med. **2**, 48 (1975).

[12] B. N. Petukhov, Yu. N. Purakhin, V. S. Georgievsky et al., Kosm. Biolog. i Med. No. **6**, 50 (1970).

[13] Yu. N. Purakhin, L. I. Kakurin, A. S. Georgievsky, B. N. Petukhov and V. M. Mikhailov, Kosm. Biolog. i Med. No. **6**, 47 (1972).

[14] V. S. Gurfinkel et al., Regulation of Posture, Moscow 1965.

Life Sciences and Space Research XIV — Akademie-Verlag, Berlin 1976

HYPERGRAVITATION AND SYMPATHO-ADRENERGIC REACTIVITY

P. GROZA, R. CÂRMACIU, E. NICOLESCU, S. CANANĂU, R. VRÂNCIANU and D. BOBIC

The "D. Danielopolu" Institute of Normal and Pathological Physiology, Bucharest, Rumania

The sympatho-adrenergic reaction of the organism subjected to hypergravitation was investigated in rats exposed to $+6G_z$. The electro- and cardiotachograms recorded telemetrically were correlated with the adrenal catecholamine content. The determinations were made in controls and in rats treated with hexamethonium and atropine administered separately or together. The rats reacted to acceleration initially by a short bradycardia followed by tachycardia. In one case there was a sustained bradycardic response.

Hexamethonium (C_6) lowered the resting heart rate, attenuated the initial bradycardia and reduced the consecutive tachycardia. In some rats during centrifugation a bradycardic and arhythmic response was found sometimes ending in cardiac arrest. Hexamethonium stored the catecholamines in the adrenals as a result of their increased concentration.

Atropin brought about resting tachycardia, disappearance of bradycardia from the very beginning of centrifugation and a more accentuated tachycardia.

Atropin and hexamethonium administrated together diminished the response to acceleration.

1. Introduction

Sympatho-adrenergic reactivity may be assessed by some visceral responses such as tachycardia, as well as by catecholamine secretion. As suggested by von Euler, an increased urinary noradrenaline excretion corresponds to a peripheral sympathetic reaction and excretion of adrenaline to an adreno-medullary response. Similar information can be obtained from the catecholamine content in the adrenals. The importance of tachycardia in readjusting the blood circulation during hypergravitation has been demonstrated [1—3]. Up to a certain limit, increase in the heart rate is proportional to hypergravity. In man psychogenic tachycardia is already present to a certain extent before beginning the experiments [3—5].

In some of the experiments, bradycardic response was noticed during the initial stage of acceleration. Grandpierre et al. [6] found a permanent bradycardia in one rat submitted to positive gravitational acceleration during a hyperbolic flight. In other experiments, some rats showed initial bradycardia during acceleration [7]. During acceleration at values close to physiological tolerance (before the onset of visual disorders), ventricular extrasystoles occur as a rule, and are particularly frequent in long-term experiments. Changes of the ECG, indicating various degrees of cardiac ischemia, were also recorded [3—5].

Tachycardia could be related to an increased tone of the sympathetic nervous system, including the adrenal medulla. The activity of the sympatho-adrenergic system could be judged from the amount of urinary catecholamines. Catecholamine concentrations in the blood and adrenal gave similar indications. In man acceleration causes an increased excretion of noradrenaline and adrenaline and an increase in plasma free fatty acids, indicating an increased sympathetic activity. As adrenaline has been previously related to emotional disturbances, this increase in urinary catecholamines was ascribed to emotional stress [8, 9]. After the administration of a beta-blocking agent (propranolol) tachycardia diminished [2].

The present paper describes a study of the rat under the influence of acceleration and of some drugs acting on the autonomic system (hexamethonium and atropine), on the heart rate and sympatho-adrenal activity assessed by the amount of catecholamines in the urine and adrenal medulla.

2. Methods

The experiments were carried out on 16 male Wistar rats (130—260 g) submitted to centrifugal acceleration ($+6G_z$). The rats were placed in appropriate containers and the heart activity was continuously registered telemetrically by electrocardiogram and cardiotachogram, with an apparatus adapted to the fast heart rate of this species. The heart rate was measured before (15—30 min), during (6—10 min), and after (20 min) acceleration. The animals were given intraperitoneally a single dose of hexamethonium (2—10 mg/kg body weight) or of atropine (50 μg/kg body weight). Some of the animals were given hexamethonium after atropine injection. The controls were given equivalent amounts of physiological saline. The catecholamine content of the adrenals was estimated by a fluorimetric method [10].

3. Results

Heart rate. In the present experiments the resting heart rate was between 300 and 420 beats/min. There were spontaneous oscillations of the heart rate occurring about 3—5 times per minute and causing a variation of about ± 10 beats per minute. They are probably related to blood pressure oscillations. They

→

Fig. 1. a, rest cardiotachogram; frequency oscillations in young rat (CTG = cardiotachogram, ECG = electrocardiogram, b.p.m. = beats per minute); b, bradycardic response at the beginning of the positive longitudinal acceleration and the tachycardic response; c, rat responding with bradycardia during acceleration; d, tachycardia in deceleration; e, rat treated with 10 mg kg^{-1} hexamethonium i.p. becoming bradycardic; f, absence of bradycardic reaction at the start and diminution of tachycardia during centrifugation after hexamethonium; g, bradycardia during centrifugation after hexamethonium; h, bradyarhythmia in some rat (lower trace; initial normal frequency); i, rat treated with hexamethonium with bradycardia produced by centrifugation who died (lower trace, normal frequency); j, atropine induced tachycardia and decreased oscillations of heart rate; k, absence of the short initial bradycardic response and the tachycardic response to acceleration in atropine-treated rat; l, absence of frequency oscillations and diminished tachycardic response in atropinized rat treated also with hexamethonium.

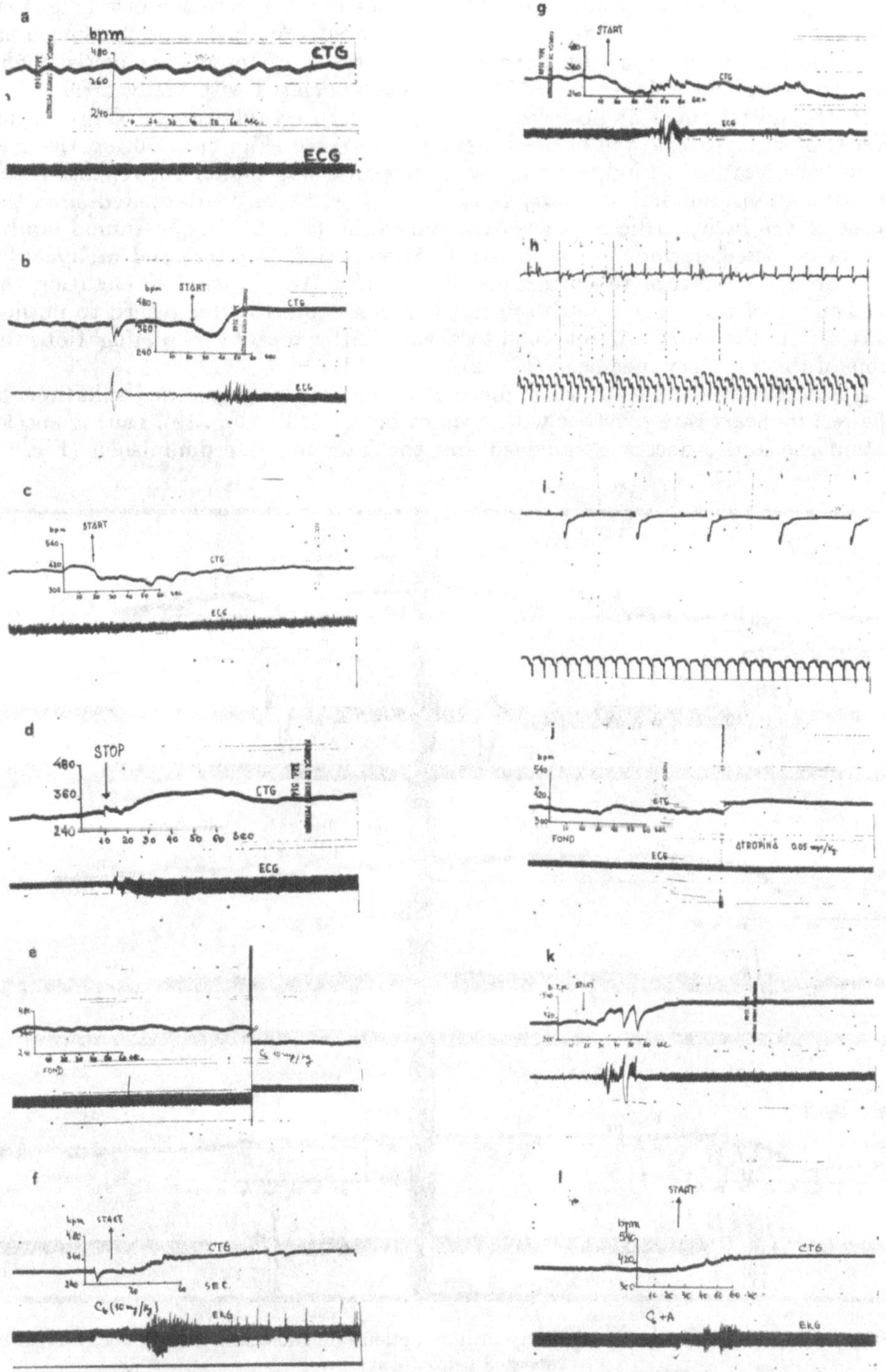
a
bpm
480
360
240
CTG
ECG
b
bpm
START
480
360
240
CTG
ECG
c
START
540
420
300
CTG
ECG
d
STOP
480
360
240
CTG
10 20 30 40 50 60 sec
ECG
e
FOND
f
bpm
START
480
360
240
CTG
30
60
SEC
EKG
g
START
CTG
ECG
h
i
j
bpm
540
420
300
CTG
FOND
ATROPINA 0.05
ECG
k
START
l
START
bpm
540
420
300
CTG
C + A
EKG

were more evident in young rats, due to their autonomic lability (Fig. 1a). Respiratory arhythmia was sometimes observed. Against this background, short periods of bradycardia developed as a result of interfering noises (light was without effect). Movement of the rats was associated with tachycardia.

In the initial stage of positive longitudinal hypergravitation, a short initial period of bradycardia was noticed lasting 15—25 sec (Fig. 1b). During the rest of the acceleration period, a tachycardic response was found. On repeating the centrifugation, the initial bradycardia was progressively attenuated and the onset of the tachycardic response occurred earlier (Fig. 2). In one animal bradycardia persisted during the whole acceleration period (Fig. 1c) and tachycardia developed after return to normal gravitation (Fig. 1d). During acceleration the oscillations of the heart rate disappeared. A few minutes after return to normal gravitation, the heart rate returned to normal. After repeated centrifugations the time of this recovery increased (Fig. 2).

Hexamethonium induced either diminution or disappearance of oscillations of the resting heart rate together with a slight bradycardia (Fig. 1e). During acceleration the initial decrease vanished and the later increase diminished (Fig. 1f)

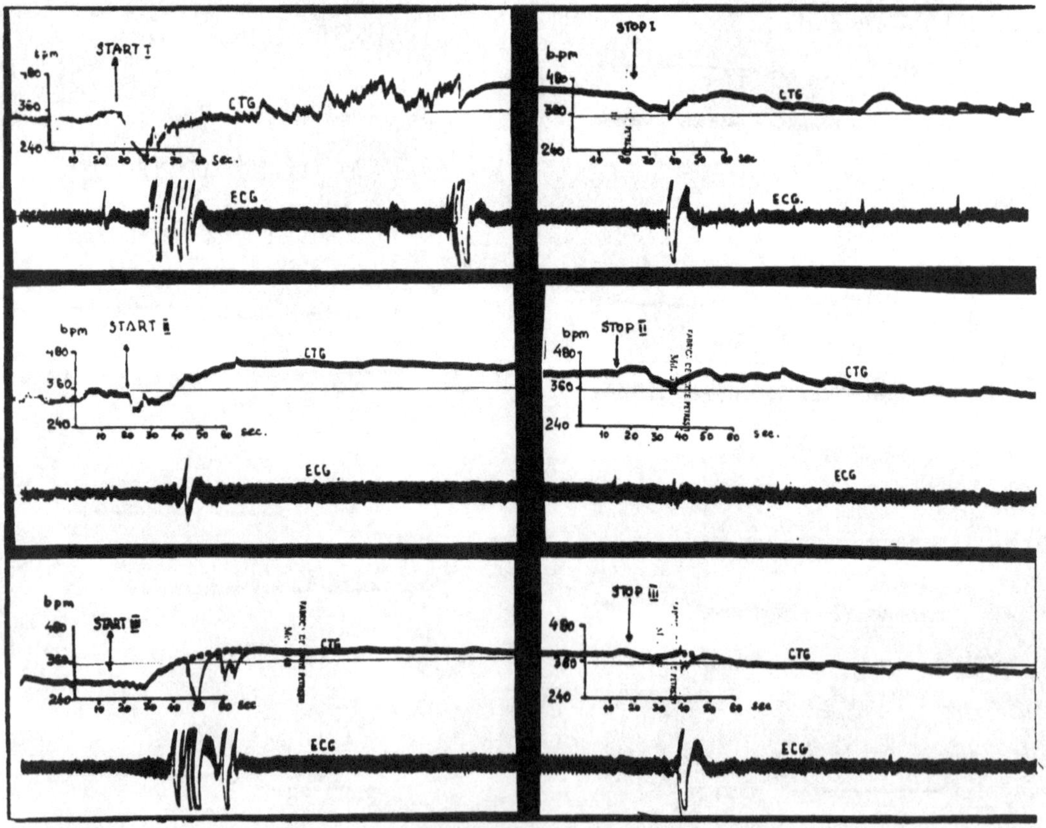

Fig. 2. Attenuation of the initial bradycardic reaction and increased time of recovery after repeated centrifugations.

or even disappeared. In some animals during centrifugation a bradycardic (Fig. 1g) and arhythmic response (Fig. 1h) were found. After an arhythmic and bradycardic response (Fig. 1i) cardiac arrest was observed in an old animal.

Atropine induced tachycardia and a decrease in the oscillations of the heart rate (Fig. 1j). The initial bradycardic response to acceleration was less marked or even absent, but the subsequent tachycardia was enhanced (Fig. 1k). After cessation of centrifugation recovery of the heart rate was slower. In atropinized

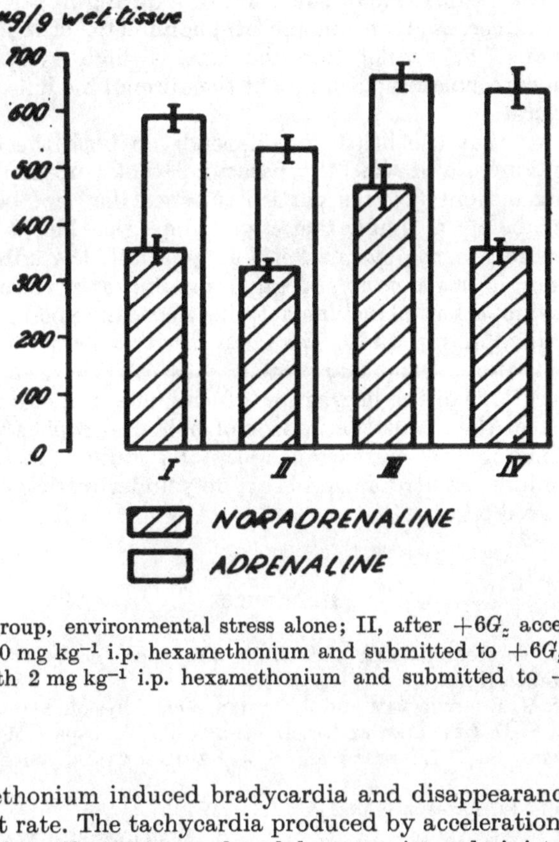

Fig. 3. I, control-group, environmental stress alone; II, after $+6G_z$ acceleration; III, rats pre-treated with 10 mg kg^{-1} i.p. hexamethonium and submitted to $+6G_z$ acceleration; IV, rats pretreated with 2 mg kg^{-1} i.p. hexamethonium and submitted to $+6G_z$ acceleration.

animals, hexamethonium induced bradycardia and disappearance of the oscillations of the heart rate. The tachycardia produced by acceleration was diminished (Fig. 1l). A similar effect was produced by atropine administered after hexamethonium.

Adrenal catecholamine concentration. Noradrenaline and adrenaline concentration in the adrenals were investigated in normal rats, in rats submitted to the environmental stress of the experiment and after $+6G_z$ acceleration for 6 minutes, with or without hexamethonium administration. Environmental stress does not produce significant noradrenaline modifications but lowers adrenaline levels. Centrifugation also brings about a further slight diminution of adrenaline. Catecholamine concentrations were higher in the adrenals after hexamethonium (Fig. 3). After centrifugation urinary catecholamine determinations showed an increase in adrenaline elimination.

4. Discussion

Besides periphera vasoconstriction and water retention, increase in the heart rate contributes to haemodynamic adjustment during hypergravitation. However a persistent bradycardia was also noticed [6]; this represents a pathological reaction. In our experiments an initial bradycardia was present which could be ascribed to the interfering action of noise, since the same reaction was obtained during the resting period. Photic stimulation had no effect on the heart rate.

Diminution of the oscillations of the heart rate during hypergravitation might be ascribed to an increased autonomic amphomimetic stimulation. The later tachycardic response to acceleration indicates a high sympathetic activity, confirmed by the catecholamine changes of the adrenal medulla, but less evident in the urinary excretion.

It is well known that the heart rate depends on both the sympathetic and parasympathetic tonus, and that the parasympathetic normally predominates. With very few exceptions [2] this particular aspect has not been investigated. Hexamethonium, being a sympathetic ganglioplegic, blocks the cholinergic synaptic transmission in the sympathetic ganglia and the adrenal medulla, as well as the cholinergic transmission of vagal impulses at myocardial level. Thus, it also has a parasympathetic blocking activity that can explain the slight bradycardic effect obtained in our study. The same effect also explains the diminution or absence of oscillations of the heart rate.

The severe alterations of the heart rate (causing death in one animal) might be attributed to lack of the sympathetic dromotropic, inotropic and trophic effect. Atropine, by blocking the muscarinic receptors, induces resting tachycardia. After hexamethonium the atropinized heart acts under resting conditions practically as a denervated organ.

References

[1] H. BJURSTEDT, G. ROSENHAMER and G. TYDÉN, paper presented at COSPAR XVI Meeting, Constance 1973.
[2] H. BJURSTEDT, G. ROSENHAMER and G. TYDÉN, Acta Physiol. Scand. **90**, 491 (1974).
[3] R. R. BURTON, S. D. LEVERETT and E. D. MICHELSON, Aerospace Med. **45**, 1115 (1974).
[4] M. J. PARKHURST, S. D. LEVERETT and S. J. SHUBROOKS JR, Aerospace Med. **43**, 708 (1972).
[5] S. J. SHUBROOKS JR, Aerospace Med. **43**, 1200 (1972).
[6] R. GRANDPIERRE, R. ANGIBOUST, R. BRICE, B. ACAILLER, P. CAZARD, G. CHATELIER, O. OLSEN and F. SORET, Revue des corps de santé, **3**, 93 (1962).
[7] P. GROZA, E. BUBUIANU and S. CANANĂU, Rev. Roum. Morph. Embriol. Physiol. **12**, 3 (1975).
[8] E. J. BECKER and F. KREUZER, Pfluger Arch. ges. Physiol. **304**, 1 (1968).
[9] McC. GOODALL, J. Clin. Invest. **41**, 197 (1962).
[10] U. S. VON EULER and F. LISHAJKO, Acta Physiol. Scand. **51**, 348 (1961).
[11] P. GROZA, S. CANANĂU, A. BORDEIANU, E. DUMITRESCU-PAPAHAGI, C. VLADESCU and E. DANIELIUC, Rev. Roum. Physiol. **20**, 157 (1974).

Life Sciences and Space Research XIV — Akademie-Verlag, Berlin 1976

A STUDY OF THE CUMULATIVE EFFECTS OF REPEATED EXPOSURES TO RADIAL ACCELERATIONS

T. N. Krupina, G. P. Mikhailovsky, A. Ya. Tizul, M. P. Kuzmin, N. I. Tsyganova and E. B. Shulzhenko

Institute for Medico-Biological Problems, Moscow USSR

After exposure to transverse accelerations of 4—10*g* changes in the vascular system of the eye and immunobiological resistance of the body persisted from 5 to 15 days. The data give evidence that the state of retinal vessels is the most informative index of the cumulative effect on the body and its recovery.

The effect of transverse accelerations on the human body has been widely investigated [1, 4—6]. Nevertheless, the influence of repeated centrifugation and the duration of the after-effect are inadequately known. No reliable criteria for the selection of optimal time intervals between exposures have been established.

In order to assess the time of the after-effect and to select optimal time intervals between repeated centrifugations, test subjects who were exposed to transverse accelerations of 4—6*g* for 120 sec, 8*g* for 60 sec and 10*g* for 20 sec were examined before and after the exposure. The examinations included the central nervous system, cardiovascular system, immunobiological reactivity, biochemical parameters of the blood and adrenal function. The investigations showed that after the exposures most of the systems developed adequate changes which were reversible, remaining within the limits of compensatory-adaptive reactions. The time of the recovery of the changes varied for different systems and depended on the value and duration of centrifugation and individual features of the test subject.

Changes in the nervous system included lability of autonomic reflexes and responses, excitation of tendon and periosteal reflexes, decline in the amplitude of brain biopotentials and increase of brain sensitivity to the changes in blood gas composition; these parameters returned to normal 24 hours after the exposure.

Electrocardiographic examinations were carried out 4—6 hours and 1, 3 and 5 days after the exposure. No clear-cut or characteristic changes in the ECG were noted as early as 4—6 hours after centrifugation. Instability of the heart rate and T-wave amplitude was the only abnormal finding, some test subjects showed an increase of the heart rate by 10—20 beats/min and decrease of the T-wave amplitude in the standard and left chest leads by 0.3—0.5 mm. The change in the T-wave amplitude persisted for 2 days, or in some cases 3—5 days after the exposure.

Mechanocardiographic studies of haemodynamics showed that an exposure to 4—6*g* acceleration was followed by an increase of the mean arterial pressure (by 20%) and cardiac index (by 51.7%) and a slight decrease of the specific peripheral resistance. These changes persisted only during the first post-exposure day. A decline in the specific peripheral resistance, and to a lesser extent of the mean arterial pressure, with an almost unchanged cardiac index remained for 7 days after an exposure to 10*g*.

Ophthalmological examinations of the test subjects demonstrated a marked reaction of the vessels of the optic nerve disc and retina: hyperaemia and ectasia of disc capillaries, moderate pericapillary oedema, and engorgement of retinal arteries and veins. Campimetric measurements showed an increase of the blind spot (physiological scotoma). Two periods can be distinguished in the changes of the vascular system of the retina: (i) the first three hours after the exposure when reactive hyperaemia predominated and (ii) the first 24 hours after the exposure when changes induced by disorders in the permeability of retinal vessels prevailed (pericapillary oedema of the retina, enlargement of the blind spot). After an exposure to 10*g* the parameters of the eye returned to normal by the 5—6th day. The size of the blind spots returned to the pre-test level one day later than the parameters of the eye. The Valsalva manoeuvre showed an increase of the blind spots even at that time. This suggested latent changes in retinal vessels. The normal size of the blind spots during the Valsalva manoeuvre is an indicator of the total recovery of the vascular system of the eye. During repeated centrifugations the recovery of the blind spot took 10—12 days, indicating the cumulative effect of transverse acceleration.

An examination of the adrenal function showed that an exposure to 4—6*g* brought about increased excretion of ketosteroids although they remained within the level of normal variation. After exposure to 8*g* the excretion of 17-ketosteroids and 17-hydroxycorticosteroids increased 2.6—3 times compared with the initial level and then decreased significantly below the control value. Excretion of the adrenal hormones returned to normal after 7—15 days after the exposure. It should however be noted that the variations in the hormonal excretion remained, as a rule, within physiological limits.

On the 2nd post-exposure day the level of properdine was 3—4 times less than the control. Then it tended to increase and reached the initial value by the 15th day in most test subjects. After an exposure to 10*g* the level of properdine returned to normal between 7 and 15 days.

Thus, our observations of different functions of the body following an exposure to acceleration give evidence that 4—6, 8 and 10*g* induce typical polymorphic changes. Vitally important functions (heart rate, respiration rate, arterial pressure) recover during the first minutes after the exposure. Changes in the vascular system of the eye, functional activity of the adrenals and immunobiologic system of the body persist for a longer time and return to the normal within 5—15 days after the exposure.

Our findings, and the data published in the literature, give grounds for believing that the most informative index of the recovery of the functions of the human body after an exposure to transverse acceleration is the state of the vessels of the eye, which may to a certain extent characterize the vascular system of the brain [2, 3] and the entire body. In addition, pronounced changes in immune parameters and adrenal function also characterize recovery processes after a prolonged exposure to radial acceleration.

References

[1] A. S. Barer, Kosm. Biolog. i Med. No. 1, 57 (1967).
[2] A. A. Kedrov and A. I. Naumenko, Problems of Physiology of Intracranial Circulation and Their Clinical Substantiation, Leningrad 1954.
[3] V. V. Klosovsky, Cerebral Circulation, Moscow 1951.
[4] A. R. Kotovskaya, S. I. Lobashkov, S. F. Simpura, P. M. Suvorov and G. F. Khlebnikov, Problems of Space Biology, Moscow, 1962, 2, 238 (1962).
[5] P. M. Suvorov, Doctorate Dissertation, Moscow 1969.
[6] E. B. Shulzhenko, Proc. IX Tsiolkovsky Lectures, Kaluga 1974 (Publ. Moscow 1975).

Life Sciences and Space Research XIV — Akademie-Verlag, Berlin 1976

HEAT EXCHANGE BETWEEN THE ORGANISM AND ENVIRONMENT UNDER CONDITIONS OF WEIGHTLESSNESS; METHODICAL APPROACH

L. Novák

Institute of Biophysics, Czechoslovak Academy of Sciences, Brno, Czechoslovakia

The spontaneous streaming of air around surfaces of warm bodies conditioned by gravitation is missing in the weightless condition. This implies a change in the thickness of the surface air layer and its interference with the heat output of an organism. The paper describes the use of an electric dynamic katathermometer (EDK) for automatic and continuous scanning of heat output and presents results of measuring basic characteristics of the surface layer under defined laboratory conditions, and their relation to heat output.

1. Introduction

The gravitational field causes a spontaneous flow of air; the heated air rises and produces, on the surface of a warm body, a layer with laminar flow designated by Büttner [1] the *surface layer*. This layer is of utmost importance for the maintenance of heat homeostasis of the organism. It determines the value of heat output by natural convection, and its thickness also determines the value of the heat output by forced convection and perspiration [2, 3]. Under conditions of weightlessness this natural regulatory mechanism of natural convection is absent. An experimental estimation of the thickness and heat conductivity coefficient of this surface layer by means of a classical method using a thermocouple is laborious and time consuming [1]. However, without such data studies on physiological mechanisms of thermoregulation of a homoiothermic organism under weightlessness are incomplete and imprecise. This problem may be solved successfully by using an electric dynamic katathermometer (EDK) whose sensor allows, in contrast to classical katathermometers, automatic and continuous investigations of values of heat output under various conditions [4, 5].

2. Methods

Experiments were carried out under standard conditions in a copper cylinder (diameter 30 cm, height 32 cm); the required temperature of the cylinder walls was maintained by means of an ultrathermostat U-10. The EDK sensor was put on a tripod placed in the middle of the cylinder together with a small fan. To determine the effect of the surface quality and body size on the values investigated, cylindrical sensors of diameters $D_1 = 2.0$ cm and $D_2 = 5.7$ cm were used with polished duralumin (emissivity 0.29) or dull black (emissivity 0.95) surfaces, respectively. The temperature of the sensor was 37 ± 0.1 °C, those of the walls and air were 15, 25 and 35 °C. The air stream velocity (0.01 to 1.5 m s^{-1}) was regulated according to a calibration curve by means of the stabilized terminal voltage of the fan. Values of heat loss by convection (H_c), the heat transfer coefficient (h_c), and the thickness of the surface layer (δ) were calculated according to the following equations:

The heat loss by convection

$$H_c = H - H_R = H - \varepsilon\sigma(T_1^4 - T_R^4) \text{ W m}^{-2} \tag{1}$$

where H = measured values of the heat output (W m^{-2}), H_R = heat output by radiation (W m^{-2}); ε = emissivity of the surface (0.29 and 0.95); σ = Stefan-Boltzmann constant = 5.668×10^{-8} (W m^{-2} deg^{-4}); T_1 = temperature of the sensor; T_R = temperature of walls (°K).

The coefficient for heat transfer by natural and forced convection

$$h_c = H_c/\Delta T \quad \text{W m}^{-2}\text{ deg}^{-1} \tag{2}$$

where

ΔT = difference between temperatures of the sensor surface and air (°C).

For the thickness of the surface layer

$$\delta = \lambda/h_c \quad \text{cm} \tag{3}$$

where

λ = heat conductivity coefficient of air 2.38 W cm^{-2} deg^{-1}.

3. Results

As shown in Table 1 the value of the heat transfer coefficient h_c increases with increasing wind velocity; however, it was lower when using a cylindrical sensor of a greater diameter. This finding coincides with the well-known fact that the diameter D of a cylindrical body is related to the formation of the surface layer

Table 1

Average Values of Heat Transfer Coefficient h_c (W m^{-2} deg^{-1}) measured on Sensors with Polished Surfaces, and Different Diameters $D_1 = 2.0$ cm, $D_2 = 5.7$ cm at $\Delta T = 2$, 12 and 22 °C and various Wind Velocity v (m s^{-1})

v (m s^{-1})		0.01	0.43	0.84	1.47
	$\bar{x}$ (W m^{-2} deg^{-1})	4.32	14.16	18.80	23.73
$D = 2.0$ cm	SE (W m^{-2} deg^{-1})	±1.18	±1.26	±0.52	±0.90
	n	6	6	6	6
	$\bar{x}$ (W m^{-2} deg^{-1})	2.84	10.93	14.67	18.27
$D = 5.7$ cm	SE (W m^{-2} deg^{-1})	±0.70	±1.07	±0.41	±0.15
	n	6	6	6	6

SE = standard error; n = number of estimates

of laminar streaming air; this results from the relationship between Nusselt's and Reynolds numbers [2, 3].

Lower values of the coefficient h_c are due to different thickness of the surface layers of laminar flowing air (Table 2). The surface layer is higher in the sensor with a greater diameter. Average values of h_c calculated from data obtained at ΔT of 2, 12 and 22 °C show the highest variation under conditions of low air

Table 2

Average Values of the Surface Layer Thickness δ (cm) measured on Sensors with Polished Surfaces, and Different Diameters $D_1 = 2.0$ cm, $D_2 = 5.7$ cm at $\Delta T = 2$, 12 and 22 °C, and various Wind Velocities v (m s^{-1})

v (m s^{-1})		0.01	0.43	0.84	1.47
	$\bar{x}$ (cm)	0.580	0.168	0.126	0.100
$D_1 = 2.0$ cm	SE (cm)	±0.17	±0.02	±0.003	±0.003
	n	6	6	6	6
	$\bar{x}$ (cm)	0.876	0.219	0.162	0.130
$D_2 = 5.7$ cm	SE (cm)	±0.24	±0.02	±0.004	±0.002
	n	6	6	6	6

SE = standard error; n = number of estimates

velocity (±25—27%). With increasing wind velocity the dependence of variation on ΔT rapidly decreases, so that at the wind velocity $v = 1.5$ m s^{-1} it is only 1%. The accuracy of EDK measurements is better than 1% and for that reason it enables one to follow accurately the dependence of the h_c value and the thickness of the surface layer δ on the temperature gradient, even under conditions of low air velocity. In calm air the mean thickness of the surface layer for the sensor with a diameter of 5.7 cm is 0.88 cm; this corresponds with values measured by means of a polished thermocouple on the surface of the human body [1].

For that reason the relationship between the average thickness δ of the surface layer and the air velocity, as measured on the sensor with the diameter of 5.7 cm, is presented in Fig. 1. A very rapid decrease in the thickness of the surface layer occurs especially in the zone of velocities of $0.01 < v < 0.5$ m s^{-1}; this is associated with a marked increase in heat output by both spontaneous and forced convection (H_c) and, therefore, by an expressed increase in the total heat output H. If a polished surface is used the value of the heat output by radiation is relatively low (Fig. 1). Within the zone of normal values of man's heat output (1—2 met* = 50—100 W m^{-2}) the total value of the heat output is markedly dependent on the air flow velocity at the value of 1 m s^{-1} at relatively low values of the temperature difference ($\Delta T = 6$ °C). Higher values of ΔT together with air flow velocity over 1 m s^{-1} exceed approximately three times the value of 50 W m^{-2}. For the sensor with a dull black surface the heat output by radiation is approximately three times higher, so that the value of 50 W m^{-2} is reached as early as at the temperature difference $\Delta T = 3$ °C and the air flow rate of 0.5 m s^{-1}. At $\Delta T = 6$ °C an air flow of 0.5 to 1 m s^{-1} increases the heat output to the level of 50—150 W m^{-2}.

* met denotes the unit of metabolic rate, defined by Gagge et al. (A. P. Gagge, A. C. Burton and H. C. Bazett, Science **94**, 428 (1941) as 50 kcal m^{-2} hr^{-1}.

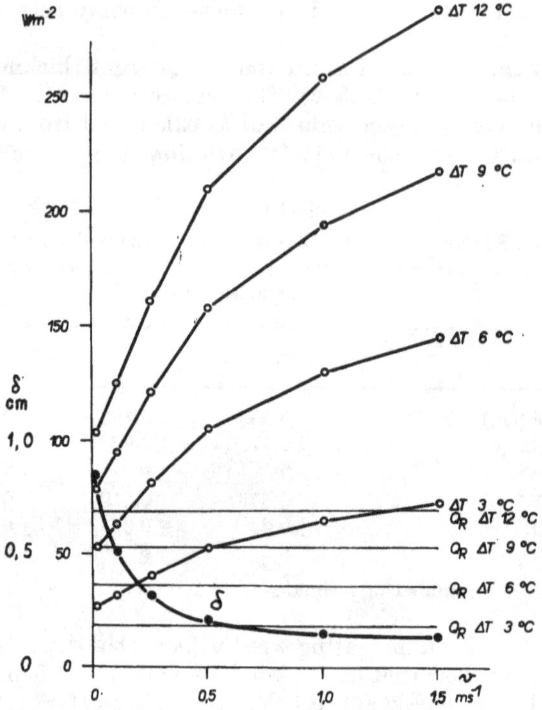

Fig. 1. Relationship between the average thickness of the surface layer δ and the air movement velocity v, its influence on the heat output by convection and radiation at various temperature differences ΔT. Q_R = radiation output. Polished sensor with the diameter 5.7 cm.

4. Discussion

The data given show that the heat output by radiation and by convection through the surface layer in the zone range 1—2 met (50—100 W m^{-2}) occurs at low values of the temperature difference ΔT and that it changes to twice or even three times these values due to increases in the air flow velocity up to 0.5 m s^{-1}. The air movement and the difference between the temperatures of the surface and the atmosphere, or the thickness of the air layer with laminar flow, can hardly be measured simultaneously within these narrow ranges by the usual equipment. When using the EDK method described, however, it is possible to obtain continuous information on the total value of the heat output and its structure. Consequences of the increasing thickness of the surface layer are shown in Fig. 2. Values of the h_c coefficient decrease markedly with increasing thickness of the surface layer in the range up to 1 cm and thereafter decrease asymptomatically to zero with increasing values of δ (e.g. for $\delta = 100$ the value of $h_c = 2.38 \times 10^{-2}$ W m^{-2} deg^{-1}). To avoid perspiration the necessary heat output from the organism has to be achieved by radiation (i.e. through the increase of ΔT). In Fig. 2 changes in the temperature difference ΔT are presented in relation to the increasing thickness of the air layer δ. As we can see with the black body the

ΔT necessary for the maintenance of the heat output of 50 W m^{-2} by radiation and conduction through an air layer of 5 cm must be approximately 8 °C, while for the sensor with a polished surface this value is approximately three times higher. If, under conditions of weightlessness, the formation of the surface layer due to a forced convection is analogous to that on earth then, if the difference in temperature is 8 °C (at which the dull black sensor radiates 50 W m^{-2} without

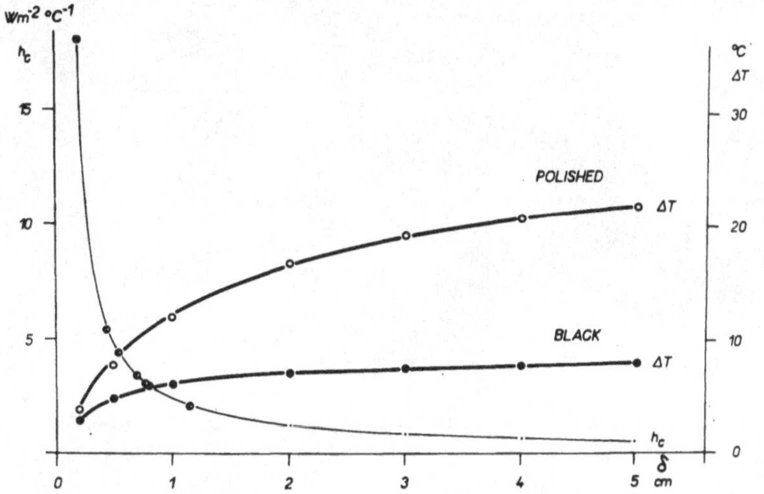

Fig. 2. Relationship between the average thickness of the surface layer δ and the temperature difference ΔT necessary for maintenance of the heat output of 50 W m^{-2} by radiation and convection from the body with polished or dull black surface.

convection), the additional forced convection with a wind velocity of 0.5 m s^{-1} at the same value of ΔT will immediately double the total heat output. With a wind velocity of 1.0 m s^{-1} it will be nearly three times as great, and this should undoubtedly influence the thermoregulatory reaction. It is the EDK, however, which enables us to measure immediately such changes induced by complex effects of many factors.

References

[1] K. Büttner, Klinische Wochenschrift **11**, 1508 (1932).

[2] J. D. Hardy, Physiological Problems in Space Exploration, Charles C. Thomas Publ. Springfield Ill. USA 1964.

[3] G. M. Rapp, in: Physiological and Behavioral Temperature Regulation, Charles C. Thomas Publ., Springfield Ill. USA 1970 (p. 55).

[4] L. Novák and J. Česnek, Kosm. Biol. i Med. **7** (2), 56 (1973).

[5] L. Novák, J. Staněk and J. Česnek, in: Depressed Metabolism and Cold Thermogenesis, Academia Praha (in press).

Planetary Quarantine

Life Sciences and Space Research XIV — Akademie-Verlag, Berlin 1976

EFFECT OF EXTREME FACTORS ON MICRO-ORGANISMS USED FOR THE CONTROL OF THE EFFECTIVENESS OF STERILIZATION

V. I. Vashkov, G. V. Scheglova, N. V. Ramkova, E. S. Zavolnaya, K. O. Fedorova and E. K. Skvortsova

All-Union Scientific Research Institute for Disinfection & Sterilization, Moscow, USSR

Survival of micro-organisms used for control of sterilization procedures was studied under conditions simulating the Martian environment (daily temperature change from $+20°$ to $-20°$ in 99.98% CO_2 + 0.02% air at 0.13×10^{-6} N m^{-2} pressure, with ultraviolet radiation spanning the whole range of the solar spectrum 300—350 MkW cm^{-2}). The test organisms were four strains of *Bacillus subtilis* and one strain of *Bacillus anthracoides*, and were inoculated on to four materials, smooth metal, porous plastic, multilayer composition material and powdered limonite. Some organisms survived on all materials, the greatest on the limonite. The resistance of the survivors to disinfectants was the same as that of the original cultures.

The possibility of existence of terrestrial micro-organisms in extreme conditions is very interesting. It is of both great theoretical and practical importance. Research workers in the USSR and USA have observed that various extreme factors of the space environment — high vacuum, low temperature, ultraviolet radiation, gaseous environment different from the terrestrial, both separately and together — do not necessarily have lethal effects on micro-organisms [1—3]. Parameters of the environment similar to those of Mars do not inhibit reproductive capacity of the micro-organisms isolated in the desert.

We have not found any data in the literature about the action of extreme conditions on micro-organisms used for the control of effectiveness of different methods of sterilization. The aim of the present work was to study survival of such micro-organisms in conditions similar to those of Mars, their reproductive capacity after exposure in these conditions and their stability to sterilizing agents. The experimental parameters were as follows: daily fluctuations of temperature from 20 °C to -20 °C, gaseous composition 99.98% CO_2 +0.02% of air; pressure 1.3×10^{-6} N m^{-2}; ultraviolet radiation with wavelength in the whole range of the solar spectrum with power 300—350 MkW cm^{-2}. These conditions were generated in the "Photostat" apparatus.

The following test microbes were selected; *Bacillus subtilis* strain 7, used for the control of gaseous sterilization (tolerance to gas mixture "OB" at 1000 mg l^{-1} and 35 °C = 120 min); *Bacillus subtilis* strain 18 used for the control of thermal sterilization (survival in dry hot air at 160 °C = 60 min, in flowing steam = 180 min); *Bacillus subtilis* strain 37 and *Bacillus anthracoides* strain 96 used for the control of sterilization by 6% hydrogen peroxide solution (tolerance for 180 to 250 min); *Bacillus subtilis* strain 157 used for the control of radiosterilization (D_{10} index = 270 krad). All the five species of microbes tested were applied to materials of four types: smooth metal, porous plastic, multilayer composition material and limonite ground into powder. Sterilized surfaces made of the selected materials (5 mm × 5 mm) and limonite were inoculated with suspensions of the microbes at a rate of 1000 microbial cells per test object. Infected samples were placed in the "Photostat" and simultaneously similar tests were performed at ambient temperatures in air in the laboratory (control). The experiment lasted for 18 days. Tests for survival were made at intervals of 3, 6, 10, 14 and 18 days. Wet swabs from the samples were made, part of the swab liquid was inoculated on to the nutrient agar, the remaining liquid was filtered through membrane filters which were then placed on the surface of nutrient agar. The data showed that the micro-organisms did not all die in the conditions of the "Photostat" though the number of viable cells was reduced 10—200 times. The greatest number of surviving cells was on the limonite, the least number on the smooth metal surface (see Table 1).

Table 1

Survival of the Micro-organisms Investigated in the "Photostat" and under Laboratory Conditions

Name of culture	Contamination*	Quantity of micro-organisms surviving after 18 days' exposure (mc/test object)							
		Smooth material (metal)		Porous material		Multilayer material		Limanite	
		Ph.	Lab.	Ph.	Lab.	Ph.	Lab.	Ph.	Lab.
Bac. subtilis strain 7	1000	5	90	1	200	42	70	108	130
Bac. subtilis strain 18	1100	1	82	1	250	34	50	52	720
Bac. subtilis strain 37	1000	3	25	1	90	17	19	20	320
Bac. subtilis strain 157	1000	3	180	1	50	20	65	77	480
Bac. anthracoides strain 96	1000	1	15	1	45	4	15	6	200

* Density of initial contamination (mc/test object). Ph. = "Photostat"; Lab. = laboratory.

Among the complex of factors acting in the "Photostat" the highest antimicrobial activity was caused by ultraviolet radiation, and therefore the spores which were exposed to UV radiation unprotected on the smooth surface died first, and the spores even if protected by only slight screening (porous surface, multilayer material, limonite) endured better the conditions of the "Photostat". It should also be noted that the number of viable micro-organisms on the porous surfaces is slightly more than we managed to isolate, because it is well known that the swabbing from these materials cannot be complete. Morphological, biochemical, cultural properties and intensity of sporeforming of the cultures which survived after the most prolonged treatment in the "Photostat" were not changed. Considering the resemblance of limonite to Martian soil, survival of the test cultures in it during 11 months was studied under laboratory conditions. The cultures under investigation remain viable in this soil for a considerable time, but survival of different microbes is not the same.

Rate of growth of the cultures after exposure in the "Photostat" and under the laboratory conditions was estimated by inoculating a broth culture and incubating for 6, 24 and 48 hours and plating on to meat-peptone agar, and also by turbidity measurements on the photoelectrocolorimeter (PEC). No difference was found. The resistance to disinfectants (flowing steam, dry hot air, the gas mixture "OB", 6% hydrogen peroxide solution, gamma-radiation) was also unchanged in the survivors after exposure to extreme conditions.

References

[1] L. K. Losina-Losinski, in: Reaction of Cells and their Biological Components on Extreme Coercions, No. 12, Moscow—Leningrad 1966.
[2] V. N. Bychenkova, in: Reaction of Cells and their Biological Components on Extreme Coercions, No. 12, Moscow—Leningrad 1966.
[3] C. A. Hagan. J. F. Codtrey and R. H. Green, Space Life Sci. 3, 108 (1971).

Life Sciences and Space Research XIV — Akademie-Verlag, Berlin 1976

STERILIZATION OF A BIOLOGICAL ANALYSER

N. V. Ramkova, A. G. Nekhorosheva, S. V. Lysenko, L. B. Chudnova, N. V. Kareev and G. V. Scheglova

All-Union Scientific Research Institute for Disinfection and Sterilization, Moscow, USSR

Experiments were made on the sterilization of a biological analyser designed for sampling the upper atmosphere. Ethylene oxide, though effective, was absorbed too strongly by nutrient medium to permit its use for the complete analyser containing nutrient medium. Gamma radiation (3.2 Mrad) was found to be satisfactory and did not impair the function of the analyser.

We report on the sterilization of a biological analyser designed for sampling micro-organisms in the air of the upper atmosphere of the earth. We evaluated both gaseous sterilization and sterilization by gamma radiation. These methods were chosen because of their proven value, both in our country and abroad, for sterilization of objects with heat sensitive elements [1—7].

Sterilization schedules for the analyser were based on test objects that had been artificially contaminated with spores of cultures highly tolerant to sterilizing agents. The density of contamination was 2—3 decimal logarithms higher than natural contamination. The test objects were placed in the sites of the analyser most unfavourable for sterilization. Effectiveness of the process was evaluated by the killing of the spores. Pure ethylene oxide was used as a sterilizing gas.

The analyser was exposed to treatment in the assembled form, both with nutrient medium and without it. Sterilization was performed at 18—23 °C, with relative humidity 80—100%, with a concentration of sterilizing gas 500 to 2000 mg l^{-1}, and time of exposure 24—64 hours. Within the above range of conditions sterility of the analyser was achieved.

It is important to study the elimination of sterilizing gas from the interior of the device and especially from the nutrient medium, because of its residual toxicity. To remove gas from the analyser after exposure we used an alternation of blowing with sterile air and pumping with a vacuum pump to a residual pressure of 0.6—1.3 N m^{-2}. The blown air was sterilized with bacterial filters.

The residual quantity of ethylene oxide in the interior of the analyser and in the nutrient medium was estimated by a chemical method. This is based on the following reactions: hydration of ethylene oxide to ethylene glycol, oxidation of ethylene glycol by iodic acid to formaldehyde, and the interaction of formaldehyde with chloric acid in the presence of sulphuric acid. The optical density of the final solution, proportional to the amount of the ethylene oxide, was measured by a photocolorimetric method.

It was discovered that ethylene oxide is absorbed by the nutrient medium in great quantities. To remove it from the medium is difficult. After maximum degassing (92 hours) the minimum residual quantity of ethylene oxide in the nutrient medium was 0.4 mg g^{-1}. Investigation were there fore made to estimate the capacity of the nutrient medium, to permit the growth of single cells of micro-organisms after sterilization by ethylene oxide. The results showed that the residual quantities of the gas are inhibitory.

Gaseous sterilization with pure ethylene oxide can thus be used only for sterilization of the interior of the analyser without nutrient medium. Sterilization of the medium in this case must be performed by some other method. It was decided to investigate sterilization of the device by gamma-rays. The high penetration of gamma-rays allows sterilization of the assembled analyser already containing nutrient medium which presented great advantages. To select the effective sterilizing dose. *Bacillus subtilis* with D_{10} index 280 krad was used as a test culture. Spores of the test culture were put on pieces of sterile polyethylene film and placed in the analyser in 10—16 sites in which the absorbed dose was minimal. Doses of $2.6 \pm 12\%$ Mrad and $2.9 \pm 12\%$ Mrad ensured complete sterilization of 60% and 85.1% of the tests respectively. A dose of $3.2 \pm 12\%$ Mrad was accepted as the sterilizing dose.

The working capacity of the analyser did not deteriorate after irradiation with 12 Mrad, which was proved by investigations performed immediately after sterilization and after six months' storage of the device.

References

[1] A. K. Astafieva, V. I. Vashkov and E. N. Nikiforova, in: Life and Science, V (1966).
[2] L. B. Chudnova, N. V. Kulkova and V. M. Tsetlin, J. Space Biol. Med. (1972).
[3] V. I. Vashkov, N. V. Kareev and L. B. Chudnova, paper presented at Int. Symp. on Disinfection and Sterilization, Moscow 1972.
[4] N. V. Ramkova, T. V. Bajbeuk-Milikova, T. L. Potapenko and R. A. Vyshegorodskaya, paper presented at Int. Symp. on Disinfection and Sterilization, Moscow 1972.
[5] N. V. Ramkova, T. L. Potapenko and R. A. Vyshegorodskaya, J. M. E. I. 8, (1974).
[6] V. R. Tyler, Amer. J. Nurs. 60, 1596 (1960).
[7] G. Schwenkler, Zbl. Bakt. Hyg. 1 Abt. Orig. B 156, 171 (1972).

Exobiology

Life Sciences and Space Research XIV — Akademie-Verlag, Berlin 1976

ON METHODS OF DETECTION OF EXTRATERRESTRIAL LIFE

A. A. IMSHENETSKY, M. D. EVDOKIMOVA and G. G. SOTNIKOV

Institute of Microbiology, USSR Academy of Sciences, Moscow, USSR

New methods have been developed for detecting microbial growth from the microflora of desert soils. The first is a polarimetric method in which the fall in optical activity due to assimilation of D-glucose is followed. Detectable changes with desert soils were seen in a few hours, and the method can be employed with small amounts of material. The second method is the release of heat from metabolizable substrates as measured by a microcalorimeter. In the presence of glucose a characteristic response from desert soils was found within 24 hours.

Two main approaches have been used to detect extraterrestrial life. American scientists have given preference to biogeochemical methods based on the determination of various gases appearing after pyrolysis of Mars soil and after adding to it a compound nutrient medium. Soviet scientists have aimed their investigation at discovering nutrient media suitable for development of microflora of desert soils and at measuring the multiplication of microbial cells, or products of their vital activity.

A simple and suitable method is for detecting microbial growth polarimetry of the culture. Dextrorotary and levoratory chemical compounds as well as optically inactive ones can be used as substances added to the medium into which the soil of a planet is to be inoculated. The presence of optically active substances in the soil is hardly probable, as even if their formation takes place in space this occurs very seldom, as is particularly proved by the absence of optically active organic substances in meteorites (carbonaceous chondrites). In the course of investigations a microbial analysis was performed on soils of deserts (Turkmenian SSR, Darvaza and Repetek regions, Nubiana desert) and with chernozem. Subsequent cultures of these soils were inoculated into a medium of the following composition: asparagin, 0.1 g; NH_4NO_3, 0.5 g; K_2HPO_4, 2.6 g; KH_2PO_4, 0.91 g; $MgSO_4 \cdot 7H_2O$, 0.3 g; $CaCl_2$, 0.01 g; NaCl, 0.1 g; mixture of trace elements, 1 ml; yeast autolysate, 5 ml; liver extract, 1 ml; distilled water, 1 l, pH 7.0. The nutrient medium was supplemented with 0.25% of D-glucose or maltose. A small quantity of carbohydrate added to the medium makes it possible to obtain a prompt response since it is quickly metabolized. The nutrient medium was poured into 700 ml Erlenmeyer flasks, 100 ml into each flask. The flasks after inoculation were exposed in a thermostat at 25°. The optical activity was determined 24, 48, 72 and 96 hours after inoculation with the help of a Roussel-Jouan polarimeter at 436 nm.

As a result of growth of micro-organisms glucose or maltose were rapidly consumed and the optical activity quickly fell. The reduction of optical activity was much slower with the inoculation of desert soil than with the inoculation of chernozem (Fig. 1).

In order to demonstrate that the fall of optical activity is correlated with the growth of micro-organisms the soil was inoculated into a glucose medium, the

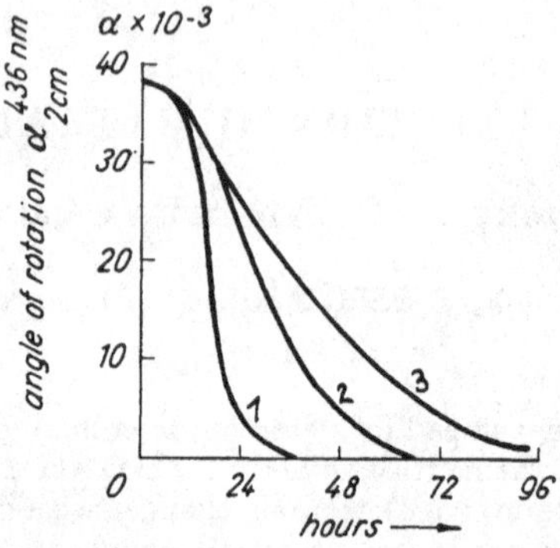

Fig. 1. The decrease of optical activity in a glucose medium as a result of development of soil microflora. 1, chernozem; 2, desert soil (Ropetek); 3, desert soil (Darvaza).

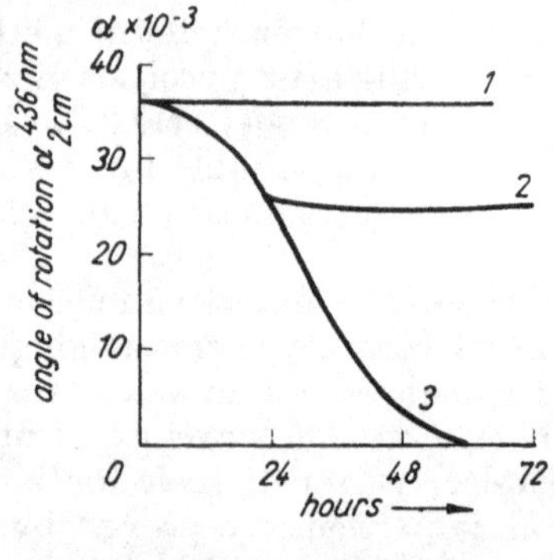

Fig. 2. Effect of formalin on decrease in optical activity of a glucose medium containing multiplying soil micro-organisms. 1, sterile medium (Control 1); 2, addition of formalin (after 24 hours), (Control 2); 3, without formalin.

first control being the flask with sterile medium and the second control being the flask supplemented with 2—3 droplets of 40% formalin per 5 ml of the medium after the fall of optical activity. The decrease in the optical activity is stopped after addition of formalin (Fig. 2).

Inoculations of desert soils in a medium containing racemic tartaric acid always gave negative results. No consumption of the dextrorotary isomer took place in this case and the nutrient medium remained optically inactive, nor was optically active (+)-tartaric acid attacked. It is quite natural to envisage the absence of microscopic fungi and bacteria assimilating (+)-tartaric acid. But it is quite possible that the soil of other deserts contains them. In any case soils containing a great variety of micro-organisms should include species assimilating these acids. Inoculation of chernozem into media containing racemic tartaric acid led to a fall in optical activity. Similar changes occurred after inoculation of the culture of microscopic fungi *Penicillium glaucum.* Fig. 3 gives an idea of changes in the optical activity after inoculation. An advantage of the polarimetric method of detection of extraterrestrial life using carbohydrate media is that the experiment requires small amounts of nutrient medium and it is possible to gain positive results from small quantities of desert soil (2—2.5 mg per 1 ml of the medium).

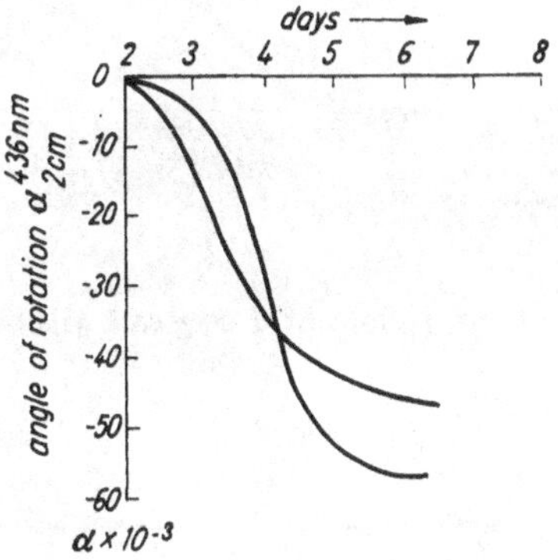

Fig. 3. The optical activity appearing in a medium containing racemic tartaric acid (as a result of consumption of dextrorotary isomer). 1, chernozem; 2, *Penicillium glaucum.*

The second method of detection of extraterrestrial life is based on estimation of heat production in cells of micro-organisms during their growth and reproduction. As is known, in some cases heat production can represent 58% of the consumed metabolites. In the process of metabolism one bacterial cell produces from 1×10^{-10} to 1×10^{-7} kcal hr^{-1}. The quantity of heat depends on several things: on the composition of the nutrient medium and on the presence in it of reductive compounds, on physiological peculiarities of micro-organisms, on the degree of aeration, etc.

Investigations were performed with soil from the Kara-Kum desert. 30 g of soil were placed into a 100 ml reaction vessel in sterile conditions. In the same vessel a pipette was placed containing the liquid whose composition has been described above. Two reaction vessels, test and control, were placed in calorimetric cells. After thermal equilibration the reaction vessel was opened and the soil was damped by the medium. A differential microcalorimeter of Calvet type was used in the investigations. The sensitivity of this microcalorimeter is about 7×10^{-6} kcal s^{-1} mm^{-1}.

In the first series of experiments the soil was moistened with the above-mentioned medium without glucose. In the second series of experiments the amount of glucose in the medium was 1%, and in the third series 5%. The amount of glucose not assimilated by micro-organisms was evaluated by polarimetry. The content of bacteria grown in the medium was determined by direct counting of cells in a Gorayev camera. With the desert soil there was a brief period of negative thermogenesis after damping it with the medium without glucose. This seems to be correlated with the moisture content of soil. Later (Fig. 4) the curve of thermogenesis slowly rises and later gradually descends.

Quite a different situation occurs when the desert soil is mixed with a nutrient medium containing 1% of glucose. The curve of thermogenesis quickly rises (Fig. 5) and reaches the maximum within a day; later the curve slowly descends

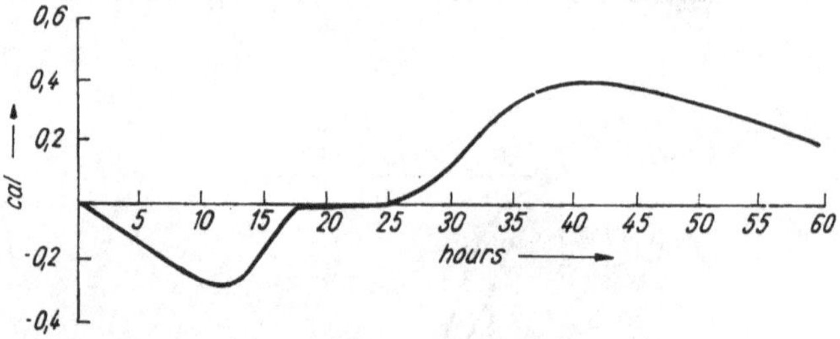

Fig. 4. The curve of heat loss by microflora of desert soil without exogenous energy source.

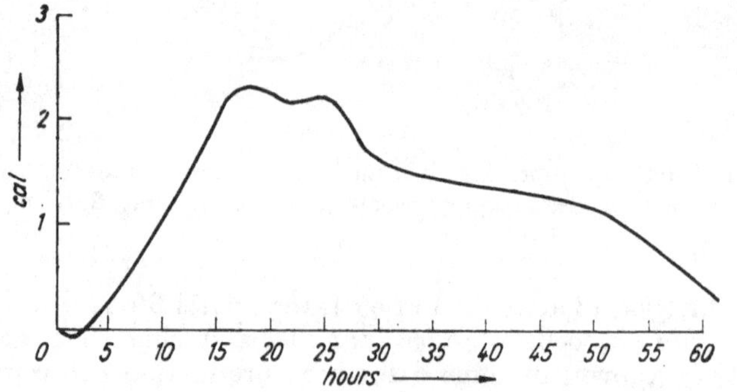

Fig. 5. The curve of heat loss during reproduction of microflora of desert soil in medium N II containing 1% of glucose.

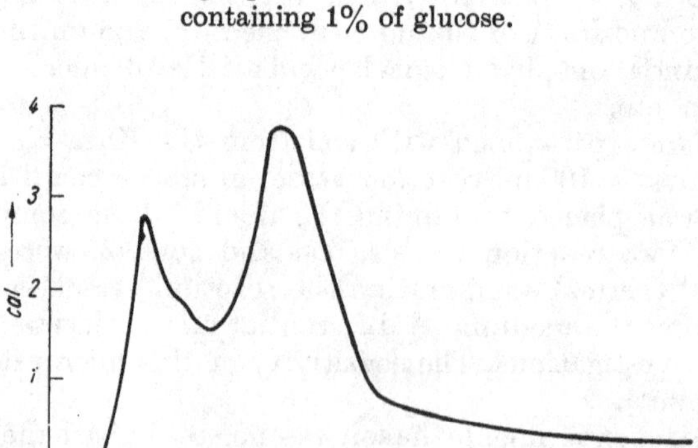

Fig. 6. The curve of heat loss during reproduction of microflora of desert soil in medium N II containing 5% of glucose.

Most data were obtained with a medium containing 5% of glucose. After 30 minutes thermogenesis rose exponentially. 10 hours later it reached 2.8 cal hr^{-1}, then some decrease in the heat production was observed, but soon after that the curve again rose and by 25 hours thermogenesis reached almost 4.0 cal hr^{-1} (Fig. 6). After this the curve falls rather quickly. How can one explain the two peaks? The reaction vessel was not supplied with air but contained some at the start: therefore it is possible that the first peak is due to aerobic microflora; the second peak may then be due to facultatively anaerobic micro-organisms and anaerobic micro-organisms utilizing residual glucose. Complete assimilation of glucose did not occur in these experiments either.

Life Sciences and Space Research XIV — Akademie-Verlag, Berlin 1976

PERFORMANCE OF FUNGI IN LOW TEMPERATURE AND HYPERSALINE ENVIRONMENTS

S. M. Siegel and T. W. Speitel

Department of Botany, University of Hawaii, Honolulu, Hawaii, USA

During the past ten years we have observed a broad array of stress capabilities in common fungi including ability to grow in aqueous ammonia and other alkaline solutions, in acids, in the presence of heavy metals, and in various salt media at low temperature.

This report is concerned primarily with (a) the performance of *Aspergillaceae* in a variety of saturated salts, (b) distinctive roles for K^+ and Rb^+ ions, and (c) the lowest temperatures at which growth in nutrient brines has been observed, namely 267 °K in as little as 14 days. We also describe a novel solid medium based upon gelatin, glycerol and water in which fungal cultures growing at 248 °K can be directly examined under oil-immersion magnification. The performance capabilities of the fungi show that tolerance or adaptability to harsh and extreme physical-chemical environments cannot be considered a unique feature of prokaryotic life forms.

Salt flats, brine pools and other natural hypersaline environments have long been recognized as real ecological niches harboring a range of biota from pseudomonad bacteria and green algae to specialized crustaceans. A notable omission in this ecological record is the fungi, although the group is known to include marine forms.

Because fungi are highly versatile metabolically and important in the recycling of organic matter in ecosystems generally, we have undertaken an extended study of their stress capabilities over the past ten years [1—5]. In addition to growth in ammonia, alkalies and acids and their ability to solubilize and accumulate metals and metalloid elements, the fungi have displayed salt tolerances comparable with or superior to those of common halophiles without the benefit of evolutionary or laboratory selection or conditioning. The ability to grow in the presence of hypertonic salt solutions also opened prospects for their use in cryobiologic investigations, at temperatures lower than the freezing point of ordinary aqueous nutrient media.

Most of these studies of fungi in severe or exotic environments have been concerned with the *Aspergillaceae*, a widespread terrestrial group without marine affinity and commonly observed on decaying vegetable foodstuffs as saprophytes, but also known as occasional animal pathogens (e.g. "aspergillosis"). The standard culture medium contains 1% each of glucose and peptone with 0.5% yeast extract and ranges in pH from 6.0 to 6.5. Such media are modified for experimental purposes by saturation with salts as desired with respect to depression of the freezing point. All cultures were set up to quintuplicate.

In early studies (cited above) it was found that *Penicillium notatum* and other forms could be cultured on freeze-thaw diurnal cycles, for example 12 hours each at 293 °K and 243 °K without marked reduction in growth rates provided they were incubated in media saturated with acetate chlorides of other salts of K, Mg or Ca. The cyclic temperature regimes left open, however, the possibility that growth took place only during the warm phase. It did show that a fungus can grow while being exposed to severe cold shock repeatedly over 30 or 60 consecutive days. Comparable capabilities were found in various species of *Penicillium* and the closely allied genus *Aspergillus.*

Table 1
Time-Temperature Relations in the Development of *Penicillium notatum* in Salt saturated Glucose-Peptone Medium

		Days required for Stage in Chloride Medium[a]				
Temperature (°K)	Stage	None	K^+	Na^+	$Na^+ + K^+$	$Na^+ + Rb^+$
297	Germination	<1	2	>120	2	3
	Macrocolony[b]	2	34	—	35	35
	Sporulation	3	62	—	67	68
279	Germination	18	18	>120	24	26
	Macrocolony	30	78	—	87	88
	Sporulation	67	100	—	110	113
266	Germination	>120	24	>120	30	32
	Macrocolony	—	88	—	104	109
	Sporulation	—	>120	—	>120	>120

[a] "$Na^+ + K^+$" and "$Na^+ + Rb^+$" indicate 0.5M K^+ or Rb^+ added to 3.5M NaCl.
[b] Colony diameter ≧ 1 mm.

Subsequently, the growth and development of *Penicillium notatum* was examined over a more modest range of continuous temperatures, 266—297 °K (Table 1). The time required for spore germination to begin, for the appearance of 1 mm colonies and for spore differentiation were noted. Saturated KCl slowed development substantially at 297 °K, but in saturated NaCl no development at all was seen during a period of 120 days. When, however, the NaCl medium was supplemented with 0.5M KCl or RbCl, the performance was nearly the same as that seen in KCl at saturation. It is of interest to note that Rb^+ with an ionic radius of 1.47 Å is closer to K^+ (1.33 Å) than the latter is to Na^+ (0.97 Å) in charge-size relations. Neither Li^+ (0.68 Å) nor Cs^+ (1.67 Å) could replace K^+ or Rb^+. Thus it seems clear that the non-performance of the organism in NaCl is an expression of Na^+ imbalance and induced K^+ requirement of unusual magnitude. With reduction in temperature, but still above freezing, the difference in developmental rates with and without salts is far smaller. That is, even a modest reduction in temperature affects rates more profoundly in the absence of salts at saturating levels. Finally, at 266 °K, the salt-free cultures are frozen and fail to develop in 120 days, whereas development through the macrocolony phase is almost as rapid as it was at a temperature level 13° higher and above the usual freezing point. The most important difference observed was the failure at the lower temperature of sporogenesis, at least within the present time limits.

One of the more striking results of this and other studies was the apparent reduction in temperature effects in saturated salt media. This effect is supported by another series of studies using mycelial growth rate data (weight basis) to calculate Arrhenius heats of activation with and without salts (Table 2). It is evident that whatever saturating salt is used, the activation energy for growth

Table 2

Salt induced Lowering of Activation Energy for Growth based upon Biomass of *Penicillium notatum* in Glucose-Peptone Medium

Saturating Salt	Temperature Range (°K)	log Increase in Biomass	E_a[a] (kcal/mole)
None	277—295	0.793[b]	16.5 ± 2.1
	265—277	—	—
KCl	277—295	0.293	6.1 ± 0.6
	265—277	0.167	4.7 ± 0.5
KOAc	277—295	0.270	5.6 ± 0.7
	265—277	0.147	4.1 ± 0.5
$Mg(OAc)_2$	277—295	0.267	5.5 ± 0.6
	265—277	0.149	4.2 ± 0.5

[a] Arrhenius heat of activation ("activation energy") ± standard error.
[b] After 10 days incubation without salts and 60 days with salts.

is always appreciably lower than in salt-free media. Thus, the presence of high salt concentrations does in fact reduce the temperature dependence of the growth process in *Penicillium*.

Exploratory studies indicated that a constant 266 °K was not the lower limit for these fungi; however, the technical problems in the observation and separation of individual small colonies in a slurry of medium and fine salt crystals are considerable. On the other hand, agar media which provide a good surface, are destroyed by freezing or by the addition of many salts. An alternative medium was devised by the dispersion of gelatin in 40—60% (v/v) aqueous glycerol, together with the usual proportions of glucose, peptone and yeast extract. When 20—40 g of gelatin is introduced rapidly into 100 ml of the glycerol-water solution, the resulting mixture remains fluid enough to pour for 5—10 minutes. When solidified in plates, this gel medium has the translucency and resiliency of 1—2% agar gels and is completely stable at temperatures of 248 °K or less. Gel surfaces were easily scanned at oil-immersion magnification by first placing a cover slip over the area to be examined followed by the usual immersion oil.

These cryogels represent severely desiccating environments and do not in general support micro-organismal growth even at room temperatures. When inoculated with suspensions of *Aspergillus niger* spores and incubated at 248 °K, the following germination responses were noted: at 30 days, 0 per 1000; at 60 days, 0.8; at 120 days, 1.3; and at 180 days, 2.5. These experiments and others concerned with the lower thermal limits of fungi are still in progress. They show the importance of including eukaryotic organisms in any consideration of the environmental capabilities of terrestrial life.

Acknowledgments

This work was supported by NASA Grant NGL 12-001-042.

References

[1] K. ROBERTS and S. M. SIEGEL, Plant Physiol. **42**, 1215 (1967).
[2] S. M. SIEGEL, Adv. Space Sci. Tech. **9**, 1 (1967).
[3] S. M. SIEGEL and C. GIUMARRO, Icarus **4**, 37 (1965).
[4] S. M. SIEGEL, T. SPEITEL and R. STOECHER, Cryobiol. **6**, 160 (1969).
[5] S. M. SIEGEL, Spaceflight **13**, 183 (1971).

Life Sciences and Space Research XIV — Akademie-Verlag, Berlin 1976

EFFECT OF SPACE FACTORS ON *Escherichia coli* B/r CELLS

H. BÜCKER, R. FACIUS, G. REITZ, C. THOMAS and H. WOLLENHAUPT

Arbeitsgruppe für Biophysikalische Weltraumforschung an der Universität Frankfurt, Frankfurt/Main, FRG

Stationary phase cells of *Escherichia coli* B/r were inactivated when they were exposed to high vacuum (10^{-6} torr). About 5% of the cells were still able to form a colony after 45 min exposure. Vacuum dried cells (0.5 torr, 120 min) show colony forming ability of 30% or more. They were inactivated to about 5% by further vacuum treatment. Vacuum treated cells showed higher permeability for various cell components. UV irradiated *E. coli* B/r cells in vacuum showed increased UV sensitivity. DNA-protein cross-links were preferentially formed in a vacuum. To obtain 63% of DNA cross linked with protein required 852 erg mm^{-2} (D_{37}) of UV irradiation in suspension and only 72 erg mm^{-2} of UV irradiation in vacuum. The protein components of DNA-protein cross-links were hydrolysed with pronase E and the amino acids directly bonding to DNA were determined. The most important amino acids concerned in DNA-protein cross-links seem to be glycine and alanine, followed by aspartic acid (asparagine), glutamic acid (glutamine) and histidine.

The sensitivity to X-rays of stationary phase *E. coli* B/r cells seems to depend on the remaining gas atmospheric in the vacuum since it varies with different pumping systems.

In a previous investigation we have shown that vacuum treatment of bacterial cells leads to damage of the cell membrane and subsequently to inactivation [1]. The inactivation of bacterial cells in a vacuum is related to their dehydration, and therefore depends on time of exposure and on the final pressure reached (Table 1). With a final pressure of 0.5 torr inactivation increases from 46% for an exposure of 10 min to about 70% for exposures of 60 min or more. A subsequent exposure of 45 min of each of these samples to a high vacuum of 3.5×10^{-7} torr leads to further inactivation up to 95%. A longer exposure to high vacuum did not cause further inactivation.

Table 1

Inactivation of Stationary Phase Cells of *E. coli* B/r after Vacuum Treatment

Final pressure (torr)	Time in vacuum (min)	Inactivation (%)
0.5	10	46.2
0.5	15	49.8
0.5	60	71.5
0.5	120	70.3
3.5×10^{-7}	45	94.6

The sensitivity to UV irradiation of 254 nm was also increased when stationary phase cells of *E. coli* B/r were irradiated in vacuum. The UV-induced damage was not photoreactivable and not reactivable, or only slightly, by excision repair [2]. However, pyrimidine dimers cannot be responsible for the increased UV sensitivity of these cells in vacuum.

We have investigated whether DNA-protein cross-links are preferentially formed in a vacuum and determined their importance. One way to assess the

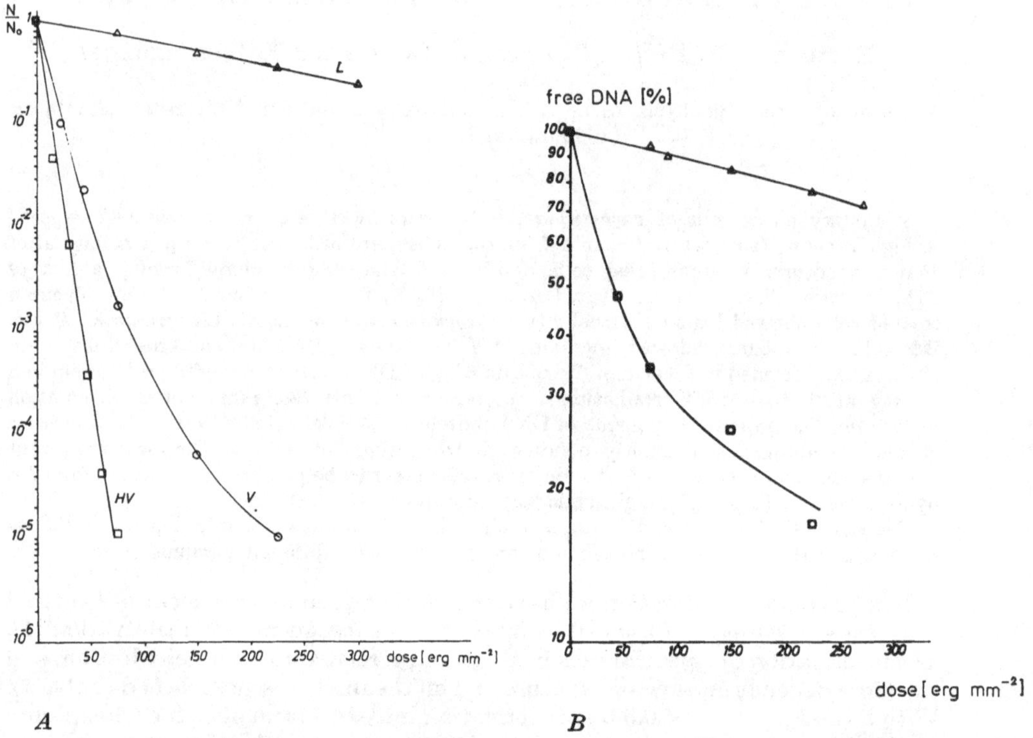

Fig. 1. A, inactivation of *E. coli* B/r cells irradiated in suspension (L), in vacuum (HV) and after vacuum treatment at the air (V); B, recovery of free DNA from the same cells.

biological importance of a given photoproduct in DNA is to determine if its yield varies with radiation exposure in the same manner as the efficiency for killing, when cells are pre-treated under various experimental conditions that affect their sensitivity to radiation. The amount of DNA that becomes cross-linked to proteins increases when stationary phase cells of *E. coli* B/r are irradiated in high vacuum (Fig. 1). The formation of DNA-protein cross-links, here expressed as the decrease in extractable DNA from the cells, may be in part responsible for the increased sensitivity of *E. coli* under these conditions. Cells irradiated at 0.5 torr or after vacuum treatment at the air, show similar yields of DNA-protein cross-links, but are not as sensitive as cells irradiated in high vacuum. The same yields were obtained with cells dried in air or in a desiccator. Other photoproducts produced only in a vacuum have also been determined in the DNA of *E. coli* cells by our group [3]. To obtain 63% of DNA cross-linked with proteins, a UV dose of 852 erg mm^{-1} (D_{37}) was required if irradiated in suspension, and only 72 erg mm^{-1} if irradiated in a vacuum.

To determine the amino acid residues which are able to cross-link with nucleic acids, the protein part of the cross-linked product was hydrolysed with pronase E, and subsequently the nucleic acids were separated both from the pronase E and the hydrolysis products, and following that the nucleic acids were hydrolysed in hydrochloric acid [4]. Amino acids were then determined by means of an amino acid analyser. No peptides were found after hydrolysis of the nucleic acids with formic acid. Pronase E hydrolysed all proteins to single amino acids.

Table 2

Amino Acids found after Enzymatic Hydrolysis of the Protein Part of the Adduct followed by Acid Hydrolysis of the Nucleic Acid Part

	After UV irradiation in vacuum		After UV irradiation in suspension	
	300 erg mm^{-2}	10^5 erg mm^{-2}	600 erg mm^{-2}	10^7 erg mm^{-2}
ala	15.54 nM	48.73 nM	13.34 nM	14.04 nM
arg	1.06	30.51	1.63	4.57
asp	5.61	30.58	7.35	10.39
cys	0	2.71	0	0.47
glu	7.69	60.09	8.55	18.71
gly	45.60	102.17	33.53	35.88
his	4.43	44.44	3.96	25.87
ile	0.43	21.15	0.88	4.14
leu	1.02	30.44	1.80	6.56
lys	2.17	39.94	2.82	8.49
met	0	38.07	0	7.35
phe	0.79	21.08	1.14	2.97
ser	3.11	3.00	1.79	2.63
thr	0.47	3.64	0.38	1.14
tyr	1.71	18.08	1.01	2.64
val	2.31	13.65	3.42	2.80
total	91.94	517.28	81.60	148.65

After this procedure we found that almost all amino acid residues in proteins are in fact able to cross-link with nucleic acids (Table 2). Only tryptophan residues are completely destroyed by the acid hydrolysis, so no information is available on them. Glycine and alanine seem to be the most important amino acids concerned in the formation of DNA-protein cross-links, because they are present in large amounts compared with the other amino acids. They are followed by aspartic acid (or asparagine), glutamic acid (or glutamine) and histidine.

These results indicate also that the photoreaction may occur through peptide radicals, which can subsequently react with nucleic acids leading to cross-links. Meybeck and Meybeck [5] have observed that the energy of the 254 nm photons is only enough to produce such radicals in glycine and alanine residues. Peptide radicals can also be formed in other amino acid residues by energy transfer from aromatic amino acid residues to the immediately neighbouring residue.

Cross-links between DNA and a large number of different cellular and exogenous compounds have become more significant in biology, because these adducts (all formed by radical mechanisms such as those found here) have been correlated with ageing processes and carcinogenesis, both of which can be accelerated by radiation.

Stationary phase cells of *E. coli* B/r T which were irradiated in a high vacuum, were also more sensitive to X-rays than those irradiated under atmospheric pressure (Fig. 2). Cells irradiated in a high vacuum produced by a turbomolecular

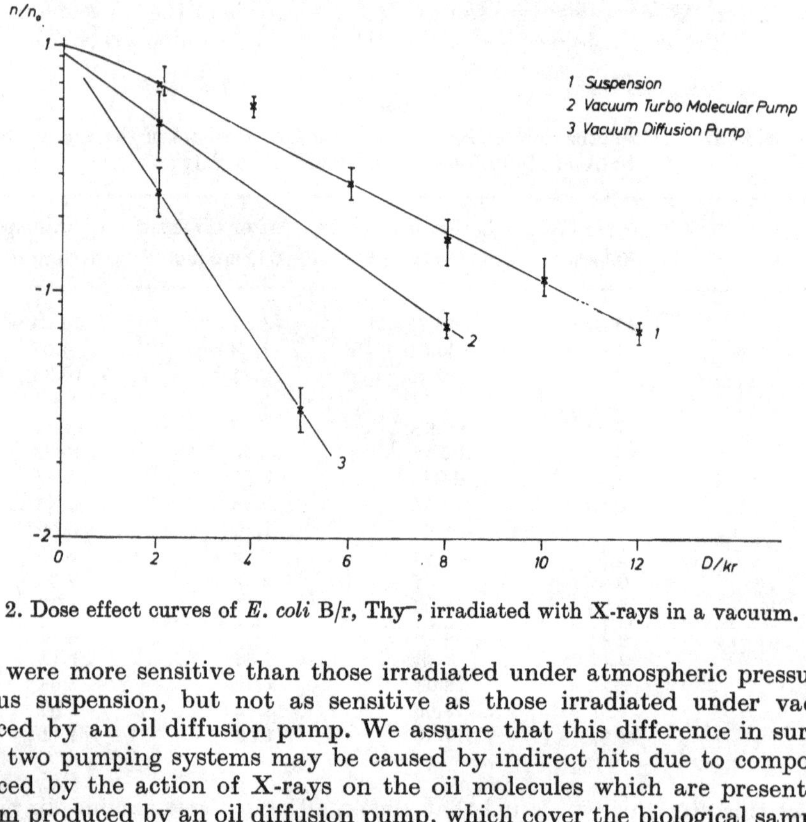

Fig. 2. Dose effect curves of *E. coli* B/r, Thy⁻, irradiated with X-rays in a vacuum.

pump were more sensitive than those irradiated under atmospheric pressure in aqueous suspension, but not as sensitive as those irradiated under vacuum produced by an oil diffusion pump. We assume that this difference in survival in the two pumping systems may be caused by indirect hits due to compounds produced by the action of X-rays on the oil molecules which are present in a vacuum produced by an oil diffusion pump, which cover the biological sample in the vacuum with a monomolecular layer.

References

[1] M. FRANKENBERG-SCHWAGER, G. TURCU, C. THOMAS, H. WOLLENHAUPT and H. BÜCKER Life Sciences and Space Research XIII, 83 (1975).
[2] H. BÜCKER and G. HORNECK, Life Sciences and Space Research VIII, 33 (1970).
[3] M. FRANKENBERG-SCHWAGER, H. BÜCKER and H. WOLLENHAUPT, Raumfahrtforschung **5**, 209 (1974).
[4] C. THOMAS et al., in: Aging, Careinogenesis and Radiation Biology, Plenum Publ. Co., New York 1976.
[5] A. MEYBECK and J. MEYBECK, Photochem. Photobiol. **6**, 365 (1967).

Life Sciences and Space Research XIV — Akademie-Verlag, Berlin 1976

ON MICRO-ORGANISMS OF THE STRATOSPHERE

A. A. IMSHENETSKY, S. V. LYSENKO, G. A. KAZAKOV and N. V. RAMKOVA

Institute of Microbiology, USSR Academy of Sciences of the USSR, Moscow, USSR

The lower parts of the biosphere are well studied since various live beings are found in oceans and at the bottom of large hollows. Contrary to this, we have no data about the upper boundaries of the biosphere. Samples were obtained with the help of specially constructed analysers which were installed in meteorological rockets and reached an altitude of 100 km.

With the help of methods completely excluding the possibility of contamination of analysers with outside microflora it became possible to prove that earth microbes carried by air currents are present in the stratosphere. At an altitude of 48—77 km *Circinella muscae, Asp. niger, Penicillium notatum* were found as well as mycobacterium and micrococcus. The correlation of these cultures with extremal factors is studied and the weight of one conidium or one cell in isolated micro-organisms is estimated. These investigations will continue.

There is information on the lower boundaries of the biosphere, since life has been detected in the deepest ocean at depths of 10—11 km (Kermadek-Tonga trench). It is more difficult to determine the upper boundaries of the biosphere as it will entail great difficulties in method.

In 1936 from a cabin lifted by the stratospheric balloon Explorer 2 to a height of 20 km a cylindrical sampling device was thrown overboard which after sampling descended by parachute. Both the sampler and parachute were first sterilized. In this way fungi and sporeforming bacteria were found at a height of 20 km [1]. In 1962—1964 stratospheric balloons inflated with helium made it possible with the help of powerful samplers to detect various micro-organisms. However, these data are not of great scientific importance since the non-sterile cover of the balloon was lying on an oat field during inflation with helium. Therefore during the flight the samplers drew in a great number of various soil micro-organisms from the balloon cover [2]. That is all we knew about the presence of micro-organisms at high altitudes.

Determination of the upper boundary of the biosphere is of great interest since this knowledge will help us to judge about the height of penetration of micro-organisms into the stratosphere, their resistance to extreme physical factors, and the content and qualitative composition of microflora in the troposphere and stratosphere.

From the methodical aspect the analysis of the stratosphere is highly complicated, as sampling must avoid penetration of extraneous microflora into the sampling device of the analyser. For this reason micro-biologists set themselves the task of designing and constructing an analyser which was installed on board meteorological rockets. A detailed description of this apparatus will be given by the designer elsewhere. The steps in its use are as follows:

1. The analyser and the nutrient medium were placed in two polyethylene bags and sealed. Then the analysers were placed into metal containers and sterilized by γ-rays (3.2—3.5 Mrad, 98.8 rad s^{-1}) for 9 hours. The sterilization affected neither the functions of the analyser nor the quality of the nutrient medium. Radioresistant spores located in 14 places inside the analyser were used in establishing the regime of γ-ray sterilization. The death of all test spores demonstrated the effectiveness of the sterilization.
2. For obtaining air samples meteorological rockets rose to a height of 100 km. In the dense strata of the atmosphere the forepart of the rocket heated to high temperatures, ensuring the death of all surface micro-organisms. At the designed altitudes the rocket cone closing the instrument compartment was discarded with the help of pyrocartridges. Sampling was made only in the stratosphere during rocket climb, which completely excluded the possibility of penetration of extraneous microflora.
3. After the cone was discarded the forepart of the rocket became the cover of the analyser which was then opened with two pyrocartridges. The analyser contained a container for the nutrient medium. The opening of the analyser permitted the precipitation of all aerosol particles including micro-organisms on the surface of the nutrient medium. Before and after sampling the film was covered with another sterile film. After sampling the instrument compartment of the rocket came down by parachute.
4. Air samples on the nutrient medium were exposed in a thermostat at 30° for 7—14 days and then analysed microbiologically. The time of appearance and the character of the colonics were recorded. The records concerned only colonies of micro-organisms precipitated on the nutrient medium during the rocket climb. After recording of colonies the micro-organisms were inoculated on additional media and studied further.

In all, six rockets with analysers were launched. In two rockets, the analysers did not work. In another case not a single microbe reached the surface of the nutrient medium after the air sampling and therefore no colonies were detected. This fact demonstrates the safety of the method of sterilization. In the other three rockets micro-organisms reached the nutrient medium and the growth of colonies was recorded. The results of microbiological analyses of the stratosphere are presented in Table 1.

The identification of colonies showed that selected micro-organisms included fungi *Circinella muscae* (Fig. 1), *Penicillium notatum* (Fig. 2), *Aspergillus niger*

Table 1

Rocket	Height during sampling (km)	No. of colonies
1	48—58	20
2	61—70	3
3	61—75	0
4	57—77	8
	77—85	0

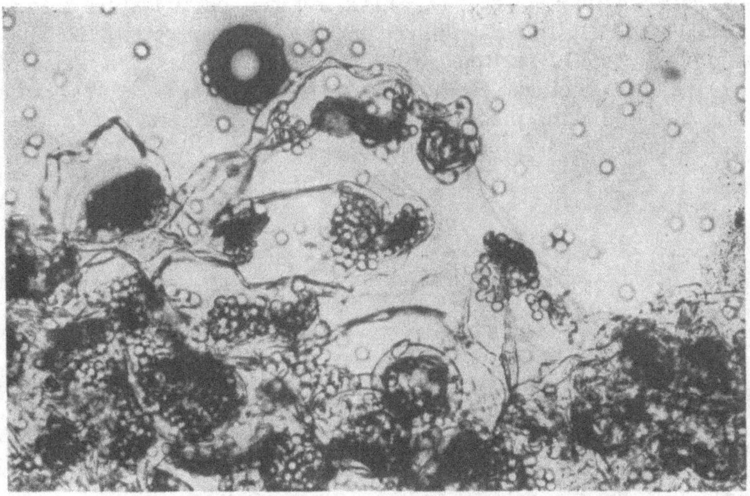

Fig. 1. Spores and mycelium *Circinella muscae* isolated from stratosphere, ×500.

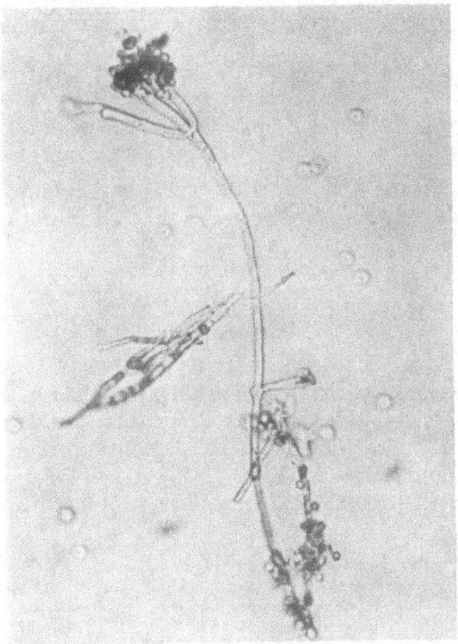

Fig. 2. *Penicillium notatum* isolated from the stratosphere, ×500.

Papulaspora anomala as well as non-sporeforming bacteria *Micrococcus albus* (Fig. 3) and *Mycobacterium luteum* (Fig. 4). Small mineral particles are known to rise high from the earth [3]. In this connection, of great interest will be a comparison of the weight of mineral particles with the weight of dried spores and conidia of fungi, or cells of bacteria, isolated from the stratosphere. We cannot reject the possibility of mycelium pieces capable of more rapid germination than conidia or spores of fungt rising to great heights.

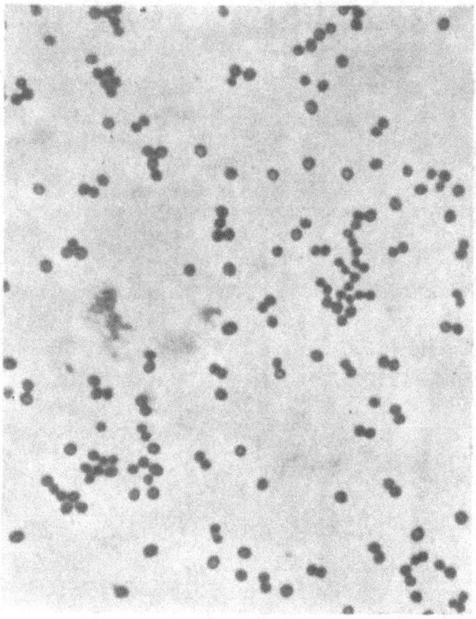

Fig. 3. Isolated culture *Mycobacterium luteum*, ×2000.

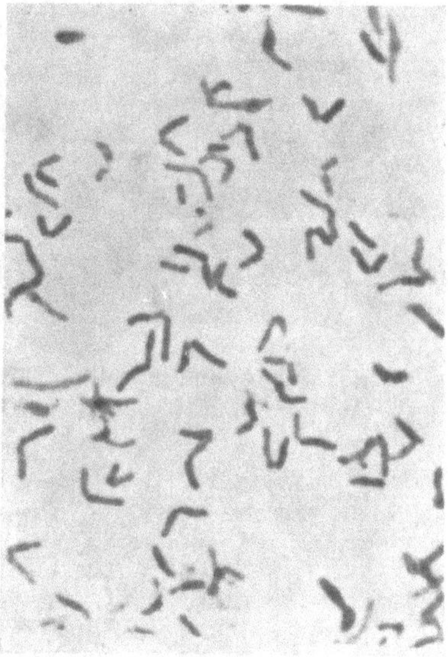

Fig. 4. Dividing *Micrococcus albus* cells isolated from the stratosphere, ×2000.

References

[1] L. A. ROGERS and F. C. MEIER, Stratosphere Flight of 1935 in the Balloon "Explorer II", National Geographic Society, US Army Air Corps 1936 (p. 276).

[2] V. W. GREENE, P. D. PEDERSON, D. A. LUNDGREN and C. A. HAGBERG, Proc. Atmospheric Biology Conf. 1964 (p. 199).

[3] B. A. BRONSHTEN and N. I. GRESHEN, Serebristye Oblaka, Nauka, Moscow 1970.

Life Sciences and Space Research XIV — Akademie-Verlag, Berlin 1976

GEOMYCOLOGY

N. J. PUERNER and S. M. SIEGEL

Department of Botany, University of Hawaii, Honolulu, Hawaii, USA

Fungi have long been known to have capabilities for reduction and alkylation of arsenate and selenate but their general capabilities for solubilizing and accumulating metallic substances have been given serious attention only in recent years. Common members of the *Aspergillaceae* cultured on boron, copper, lead and other metals or oxides can solubilize and concentrate the elements or their compounds. To account for bio-solubilization of the metals, we have set up a model study, incubating selected metals, e.g. mercury, in solutions of various metabolites including L-lysine and citric acid. Results of 100—300 days incubation showed that many metals can in fact be readily solubilized, and in some cases more effectively at pH 6—7 than at pH 1.5—2.

Established microbiological interactions upon or within the lithosphere include surface degradation of igneous rocks by lichens, oxidation and deposition of iron, manganese and sulfur by bacteria, and conversions of inorganic arsenic, selenium, tellurium and mercury to organic derivatives [1—7]. The utility of selective metal accumulation in higher plants has been recognized in geobotanical prospecting [8], but such capabilities have only recently been recognized in fungi, particularly members of the *Aspergillaceae* [9, 10].

Species of *Penicillium* and *Aspergillus* were grown in standard glucose-peptone media at 24—25 °C. These media were either controls or supplemented with metals or metal oxides as specified below. In the cultures supplied with B, Al, Bi, HgO and PbO, the elements were 100 to 150 mesh powders, and the oxides, even finer, were sealed in dialysis bags floated in the medium to avoid errors because of adhesion to fungal filaments. The other elements were presented in thin sheet or foil form with shallow layers of medium so that contact could be made with the fungal mycelium. Cultures were inoculated with conidiospores and, in all results presented here, incubated for 30 days. Analyses for the metal were carried out on both filtered medium and washed mycelium using emission spectrography or atomic absorption spectrophotometry (Hg, Cd, Cu, Pb).

Model experiments using aqueous solutions of selected metabolites were also set up using a variety of metals or compounds. Of these, trials with Cu, pyrite ($CuFeS_2$), Hg and cinnabar (HgS) are presented here. All analyses were carried out using atomic absorption. All experiments were duplicated and analyses run in duplicates or triplicate.

Penicillium was grown in the presence of elements representing four periodic groups and a range of redox and metallic properties (Table 1). In the case of B,

Table 1

Element Accumulation in Cultures of *Penicillium* and *Aspergillus*

Organism	Element	Element Analysis (ppm)			% Dry wt
		Medium[a]	Organism[b]	Ratio	
Penicillium notatum	B°	≪10	3200	≪320	1.6
	H_3BO_3	1000	4000	4	2.0
	Al°	ca. 100	10600	106	5.3
	Zn°	28	4600	164	2.3
	Cd°	12	1000	83	0.5
	Bi°	ca. 10	600	60	0.3
	Cu°	132	16000	121	8.0
Aspergillus clavatus	Hg°	2	128	64	0.06
	HgO	234	2680	11	1.34
	PbO	32	1080	34	0.54

[a] In filtered culture medium after 30 days.
[b] In washed mycelium, fresh wt after 30 days.
° chemical notation for elemental or zero-valent form.

the element and highly stable oxidized form H_3BO_3 were compared. The acid saturated the medium (0.2M) and constituted a stress condition in itself. In spite of its well-known refractory character, B was better solubilized and concentrated over 300-fold by the organism. Evidence was also obtained suggesting that one mode of B assimilation may take place via synthesis of boranes. The two concentration parameters of interest are the ratio, based upon wet weight: aqueous solution and percentage of total dry matter. Excluding H_3BO_3, the concentration ratios ranked, disregarding lesser difference:

$$B > Zn, Cu, Al > Cd, Bi.$$

On a dry matter basis, the ranking differed:

$$Cu > Al > Zn, B > Cd, Bi.$$

This suggests the complex nature of the interactions involving solubilization, uptake, intracellular deposition and cytotoxic limits. The uptake of Cu by *Penicillium* was evident prior to quantitative analysis in the distinct blue color of the mycelium and the massive precipitation in some cells of CuS upon exposure to ammonium sulfide (Fig. 1). Etch marks on the Cu surface followed hyphal filament lines [10]. The actual Cu content, 8% or 1.4 mmoles g^{-1}, is extraordinary. Experiments with *Aspergillus* showed that highly toxic Hg and Pb can also be rendered soluble and absorbed.

It was noted that the concentrations of some elements in the spent medium exceeded normal solubility levels. Thus, dissolved Cu at 132 ppm was about 66-fold higher than in sterile nutrient of the same age and Hg at 2 ppm was 80-fold greater. These enhanced concentrations in media after 30 days of culture suggested that fungal metabolites be considered a factor in metal solubilization. Accordingly, experiments were initiated as models using two elements and a

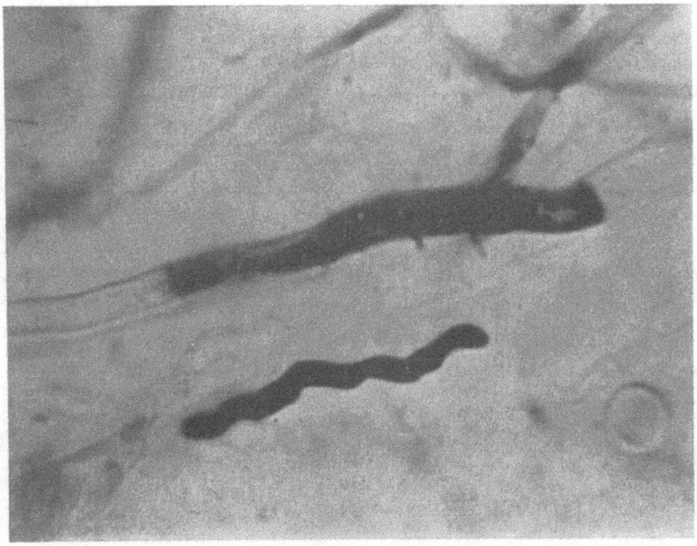

Fig. 1. Hyphal cells of *Penicillium notatum* grown on copper after washing and exposure to ammonium sulfide. The dense black CuS is evident within some cells. Photographed at 760×.

Table 2

Effect of Metabolites on the Solubility of Mercury and Mercury and Cinnabar (HgS)

Solute (100 mM)	pH	Hg in Solution (100 days)			
		Hg°		HgS	
		μM	solute/water	μM	solute/water
Water	5.5	0.12	—	0.01	—
Ammonium chloride	7.0	20	167	0.35	350
D-glucoseamine	5.5	29	242	0.42	42
L-lysine	6.5	28	233	1.56	156
L-asparagine	7.0	118	983	—	—
α-ketoglutaric	1.5	24	200	—	—
Citric acid	1.5	24	200	0.60	60

native mineral derivative of each (Tables 2 and 3). It is obvious from the calculated ratios of solute to water that ammonium chloride and a variety of organic compounds enhance the solubility not only of the elements but also of their high refractory sulfides. Evidence for selectivity is provided by the differences in the efficacy of solutes toward Hg, HgS, Cu and FeS_2 respectively, and in the differential solubilization of Cu and Fe from pyrites.

On the basis of the biological and chemical data presented here, we suggest that more attention be given to fungi as participants in or initiators of transformations of primary rock surfaces and as factors in biogeochemical cycling of inorganic substances.

Table 3
Effect of Metabolites on the Solubility of Copper and Pyrites ($CuFeS_2$)

Solute (100 mM)	pH	Cu in Solution (300 days)	
		Cu° (μM)	Pyrite (μM)
Water	5.5	20	8.0 (180)[a]
NH_4Cl	7.0	3000	15.8 (666)
D-glucoseamine	5.5	2000	15.8 (1260)
C-lysine	6.5	1120	17.6 (810)
Litric acid	1.5	650	14.4 (378)

[a] (180) denotes 180 μM Fe in solution containing 8 μM Cu, etc.

Acknowledgments

This work was supported by NASA Grant NGL 12-001-042.

References

[1] M. ALEXANDER, Introduction to Soil Microbiology, John Wiley, New York **1961**.
[2] F. CHALLENGER, Quart. Rev. Chem. Soc. Lond. **9**, 755 (1955).
[3] S. JENSEN and A. JERNELÖV, Nature **223**, 753 (1969).
[4] K. J. PUCKETT, E. NIEBOER, M. J. GORYNSKI and D. H. S. RICHARDSON, New Phytol. **72**, 329 (1973).
[5] S. M. SIEGEL, Environ. Biol. Med. **2**, 19 (1973).
[6] M. E. WILLIAMS and E. D. RUDOLPH, Mycologia **66**, 648 (1974).
[7] S. M. SIEGEL, B. Z. SIEGEL, N. J. PUERNER, T. SPEITEL and F. THORARINSSON, Water, Air and Soil Pollution, **1**, 212 (1973).
[8] H. CANNON, Science **132**, 591 (1960).
[9] K. ROBERTS and S. M. SIEGEL, Plant Physiol. **42**, 1215 (1967).
[10] M. R. D. SEAWARD, Lichenologist **5**, 423 (1973).

Index of Authors

FSC
www.fsc.org
MIX
Papier | Fördert gute Waldnutzung
FSC® C083411

Kolibri 360 GmbH
Ferdinand-Jühlke-Straße 7
99095 Erfurt, Deutschland
produktsicherheit@kolibri360.de